Wearable Electronics and Embedded Computing Systems for Biomedical Applications

Special Issue Editors

Enzo Pasquale Scilingo
Gaetano Valenza

MDPI

Special Issue Editors
Enzo Pasquale Scilingo   Gaetano Valenza
University of Pisa   University of Pisa
Italy   Italy

Editorial Office
MDPI AG
St. Alban-Anlage 66
Basel, Switzerland

This edition is a reprint of the Special Issue published online in the open access journal *Electronics* (ISSN 2079-9292) from 2016–2017 (available at: http://www.mdpi.com/journal/electronics/special_issues/weecs).

For citation purposes, cite each article independently as indicated on the article page online and as indicated below:

Author 1; Author 2; Author 3 etc. Article title. *Journal Name*. **Year**. Article number/page range.

ISBN 978-3-03842-386-7 (Pbk)
ISBN 978-3-03842-387-4 (PDF)

Table of Contents

About the Guest Editors

Enzo Pasquale Scilingo, Ph.D., is an Associate Professor in Electronic and Information Bioengineering at the University of Pisa, Italy. He received the Laurea Degree in Electronic Engineering from the University of Pisa and his Ph.D. degree in Bioengineering from the University of Milan, Italy, in 1995 and 1998 respectively. For two years he was postdoctoral fellow with the Italian National Research Council and, for two years, a post-doctoral fellow at the Information Engineering Department of the University of Pisa. Currently, he is pursuing his research work mainly at the Research Center "E. Piaggio". He has several teaching activities, he is supervisor of several Ph.D. students and is leading the laboratory Biolab at the Information Engineering Department. He coordinated a European project EC-FP7-ICT-247777 "PSYCHE-Personalised monitoring SYstems for Care in mental Health", and he is currently coordinating the European project H2020-PHC-2015-689691 NEVERMIND—NEurobehavioural predictiVE and peRsonalised Modelling of depressIve symptoms duriNg primary somatic Diseases with ICT-enabled self-management procedures. His main research interests are in wearable monitoring systems, human–computer interfaces, biomedical and biomechanical signal processing, modelling, control and instrumentation. He is author of more than 150 papers in peer-reviewed journals, contributions to international conferences and chapters in international books. He is co-author of two books edited by Springer. He is currently serving as a reviewer to many international journals and as a member of the Program and Scientific Committees of annual international conferences. He is a guest associate editor of the *Frontiers in Neuroengineering* journal, guest editor of a Special Issue in the *Journal of Biomedical and Health Informatics* Sensor Informatics for Managing Mental Health, 2015, and is an associate editor of *ETRI* Journal, *Electronics, Complexity* journal, and PLOS ONE.

Gaetano Valenza, M.Eng., Ph.D., is currently an Assistant Professor of Bioengineering at the University of Pisa, Pisa, Italy. In 2009, he started working at the Bioengineering and Robotics Research Centre "E. Piaggio" in Pisa and, in 2011, he joined the Neuro-Cardiovascular Signal Processing Unit within the Neuroscience Statistics Research Laboratory at the Massachusetts Institute of Technology, Cambridge, USA. In 2013, he received his Ph.D. degree in Automation, Robotics, and Bioengineering from the University of Pisa and, in the same year, was appointed as a Research Fellow at Harvard Medical School/ Massachusetts General Hospital, Boston, USA. His research interests include statistical and nonlinear biomedical signal and image processing, cardiovascular and neural modeling, and wearable systems for physiological monitoring. Applications of his research include the assessment of autonomic nervous system activity on cardiovascular control, brain–heart interactions, affective computing, assessment of mood and mental/neurological disorders. He is the author of more than 100 international scientific contributions in these fields, published in peer-reviewed international journals, conference proceedings, books and book chapters, and is an official reviewer of more than sixty international scientific journals, and research funding agencies. He has been involved in several international research projects, and currently is the scientific co-coordinator of the European collaborative project H2020-PHC-2015-689691-NEVERMIND. Dr. Valenza has been guest editor of several international scientific journals, and is currently member of the editorial board of PLOS ONE, of *Complexity,* and of the *Nature*'s journal Scientific Reports.

electronics

MDPI

Editorial

Recent Advances on Wearable Electronics and Embedded Computing Systems for Biomedical Applications

Enzo Pasquale Scilingo and Gaetano Valenza *

Bioengineering and Robotics Research Centre "E. Piaggio", and Department of Information Engineering, School of Engineering, University of Pisa, Largo Lucio Lazzarino 1, 56122 Pisa, Italy; e.scilingo@ing.unipi.it
* Correspondence: g.valenza@ieee.org

Academic Editor: Mostafa Bassiouni
Received: 12 January 2017; Accepted: 16 January 2017; Published: 24 January 2017

1. Introduction

The application of wearable electronics in the biomedical research and commercial fields has been gaining great interest over the last several decades. Small-sized, lightweight monitoring systems with low-power consumption and, of course, wearability allow for the collection of physiological and behavioral data in ecological scenarios (e.g., at home, during daily activities or sleep, during specific tasks, etc.) with a minimal discomfort for the end user [1]. As a result, outpatient-monitoring care can be associated with improved quality of life, especially for patients with chronic disease, possibly preventing unnecessary hospitalizations and reducing direct and indirect healthcare costs.

To this extent, research efforts have been focusing on the development of innovative sensors (e.g., smart-textile or contactless electrodes) and sensing platforms, as well as effective algorithms for embedded signal processing and data mining. Furthermore, significant endeavors have been related to small-scale integration of analog/digital sensor signal conditioning and energy harvesting, especially in the case of wireless body area/sensor networks.

The high impact of wearable technology in the frame of multidisciplinary scientific research is also confirmed by the significant number of studies published throughout the last several decades. By searching the keywords "wearable", "monitoring", and "human" in the Scopus database, taking into account article title, abstract, and keywords, a total of 2531 entries have been found, starting with less than 10 articles per year before 2001 and reaching more than 200 articles per year since 2011.

2. The Present Special Issue

In the frame of wearable electronics and embedded computing systems for biomedical applications, this Special Issue of Electronics includes a total of 14 papers, including two review papers and twelve research articles [2–15]. The issue spans a wide range of topics, including smart sensing footwear systems, human-body hydration monitoring, textile-based ECG monitoring in human and horses, biotelemetry and telemedicine, support for surgical navigation, wearable autonomic nervous system activity monitoring, and hand posture and tactile pressure sensing [2–15].

More specifically, Carbonaro et al. [7] propose a sensorized shoe for gait analysis. The shoe has built-in force sensors and a triaxial accelerometer, and is able to transmit sensor data to the smartphone through a wireless connection. Experimental results confirmed a reliable detection of the gait phases. Boehm et al. [5] propose a sensorized T-shirt able to acquire a long-term, multichannel electrocardiogram (ECG) with active electrodes, therefore avoiding the use of adhesive gel electrodes. Experimental results validated the proposed wearable monitoring system as compared with a commercial Holter ECG in healthy volunteers during movement phases of lying down, sitting, and walking. De Marcellis et al. [9] propose a pulsed coding technique based on optical ultra-wideband modulation for wireless implantable biotelemetry systems with low power consumption. The overall architecture implementing this optical modulation technique employs sub-nanosecond pulsed laser as the data transmitter and small sensitive area photodiode as the data receiver. Guidi et al. [12] propose a textile-based, wearable system for heart rate variability (HRV) monitoring in humans and animals, aiming to study human–horse interaction. Experimental results compared the performance of the proposed wearable system with a standard system in terms of amount of movement artifact. A support vector machine classifier showed the discrimination of three distinct real human–animal interaction levels.

Farooq et al. [10] present a method for the automatic quantification of chewing episodes captured by a piezoelectric sensor system. Experimental results were related to the estimation of the number of chews as compared to manually annotated chewing segments, and an artificial neural network-based automatic classification of "food intake" or "no intake" classes. In the context of surgical navigation systems, Cutolo et al. [8] propose an algorithm suitable for wearable stereoscopic augmented reality (AR) video see-through system. The video-based tracking relies on stereo localization of three monochromatic markers rigidly constrained to the scene. This approach provides a viable solution for the implementation of wearable AR-based surgical navigation systems. Caldara et al. [6] developed a potentially implantable blood pressure telemetry system, based on an active Radio-Frequency IDentification (RFID) tag, aiming to continuously measure the average systolic and diastolic blood pressure of small/medium animals. RFID energy harvesting has also been investigated. The authors present an experimental laboratory characterization and in vivo tests. Greco et al. [11] propose a wearable system for monitoring the electrodermal activity (EDA) signals during emotional elicitation. EDA was studied at different frequency sources through data gathered from healthy subjects undergoing visual affective elicitations. The authors conclude that the frequency of the external electrical source affects the accuracy of arousal recognition. Saponara et al. [15] present a scalable remote model for telemedicine scenarios using wireless biomedical sensors, an embedded local unit (gateway) for sensor data acquisition/processing/communication, and a remote e-Health service center. The use of a mix of commercially available sensors and new custom-designed ones was also presented. Bianchi et al. [4] propose an integrated sensing glove combining a low number of knitted piezoresistive fabrics to reconstruct both hand posture and tactile pressure sensing. To this end, a priori information of synergistic coordination patterns in grasping tasks was employed. In the frame of a wireless body area network, Liao et al. [14] derive an analytical and accurate 2.45 GHz model based on a 3D heterogeneous human body model. The proposed approach outperforms other modulation techniques, enabling the support of a 30 Mbps data transmission rate up to 1.6 m and affording more reliable communication links when the transmitter power is increased. Finally, Asogwa et al. [2] propose a non-intrusive method for tracking hydration rates with a resolution of 100 mL of water. The authors state that the real-time changes in galvanic coupled intrabody signal attenuation can be integrated into wearable electronic devices to evaluate body fluid levels on a particular area of interest and can aid the diagnosis and treatment of fluid disorders such as lymphoedema.

Concerning the review articles included in the special issue, Hegde et al. [13] compare footwear-based wearable systems, focusing on embedded sensors and electronics. This review article describes key application scenarios (including gait monitoring, plantar pressure measurement, posture and activity classification, body weight and energy expenditure estimation, biofeedback, navigation, and fall risk applications), utilizing footwear-based systems with critical discussion on their merits. Furthermore, energy-harvesting issues are also discussed. Bianchi et al. [3] review fabric-based approaches for the development of wearable haptic systems. Particularly, some examples of fabric-based systems that can be applied to different body locations and can elicit different haptic perceptions are presented, along with critical perspective and future developments of this approach.

3. Conclusions and Prospective Future Research Directions

The topic of wearable electronics and related embedded computing has been investigated for more than two decades, and has been exploited for a huge variety of biomedical applications. To this end, reliable solutions to collect informative, possibly long-term psycho-physiological, behavioral, and biomechanical data in ecological scenarios have been achieved. Overall, considering the sensor technology, textile-based monitoring, as well as smart-watches for cardiovascular and activity monitoring, represents a milestone for the development of these systems, maximizing comfort and usability for the end-user. Of note, a recent focus on wearable monitoring applications for animals has been successfully pursued.

Nevertheless, besides substantial benefits and widespread use in many research fields, wearable systems may be seriously affected by movement artifacts. More generally, many wearable monitoring systems have not entered the biomedical market due to certification issues and related high costs.

Much more effort is thus required to ensure that wearable monitoring systems for biomedical applications reach a proper level of reliability and compliance with strict local regulations.

From a technological point of view, we envisage that future research directions will be directed toward contactless monitoring systems (e.g., UWB).

Acknowledgments: First of all, we would like to thank all authors of this Special Issue for their excellent contributions. We also would like to thank the reviewers who spent considerable time in the reviewing process and who made extremely valuable suggestions to improve the quality of the submitted papers. We are also very grateful to Mostafa Bassiouni, the editor-in-chief, for giving us the opportunity to guest-edit the Special Issue, and the entire staff of the Editorial Office of Electronics for our pleasant collaboration. Finally, we acknowledge partial financial support from the European Commission under Horizon 2020, grant no. 689691 NEVERMIND (NEurobehavioural predictiVE and peRsonalised Modelling of depressIve symptoms duriNg primary somatic Diseases with ICT-enabled self-management procedures).

Conflicts of Interest: The authors declare no conflict of interest.

References

1. Van Laerhoven, K.; Lo, B.P.; Ng, J.W.; Thiemjarus, S.; King, R.; Kwan, S.; Gellersen, H.; Sloman, M.; Wells, O.; Needham, P.; et al. Medical healthcare monitoring with wearable and implantable sensors. In Proceedings of the 3rd International Workshop on Ubiquitous Computing for Healthcare Applications, Nottingham, UK, 6–7 September 2004.
2. Asogwa, C.O.; Collins, S.F.; Mclaughlin, P.; Lai, D.T. A Galvanic Coupling Method for Assessing Hydration Rates. *Electronics* **2016**, *5*, 39. [CrossRef]
3. Bianchi, M. A Fabric-Based Approach for Wearable Haptics. *Electronics* **2016**, *5*, 44. [CrossRef]
4. Bianchi, M.; Haschke, R.; Büscher, G.; Ciotti, S.; Carbonaro, N.; Tognetti, A. A Multi-Modal Sensing Glove for Human Manual-Interaction Studies. *Electronics* **2016**, *5*, 42. [CrossRef]
5. Boehm, A.; Yu, X.; Neu, W.; Leonhardt, S.; Teichmann, D. A Novel 12-Lead ECG T-Shirt with Active Electrodes. *Electronics* **2016**, *5*, 75. [CrossRef]
6. Caldara, M.; Nodari, B.; Re, V.; Bonandrini, B. Miniaturized Blood Pressure Telemetry System with RFID Interface. *Electronics* **2016**, *5*, 51. [CrossRef]
7. Carbonaro, N.; Lorussi, F.; Tognetti, A. Assessment of a Smart Sensing Shoe for Gait Phase Detection in Level Walking. *Electronics* **2016**, *5*, 78. [CrossRef]
8. Cutolo, F.; Freschi, C.; Mascioli, S.; Parchi, P.D.; Ferrari, M.; Ferrari, V. Robust and Accurate Algorithm for Wearable Stereoscopic Augmented Reality with Three Indistinguishable Markers. *Electronics* **2016**, *5*, 59. [CrossRef]
9. De Marcellis, A.; Palange, E.; Nubile, L.; Faccio, M.; Di Patrizio Stanchieri, G.; Constandinou, T.G. A pulsed coding technique based on optical UWB modulation for high data rate low power wireless implantable biotelemetry. *Electronics* **2016**, *5*, 69. [CrossRef]
10. Farooq, M.; Sazonov, E. Automatic Measurement of Chew Count and Chewing Rate during Food Intake. *Electronics* **2016**, *5*, 62. [CrossRef]

11. Greco, A.; Lanata, A.; Citi, L.; Vanello, N.; Valenza, G.; Scilingo, E.P. Skin Admittance Measurement for Emotion Recognition: A Study over Frequency Sweep. *Electronics* **2016**, *5*, 46. [CrossRef]
12. Guidi, A.; Lanata, A.; Baragli, P.; Valenza, G.; Scilingo, E.P. A Wearable System for the Evaluation of the Human-Horse Interaction: A Preliminary Study. *Electronics* **2016**, *5*, 63. [CrossRef]
13. Hegde, N.; Bries, M.; Sazonov, E. A Comparative Review of Footwear-Based Wearable Systems. *Electronics* **2016**, *5*, 48. [CrossRef]
14. Liao, Y.; Leeson, M.S.; Higgins, M.D.; Bai, C. Analysis of In-to-Out Wireless Body Area Network Systems: Towards QoS-Aware Health Internet of Things Applications. *Electronics* **2016**, *5*, 38. [CrossRef]
15. Saponara, S.; Donati, M.; Fanucci, L.; Celli, A. An Embedded sensing and communication platform, and a healthcare model for remote monitoring of chronic diseases. *Electronics* **2016**, *5*, 47. [CrossRef]

![electronics logo] *electronics*

MDPI

Article

Assessment of a Smart Sensing Shoe for Gait Phase Detection in Level Walking

Nicola Carbonaro [1], Federico Lorussi [1] and Alessandro Tognetti [1,2,*]

[1] Research Center "E. Piaggio", University of Pisa, Largo Lucio Lazzarino 1, 56126 Pisa, Italy; nicola.carbonaro@centropiaggio.unipi.it (N.C.); f.lorussi@ing.unipi.it (F.L.)
[2] Information Engineering Department, University of Pisa, via G. Caruso 16, 56122 Pisa, Italy
* Correspondence: a.tognetti@centropiaggio.unipi.it

Academic Editor: Enzo Pasquale Scilingo; Mostafa Bassiouni
Received: 15 August 2016; Accepted: 4 November 2016; Published: 16 November 2016

Abstract: Gait analysis and more specifically ambulatory monitoring of temporal and spatial gait parameters may open relevant fields of applications in activity tracking, sports and also in the assessment and treatment of specific diseases. Wearable technology can boost this scenario by spreading the adoption of monitoring systems to a wide set of healthy users or patients. In this context, we assessed a recently developed commercial smart shoe—the *FootMoov*—for automatic gait phase detection in level walking. FootMoov has built-in force sensors and a triaxial accelerometer and is able to transmit the sensor data to the smartphone through a wireless connection. We developed a dedicated gait phase detection algorithm relying both on force and inertial information. We tested the smart shoe on ten healthy subjects in free level walking conditions and in a laboratory setting in comparison with an optical motion capture system. Results confirmed a reliable detection of the gait phases. The maximum error committed, on the order of 44.7 ms, is comparable with previous studies. Our results confirmed the possibility to exploit consumer wearable devices to extract relevant parameters to improve the subject health or to better manage his/her progressions.

Keywords: wearable technology; gait phase; gait cycle, accelerometers; force sensors; walking; smart shoe

1. Introduction

In the last decade, many groups have carried out research and development on wearable electronics and sensors for unobtrusive, ambulatory and daily-life monitoring of human subjects. The results obtained have shown the possibility to use personal wearable devices to assist and support chronic patients [1–7], elderly people [8,9], emergency operators [10,11] and also healthy subjects for sports, wellness and prevention [12–14]. At the same time, the wearable technology market has exploded and is expected to further increase over the next few years, as proved by the growing interest of big players such as Google, Apple and Samsung.

The current trend is to augment objects worn on the body—e.g., watches, glasses, bracelets—with information and communications technology (ICT) to enable a bi-directional data exchange with a smartphone. These wearable devices or simply *wearables* have been initially conceived as technological *gadgets* but have the potential to support the user in the self-management of his/her health and wellness. Indeed, smart bracelets and/or watches can include physiological (e.g., photoplethysmography, electrodermal activity) or inertial (accelerometers, gyroscopes) sensors able to perform real-time monitoring of subject's health parameters and movement/physical activity. Recent studies have reported the first attempts to employ smart watches/bracelets in e-health applications [15–19] and many more are expected in the years to come.

Another trend, less explored but not less promising, is to embed ICT devices inside the shoe. The shoe is the ideal place to integrate sensors and communications technology: it has enough space for the micro-devices and it is the object that every one wears for most of the day. This last aspect is the key factor to enable user's acceptance, as the user does not have to wear additional items and the technology can be completely *hidden* and *transparent* for him/her. The integration of inertial and force sensors in the shoe may enable a wide number of applications, ranging from simple activity/fitness tracking (e.g., activity classification, step count, burned calories) fragile people assistance (e.g., fall detection, pedestrian navigation) to complex biomedical assessment and gait analysis.

Gait analysis is the study of human locomotion, and it is used to assess and treat patients with conditions affecting their walking activity [20]. The way we walk consists of consecutive *gait cycles*. Each gait cycle includes a predefined sequence of phases (Heel-strike HS, Stance ST, Heel-off HO, Swing SW; see Appendix A for a reference on the adopted terminology). Both *temporal* and *spatial* gait parameters are important to assess a disease and/or a traumatic event, and also to define and optimize the treatment (e.g., rehabilitation, physical therapy). The temporal and spatial gait parameters are used in many biomedical and e-health applications, such as assessment of the recovery in stroke patients [21,22] and gait-cycle-based control of functional electrical stimulators (FES) for *drop foot* compensation [23–25]. In the robotic rehabilitation field, the quantitative evaluation of the gait parameters allows the quantification of the improvements of the gait patterns [26]. In addition, the spatial gait parameters, such as stride length, can be associated with fall risk [27], or elaborated for foot motion localization in applications such as emergency operator rescue and pedestrian navigation [28–31].

The reviews from Rueterbories et al. [32] and from Taborri et al. [33] provide a full overview of the gait phase detection methods and technologies. Many research works performed gait phase detection through inertial sensors (accelerometers, gyroscopes, inertial measurement units - IMU) applied to different body segments (pelvis, thigh, shank, foot) [34–38]. As a recent relevant example, the work of Van Nguyen et al. [39] focuses on an IMU sensor for an accurate estimation of foot position, velocity and attitude. Many other works were focused on *on-foot* sensors. Most of the *on-foot* systems dealt with force based methods, employing foot-switches or force sensitive resistors (FSRs) to measure the body/ground force interaction [40–42]. Force based methods have reliable performance, but are unable to discriminate walking activities from load changes (i.e., in foot drop control, this implies the user turning off the detection/control system at the end of the walking activity). To solve this issue, Pappas et al. [43] combined force and inertial sensors to obtain a reliable gait phase detection system, able to discriminate walking activities from load shifting in static tasks. They employed three FSRs (one under the heel and two under the fore-foot region) and a gyroscope attached to the back side of the shoe (above the heel). Force sensors detected the foot loading/unloading, while the gyroscope estimated the foot inclination and rotational velocity. In their work, Pappas et al. [43] detected four gait phases through a state machine whose transitions were governed by a rule based algorithm applying predefined thresholds on the parameters extracted from the sensors (foot loading/unloading, inclination, rotational velocity). Other examples of force and inertial sensors combination can be found in [44,45]. As underlined in [33], gait phase partitioning algorithms can be divided into three main classes. *Threshold-based* methods—such as the above reported from Pappas et al. [43]—which apply predefined or adaptive thresholds to the sensor signals, are simple and are often suitable for integration in embedded systems. *Machine learning* methods and in particular Hidden Markov Models (HMM) are more complex than threshold based methods but have shown improved performance in gait phase detection. As a relevant example, Mannini and Sabatini [37] developed an HMM classifier which detected four gait phases from an uni-axial foot-mounted gyroscope, and achieved significantly better performace than the threshold-based method applied to the same dataset. A more recent trend is to apply hierarchical decision to the output of two or more HMMs. As reported in the work from Taborri et al. [38], *hierarchical* methods provide excellent performance and are compatible with real-time implementation.

In-shoe sensor systems have been commonly developed for real-time detection of gait parameters and walking patterns with applications in the assessment of specific foot pathologies (e.g., flat foot [46], diabetic foot [47]) and posture/activity recognition for healthy subjects [48–50] or people affected by neurological conditions such as stroke [51,52] or celebral palsy [26,53]. In-shoe sensor systems are generally based on movable pressure sensing insoles and/or inertial sensors, combined with an external electronic module (signal acquisition/pre-elaboration, data transmission). Examples of in-shoe systems can be found in the works from Edgar et al. [41] and Bae et al. [54].

Commercial products are limited to professional instruments for clinical evaluation and to a few consumer devices for sports and training of healthy users. Professional products include the F-scan [55] (Tekscan Inc., Boston, MA, USA) and the Pedar [56] (Novel Inc., Munich, Germany) systems that are sensing insoles for the monitoring of dynamic temporal and spatial pressure distributions. Example applications of professional products are gait stability analysis [57], gait phase detection [58] and analysis of the gait characteristics during running [59]. Despite the reliable performance and the high spatial resolution, the professional systems are not suitable for long-term monitoring in daily life conditions: both systems are expensive (on the order of several kEuros) and use electrical wires to connect the insole to the waist-worn acquisition system. The consumer products are generally made by applying an external measurement and transmission unit to a dedicated shoe (e.g., *Adidas miCoach*, *Nike + iPod*). The common aspect is the reduced number of sensors (typically only one inertial sensor) and the interaction with the smartphone in which a dedicated app can deliver special information to the users (e.g., workout time, velocity, distance travelled, calories), engaging them for reaching higher performance during physical activity.

In the current paper, we assessed a recently developed commercial smart shoe for automatic gait phase detection in level walking. The prototype we employed is the *FootMoov* smart shoe [60] produced by the Italian shoe factory Carlos srl (Fucecchio, Firenze, Italy). FootMoov was originally designed as a mobile game controller and for simple physical activity training and coaching. FootMoov can be interfaced with the smartphone through a WiFi connection and has built-in force sensors (heel and forefoot), triaxial accelerometer (forefoot), chargeable battery and acquisition/transmission module. We chose to assess and develop the gait phase detection algorithm for FootMoov since, unlike the current professional and consumer products, all of the hardware—including sensors and electronics—is integrated *inside* the shoe, and no external modules or application of additional parts are needed.

On the other side, FootMoov is a *consumer* product. Thus the sensor number and locations are not optimized for gait analysis. In addition, the precise sensor locations and orientation inside the shoe is unknown. In accordance with the current trends in wearable technology, the aim of this work is to show the possibility to employ a low cost smart shoe as a tool for biomedical and e-health applications, allowing the continuous and long-term monitoring of users/patients in daily life. To the best of our knowledge, the assessment of such a wearable product does not exist in the current literature.

In particular, we assessed the smart shoe for the real-time quantification of the temporal parameters of gait. We developed a dedicated gait phase detection algorithm that combines the information extracted from the force and accelerometer sensors of the FootMoov smart shoe. In a first experiment, we collected data on ten healthy subjects in free level walking conditions to perform a preliminary and qualitative assessment of the system and algorithm performance. In this test, the algorithm recognized 5925 strides over the total amount of 6000 strides (98.7%). In a second test, we performed a quantitative evaluation of the system in comparison with a reference gait phase signal obtained by an optical motion capture instrument. We evaluated the time difference in the onset of the detected gait phases with respect to the reference (mean error of 44.7 ms) and the error in the estimation of the single phase durations (minimum error of 0.036 s for Heel Strike , with a maximum error of 0.11 s for Heel Off).

2. Material and Methods

2.1. FootMoov

FootMoov is an innovative smart shoe with sensors and an electronic unit fully integrated inside the footwear, below the insole. The wireless (WiFi) connection enables the use of dedicated smartphone or tablet apps for the acquisition and elaboration of the sensor data.

The smart shoe includes two force sensors and one triaxial accelerometer. As shown in Figure 1, the force sensors are located approximately in the center of the heel and forefoot regions, while the accelerometer is positioned below the insole in correspondence of the shoe tip. FootMoov has a built-in battery, chargeable through a mini-USB connector placed in the rear part of the shoe, below the turn on/off switch. Sensors and electronics are integrated in the right shoe only. Inclination and foot movement could be estimated from accelerometer data, while the foot mechanical interaction with the ground can be extracted from the force sensors.

Figure 1. The FootMoov prototype. The green arrows indicate the location of the heel and forefoot force sensors (output F_h and F_f respectively). The local reference frame of the calibrated accelerometer (outputs ax, ay, az) is reported in red (x_a, y_a, z_a). In the rear part of the shoe it is possible to see the on/off switch and the mini-USB connector for battery charging.

The accelerometer integrated in the FootMoov system is a tri-axial digital sensor with low power consumption, ultra-compact dimension and 12-bit resolution for a dynamic full scale range of ± 2 g. Force sensors are analogue force sensitive resistors (FSRs) in which a variation of the electrical resistance is generated when a pressure in the sensing area is applied. The output characteristic of the FSRs is inversely proportional to force. When the sensor is unloaded the resistance is higher than 2 MΩ. Then, when applied load increases, the electrical resistance decreases. An inverting analog circuit amplifies the FSR output before it is digitally converted. The core of the FootMoov electronics is a low-power and low-cost microcontroller. This device manages sensor data acquisition and wireless transmission. In particular, the microcontroller provides the analog-to-digital conversion of force sensors signal and the digital I/O ports for the acquisition of the digital accelerometer data. A WiFi module is integrated in the FootMoov hardware for the transmission of the sensor data packet to remote devices such as smartphones, tablets or PCs. The transmitted packet is composed of the timestamp (microcontroller internal time reference), the accelerometer data (expressed in logical values) and by two more values related to the front and rear force sensor, respectively (logical values corresponding to the converted force signals). By knowing the packet structure and its transmission protocol (i.e., information supplied by the FootMoov producer), we developed a dedicated app to allow the wireless connection and data exchange with FootMoov. This app is able to retrieve, store and visualize in real-time the sensor signals. The app is based on the TCP/IP protocol in which the FootMoov shoes act as the server while the mobile devices (smartphones, tablets or PCs) are the clients. The client (user) sends a query to the server (FootMoov system) and once the server answers through the client IP address, the connection is established. Then, a "Start" button turns on, and the user can launch the data acquisition session. FootMoov samples and streams the logical values of the converted sensors data to the smartphone (25 Hz).

2.2. Gait Phase Detection

To develop our gait phase detection algorithm, we started from a revision of the existing methods. Since FootMoov is endowed with force and inertial sensors (see Section 2.1), we were inspired to create our gait detection algorithm from the one of Pappas et al. [43]. In particular, we employed a similar state machine with four states corresponding to the four gait phases described in Section 1. As described in Section 2.2.1, we pre-processed the FootMoov sensor data to obtain the quantities for the rule based transitions between the machine states. More specifically, we calibrated and processed the triaxial accelerometer signal to extract the foot inclination (the yaw angle ψ) and the inclination velocity (the yaw time derivative ψ'), and we acquired the heel and fore-foot loading signals from the FootMoov force sensors (F_h and F_f respectively). Finally, Section 2.2.2 describes the gait phase detection algorithm applied to pre-processed FootMoov signals.

2.2.1. Signal Pre-Processing

We pre-processed the accelerometer signal to estimate the foot inclination (yaw angle ψ), required by the phase detection algorithm described in Section 2.2.2. We converted the raw accelerometer signals from logical values to the measured acceleration expressed in units of g. Note that the orientation of the accelerometer inside the shoe is unknown and not necessarily aligned with the reference frame reported in Figure 1 (i.e., due to internal shoe conformation and/or fabrication tolerance). This aspect implies an offset in the detected inclination (i.e., the inclination is not zero when the shoe lies on an horizontal plane). To avoid the inclination offset, we conceived a dedicated calibration phase. The accelerometer calibration is a two-step procedure derived from the one described in our previous work [61]. In the first step, we measured the accelerometer output when the user was standing upright in a natural position to align the z-axis of the accelerometer with the axis Z_a (in the upright position, Z_a is supposed parallel to the absolute vertical, see Figure 1). In a second step, starting from the same up-right position, asked the subject to perform a simple foot movement (three consecutive dorsi-flexions of the ankle). This second measurement allows us to obtain the final alignment by applying the transformation (rotation along Z_a) that minimizes the variation of the accelerometer x component. Triaxial accelerometers measure both inertial acceleration and local gravity. In static conditions, only the gravity is present and the inclination of the accelerometer with respect to the vertical is known. In these conditions, the Euler angle ψ (ZYX convention [62]) can be obtained in terms of the accelerometer components:

$$\psi = atan2(ay, az),\tag{1}$$

where $atan2$ is the four quadrant inverse tangent and ax, ay, and az are the calibrated accelerometer components expressed in units of g. In dynamic conditions, the estimation of ψ by the accelerometer components (Equation (1)) is not reliable due to the effect of the inertial acceleration. It is well known that the inclination estimation error increases as the activity intensity increases (e.g., running, jumping). In literature, to overcome this issue, low pass filtering with very low cut-off frequencies [63] or complex Kalman filter based techniques [64] were applied. These techniques can introduce delays that are not compatible with gait phase detection. Considering the accelerometer inside the FootMoov prototype (Figure 1), the yaw angle represents the rotation angle of the foot around the accelerometer x-axis (positive for anti-clockwise rotations). In our application, we can suppose having quasi-static conditions when the foot is in contact with the ground (from the Heel-strike to the end of the Heel-off phase) and dynamic conditions when the foot is flying forward (Swing phase). As we will describe in Section 2.2.2, our gait phase detection algorithm exploits the inclination information to detect the transitions between Heel-strike to Stance and Stance to Heel-off. In these situations, we can consider the accelerometer in quasi-static conditions and directly apply Equation (1) to the calibrated accelerometer components to obtain the yaw angle (ψ).

For the force sensors, we simply scaled the outputs (logical values) to obtain the F_h and F_f signals with the following characteristics: (i) $F_h = 0$ and $F_f = 0$ when the sensors are unloaded and (ii) $F_h \approx 1$ and $F_f \approx 1$ when the sensors are fully loaded. We determined the scale factor after a preliminary experimental session in our laboratory. We recorded data from three subjects requested to walk for 60 s. In this preliminary walking trial, we evaluated the scale factors as the mean of the relative maximum of the force signals ($S_{Fh} = 1405$ and $S_{Ff} = 2003$ for the heel and front sensors, respectively). We also evaluated the baselines of F_h and F_f as the sum of the mean and the standard deviation of the un-loaded sensors ($b_{F_h} = 0.008$ and $b_{F_f} = 0.005$ for the heel and front sensors, respectively).

2.2.2. Detection Algorithm

The state machine is reported in Figure 2. Four state transitions (E_i, $i = 1, ..., 4$) are allowed and correspond to the gait events of the normal walking.

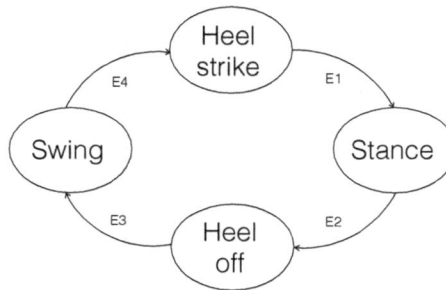

Figure 2. Gait phase detection algorithm.

The state transitions are governed by the following rules:

- E1 (Heel-strike to Stance): in the Heel-strike state, Stance is detected when both heel and fore-foot sensors are loaded or the foot rotational velocity is close to zero ([$F_h > th_{Fh}$ AND $F_f > th_{Ff}$] OR $|\psi'| < \delta_{\psi'}$);
- E2 (Stance to Heel-off): in the Stance state, Heel-off is detected if the heel sensor is unloaded and the foot inclination angle (ψ) exceeds a certain threshold ($F_h < th_{Fh}$ AND $\psi > th_{\psi}$);
- E3 (Heel-off to Swing): in the Heel-off state, Swing is recognized if heel and fore-foot are unloaded and the rotational velocity turns from positive to negative ([$F_h < th_{Fh}$ AND $F_f < th_{Ff}$] AND $\psi' < 0$);
- E4 (Swing to Heel-strike): in the Swing state, Heel-strike is detected as the heel touches the ground and the heel sensor is loaded ($F_h > th_{Fh}$).

We also evaluated the stride period as the temporal distance between the onset of two consecutive Heel-strike phases and, accordingly, the cadence as the number of strides per minute. The state machine has the same working frequency of the FootMoov sampling rate (25 Hz).

We set the thresholds of the force sensors as two times their baselines (see Section 2.2.1). We fixed $th_{\psi} = 0.035$ rad and $\delta_{\psi'} = 0.52$ rad/s in accordance with the literature values from [43,65]. The 60 s preliminary walking trial was also useful to verify these parameters.

2.3. Experiments

To assess the FootMoov system and the gait phase detection algorithm, we conceived two different experimental tests.

2.3.1. Free Walking Experiment

The first test aimed at a qualitative assessment of the prototype and the gait detection algorithm in daily life level walking conditions. Ten healthy subjects were recruited and were asked to walk on level ground. The subjects walked in outdoor conditions in an open space free of obstacles. The subjects had no physical or neurological impairments that affect the characteristics of their walking activity. The subject characteristics were different for age, weight and height, and are reported in Table 1. They were instructed to choose the cadence of a normal walking activity. We requested the subjects to wear the FootMoov shoes and to execute a fixed number of strides (150 strides) repeated for four times. During each session, number of strides, distance traveled and time elapsed were manually stored as useful information for the evaluation of the system and the algorithm developed. Through the app described in Section 2.1, we collected the FootMoov raw data (force sensors and accelerometer) that were transmitted and stored in the smartphone that was kept in the subject's pocket during the walking trials. Data were elaborated off-line in Matlab by applying the pre-processing algorithms of Section 2.2.1 and the gait phase detection algorithm of Section 2.2.2. Here—to obtain a preliminary qualitative evaluation—we compared the number of strides detected by the algorithm (i.e., as occurrence of two consecutive heel strike phases) with the total number of strides (150 for each trial).

Table 1. User characteristics.

Age (Mean ± Std) [Years]	Weight (Mean ± Std) [kg]	Height (Mean ± Std) [m]
26.75 ± 3.1	76.875 ± 8.5	1.768 ± 0.06

2.3.2. Motion Capture Experiment

The second test aimed at a quantitative evaluation of the smart shoe and the detection algorithm. We evaluated the FootMoov prototype and our gait phase detection algorithm in comparison with a reference gait phase signal obtained through an optical motion capture device. We adopted a four camera optical motion capture system (Smart DX100 produced by BTS Bioengineering [66]). Two passive markers were applied to the FootMoov shoe, as shown in Figure 3, in correspondence with the tip and heel area. The absolute position of the markers was acquired by the BTS system with a working frequency of 100 Hz. According to [43], we built the gait phase reference signal by applying a rule based algorithm to the vertical positions of the heel and tip markers. Subjects were asked to wear the FootMoov shoes (with the passive markers on) and to walk within the workspace of the cameras (see Figure 3). At the same time, raw sensor data coming from the FootMoov system were acquired and stored in the smartphone. The space used for this experiment was about 5 meters long and 2 meters wide, and the subjects repeated the session four times. Data were again elaborated off-line (pre-processing: Section 2.2.1, gait phase detection: Section 2.2.2) and were synchronised and compared to the reference gait signal.

To synchronize the two acquisition systems (FootMoov and BTS motion capture), we generated a reference signal to determine the time period in which the trial session occurred. The reference signal is a step waveform signal (TTL value) which is "High" during the acquisition period, and "Low" otherwise. This signal, activated by the smartphone app, is used to manage the BTS system through a specific port devoted to external signal acquisition. In particular, we used this step waveform signal as a trigger for the control of the recording session of the motion capture system. Moreover, to double-check the correct data alignment, the subject was requested to perform a particular foot movement at the beginning of each session. This movement generates a significant variation in both sensor data and marker signals, easily recognizable as a session starting point.

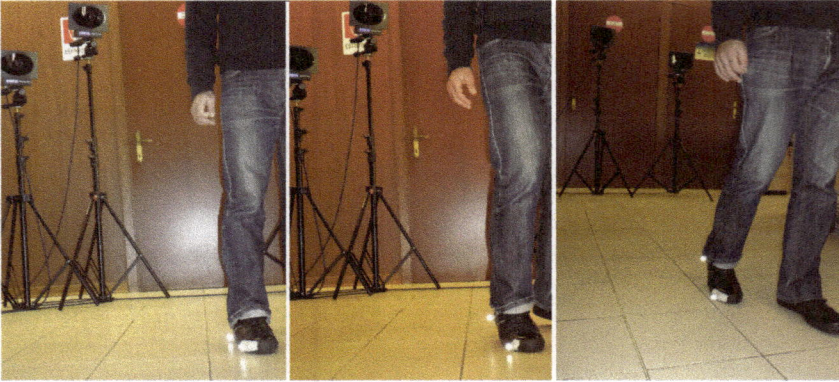

Figure 3. Subject performing the trials within the motion capture (BTS system) workspace. The two markers, attached to the right FootMoov shoe at the tip and heel area, are tracked by the BTS system during the walking activity.

The purpose of this experiment was to test FootMoov and our gait phase detection algorithm in terms of duration of the single gait phases and time delay of their onset. We evaluated the mean and the standard deviation of the durations of each gait phase (HS, ST, HO, SW) and compared the values obtained with the reference phase duration statistics. In addition, we calculated the error (e_{ij}) expressed as the phase duration difference between our algorithm (FM) and the reference (REF): $e_{ij} = \Delta T_{ij}^{FM} - \Delta T_{ij}^{REF}$, where ΔT is the phase duration and i indicates the gait phase (HS, ST, HO or SW) of the $j - th$ step. The mean and the standard deviation of the errors (e_{ij} sequences) were also evaluated. We performed a statistical analysis on the e_{ij} sequences to verify, for each phase, the null hypothesis that *the error in gait phase duration is a zero mean random variable*. In particular, we performed a *t*-test on each e_{ij} sequence. The significance level was fixed at $\alpha = 0.05$. For each gait phase, the corresponding *p*-value of the different samples have been computed. Calculations were performed by using the function *t*-test included in Matlab® (Mathworks, Massachusetts, MA, USA).

3. Results and Discussion

3.1. Free Walking Experiment

Figure 4 shows a typical output of the gait phase detection algorithm for two subjects performing the evaluation test in the outdoors, and in free walking conditions (first test described in Section 2.3.1). The force and accelerometer signals (F_h, F_f, ax, ay, az) and the detected gait phase signal are reported. The gait phase signal has four levels corresponding to the four gait phases: (1) Heel-strike; (2) Stance; (3) Heel-off; and (4) Swing. We evaluated the number of strides recognised in comparison with the effective number of strides (150 strides, performed by 10 users for four times each, for a total of 6000 steps registered). The algorithm recognized 5925 strides over the total amount of 6000 strides (98.7%). In all the detected steps, we observed the correct sequence of the four gait phases. Despite this being a preliminary and qualitative evaluation (i.e., no comparison with a reference system), it is important to underline that a reliable stride count may be important to estimate the distance travelled, as demonstrated by the study of Truong et al. [67]. Note that, even if we performed an off-line processing, our algorithm can be suitable for real-time implementation (i.e., mandatory for applications such as the drop foot control). To this aim, we estimated the computational load as the mean time spent by the algorithm (both pre-processing and gait phase detection) to estimate the current state. The value obtained was 0.9 ms and is compatible with real-time implementation.

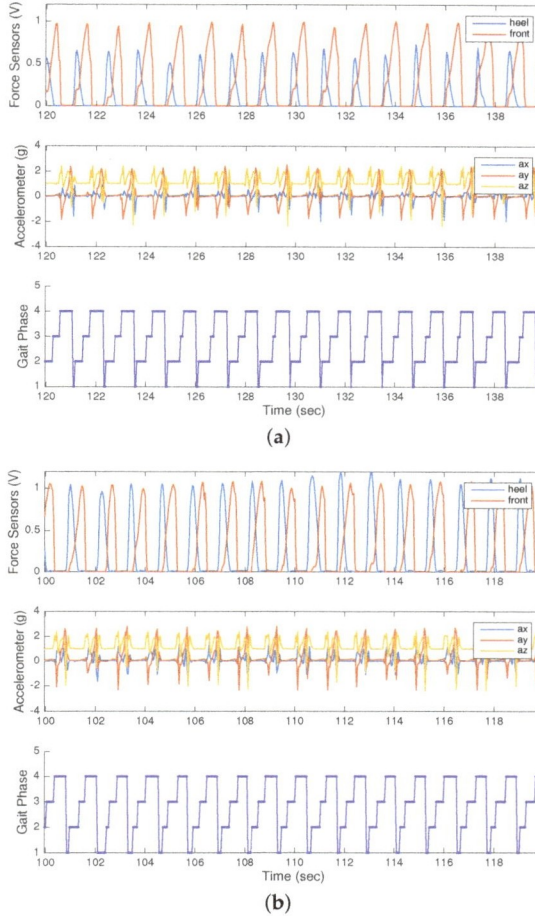

(a)

(b)

Figure 4. Typical output of the gait phase detection algorithm for two different subjects. Each subfigure - (**a**) and (**b**) - reports the sensor signal and the gait phase detection for one subject. In the top graphs, the force value F_h and F_f are reported. The middle figures report the calibrated accelerometer components (ax, ay, az). The bottom traces are related to the detected gait phases.

3.2. Motion Capture Experiment

Figure 5 shows a typical comparison between the FootMoov gait phase detection and the reference signal obtained by the optical motion capture system (second test described in Section 2.3.2).

For a first quantitative evaluation, we calculated the average time delay that affected the onset of the gait phase detection. The average delay among the four phases was 44.7 ms. This delay was comparable with the results of previous studies such as [43]. Note that the low sample frequency (25 Hz) of Footmoov introduces a time resolution of 40 ms, comparable with the error committed.

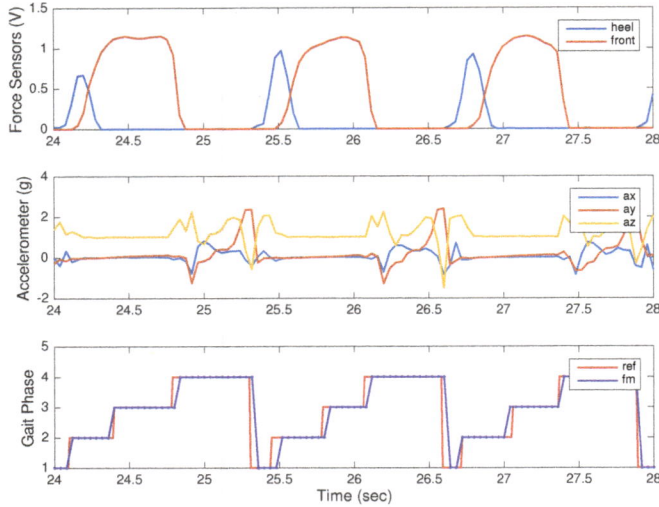

Figure 5. Comparison of our gait phase detection algorithm and the reference gait phase signal. In the top graph, the force value F_h and F_f are reported. The middle figure reports the calibrated accelerometer components (ax, ay, az). In the bottom figure, the reference gait phase signal is reported in red while the signal obtained by FootMoov elaborated with our detection algorithm is reported in blue.

In addition, Table 2 summarises the performance of the sensing shoe and our gait detection algorithm in terms of duration of the single gait phases detected (for each phase, we extracted the mean and the standard deviations of REF and FM and the related error as defined in Section 2.3.2). For all the four phases, the *t*-test performed on the e_{ij} sequences indicated that the null hypothesis (i.e., *the error sequence distribution has zero mean*) cannot be rejected. The corresponding *p*-values are reported in Table 2. The mean values of the detected and the reference phase durations are comparable. The minimum error (mean + standard deviation) is in the HS phase (0.036 s) while the maximum is in the HO phase (0.11 s). These errors, always lower than 3 times the time resolution, may be due to the fixed system of thresholds (defined in Section 2.2.2) that could not be tailored to the specific characteristics of all the subjects.

Table 2. Mean and standard deviation of the duration of the gait phases (REF and FM), mean and standard deviation of errors (e_{ij}) and *p*-values of the statistical evaluation.

		Mean [s]	Std [s]	*p*
HS	REF	0.125	0.013	
	FM	0.117	0.03	
	e	0.008	0.028	0.3
ST	REF	0.350	0.057	
	FM	0.310	0.098	
	e	0.039	0.065	0.07
HO	REF	0.293	0.061	
	FM	0.249	0.099	
	e	0.044	0.07	0.12
SW	REF	0.535	0.043	
	FM	0.514	0.036	
	e	0.021	0.050	0.25

3.3. Study Limitations

Despite the promising performance, several limitations of our work should be mentioned. First, we tested our algorithm for level walking conditions only, without considering other possible typical situations such as stair ascending/descending or walking on irregular ground (i.e., in stair ascending/descending the first contact after the swing phase is not likely to be on the heel). The second limitation is that we tested the system and the algorithm with healthy subjects with normal walking styles (pathological gaits can have different sensor signatures, difficult to be detected by the algorithm). However, our study was conceived as a preliminary evaluation of the commercial wearable technology and as a test for our algorithms to be used in daily life conditions for biomedical applications, but a more intensive testing phase with different walking conditions would be needed. We expect that both issues could be solved by adding new possible transitions between the machine states. Another possible solution would be to study and develop a system of adaptive thresholds that can be tailored to the particular subject's physical characteristic walking style or pathology. The last limitation is related to the reference signal obtained with the optical system. A validation against a force platform would be more solid and will be considered in future works.

4. Conclusions

In this paper, we have reported the assessment of a consumer smart sensing shoe for the detection of gait phases during walking activity. We developed a gait phase detection algorithm that fuses data of inertial and force sensors built-in the FootMoov smart shoe. We tested the prototype and the detection algorithm in free level walking conditions on ten healthy subjects and in a laboratory setting in comparison with an optical motion capture system. The combination of force sensors and accelerometers provided a reliable detection of the gait phases and made it possible to discriminate walking activity from load shifting in static tasks. As a preliminary and qualitative result, the algorithm, tested in free walking conditions, recognised 5925 strides over a total amount of 6000 strides (98.7%). To obtain a quantitative evaluation of the performance of the wearable system and the detection technique, we compared the gait phase signal with a reference signal obtained by an optical motion capture instrument. First of all, we extracted the time difference between our gait phase signal and the reference. We achieved a mean delay of 44.7 ms, comparable to previous study in this field, and mostly limited by the low sampling frequency of the smart shoe (25 Hz). In terms of possible applications, the achieved performance may be compatible with real time FES drop foot compensation (errors below 70 ms are considered sufficiently small [68]). We also assessed the reliability of the temporal duration of the gait phases we detected. The mean durations of the single gait phases are comparable.

In conclusion, we have demonstrated the possibility to use a consumer and low cost wearable device for the estimation of temporal parameters of gait. The demonstrated performance and the characteristics of the prototype—that is a conventional shoe that can be worn without any discomfort for the user—may boost the application of such technology in many fields. For example, it may be possible to monitor or prevent specific gait conditions or to manage and coach the recovery from a physical or neurological pathology. Another possible application may be to use the smart shoe as a feedback device to train healthy subjects or athletes in the optimisation of their walking/running behaviour. Future works will be devoted to extensive tests on a wider number of subjects (both normal and pathological walkers) with different walking conditions. The possibility to develop a system of subject-specific or pathology-specific thresholds will also be considered. We will also study the detection of spatial gait parameters to complete the set of relevant parameters that can be extracted by the wearable system.

Acknowledgments: The authors thank Luigi Campigli, head of *FootMoov*, for his kind and competent assistance.

Author Contributions: N.C., A.T. performed the data acquisition; N.C., F.L., A.T. reviewed the methods and developed the gait detection algorithm; N.C., A.T. elaborated on the data; N.C., A.T. drafted the manuscript; N.C., F.L., A.T. revised and finalized the manuscript.

Conflicts of Interest: The authors declare no conflict of interest.

Appendix A. Temporal and Spatial Gait Parameters

According to the definition from Pappas et al. [43], the normal gait cycle can be divided in four consecutive periods:

- *Heel-strike*: starts with the initial contact of the heel and ends when the entire foot is on the ground;
- *Stance*: the entire foot is in contact with the ground;
- *Heel-off*: the frontal part of the foot touches the ground while the heel is above the ground;
- *Swing*: the foot is not in contact with the ground and moves forward.

The *gait events* are defined as the transitions from one phase to the next. The *walking cycle* is the period from *heel strike* to *heel strike* of the same foot [69].

The main *temporal* gait parameters are: (i) the *stride period*: the time from two consecutive ground contacts of the same foot; (ii) the *step period*: the time from two consecutive ground contacts of different feet; (iii) the duration of the gait phases (Heel-strike, Stance, Heel-off, Swing), normally expressed in percentages to facilitate the comparison between subjects; and (iv) the *cadence*: the number of strides in a minute. The main *spatial* gait parameters are: (i) the *stride length*: distance between two consecutive ground contacts of the same foot; and (ii) the *step length*: distance between two consecutive ground contacts of the different feet. Note that if the step length and the cadence are known, it is possible to determine the step velocity.

References

1. Amft, O.; Habetha, J. Smart medical textiles for monitoring patients with heart conditions. In *Smart Textiles for Medicine and Healthcare Materials, Systems and Applications*; Woodhead Publishing Limited: Cambridge, UK, 2007; pp. 275–301.
2. Valenza, G.; Nardelli, M.; Lanata, A.; Gentili, C.; Bertschy, G.; Paradiso, R.; Scilingo, E.P. Wearable monitoring for mood recognition in bipolar disorder based on history-dependent long-term heart rate variability analysis. *IEEE J. Biomed. Health Inform.* **2014**, *18*, 1625–1635.
3. Carbonaro, N.; Anania, G.; Mura, G.; Tesconi, M.; Tognetti, A.; Zupone, G.; de Rossi, D. Wearable biomonitoring system for stress management: A preliminary study on robust ECG signal processing. In Proceddings of the 2011 IEEE International Symposium on a World of Wireless, Mobile and Multimedia Networks (WoWMoM), Lucca, Italy, 20–24 June 2011.
4. Patel, S.; Park, H.; Bonato, P.; Chan, L.; Rodgers, M. A review of wearable sensors and systems with application in rehabilitation. *J. Neuroeng. Rehabil.* **2012**, *9*, 32.
5. Tognetti, A.; Lorussi, F.; Dalle Mura, G.; Carbonaro, N.; Pacelli, M.; Paradiso, R.; de Rossi, D. New generation of wearable goniometers for motion capture systems. *J. Neuroeng. Rehabil.* **2014**, *11*, doi:10.1186/1743-0003-11-56.
6. Dalle Mura, G.; Lorussi, F.; Tognetti, A.; Anania, G.; Carbonaro, N.; Pacelli, M.; Paradiso, R.; de Rossi, D. Piezoresistive goniometer network for sensing gloves. In Proceddings of the XIII Mediterranean Conference on Medical and Biological Engineering and Computing 2013, Seville, Spain, 25–28 September 2013; Volume 41, pp. 1547–1550.
7. Carbonaro, N.; Dalle Mura, G.; Lorussi, F.; Paradiso, R.; De Rossi, D.; Tognetti, A. Exploiting wearable goniometer technology for motion sensing gloves. *IEEE J. Biomed. Health Inform.* **2014**, *18*, 1788–1795.
8. Wagner, F.; Basran, J.; Dal Bello-Haas, V. A review of monitoring technology for use with older adults. *J. Geriatr. Phys. Ther.* **2012**, *35*, 28–34.
9. Baig, M.M.; Gholamhosseini, H.; Connolly, M.J. A comprehensive survey of wearable and wireless ECG monitoring systems for older adults. *Med. Biol. Eng. Comput.* **2013**, *51*, 485–495.
10. Bonfiglio, A.; Carbonaro, N.; Chuzel, C.; Curone, D.; Dudnik, G.; Germagnoli, F.; Hatherall, D.; Koller, J.; Lanier, T.; Loriga, G.; et al. Managing catastrophic events by wearable mobile systems. In Proceedings of the Mobile Response: First International Workshop on Mobile Information Technology for Emergency Response, Mobile Response 2007, Sankt Augustin, Germany, 22–23 February 2007; Revised Selected Papers; Lecture Notes in Computer Science; Springer: Berlin/Heidelberg, Germany, 2007; pp. 95–105.

11. Klann, M. Tactical Navigation Support for Firefighters: The LifeNet Ad-Hoc Sensor-Network and Wearable System. In Proceedings of the Mobile Response: Second International Workshop on Mobile Information Technology for Emergency Response, MobileResponse 2008, Bonn, Germany, 29–30 May 2008; Revised Selected Papers; Lecture Notes in Computer Science; Springer: Berlin/Heidelberg, Germany, 2009; pp. 41–56.

12. Tesconi, M.; Tognetti, A.; Scilingo, E.; Zupone, G.; Carbonaro, N.; De Rossi, D.; Castellini, E.; Marella, M. Wearable sensorized system for analyzing the lower limb movement during rowing activity. In Proceedings of the IEEE International Symposium on Industrial Electronics 2007, Vigo, Spain, 4–7 June 2007; pp. 2793–2796.

13. Chambers, R.; Gabbett, T.J.; Cole, M.H.; Beard, A. The use of wearable microsensors to quantify sport-specific movements. *Sports Med.* **2015**, *45*, 1065–1081.

14. Redfern, J. The Evolution of Physical Activity and Recommendations for Patients with Coronary Heart Disease. *Heart Lung Circ.* **2016**, *25*, 759–764.

15. Arberet, S.; Lemay, M.; Renevey, P.; Sola, J.; Grossenbacher, O.; Andries, D.; Sartori, C.; Bertschi, M. Photoplethysmography-based ambulatory heartbeat monitoring embedded into a dedicated bracelet. *Comput. Cardiol.* **2013**, *40*, 935–938.

16. Hataji, O.; Kobayashi, T.; Gabazza, E. Smart watch for monitoring physical activity in patients with chronic obstructive pulmonary disease. *Respir. Investig.* **2016**, *54*, 294–295.

17. Wile, D.; Ranawaya, R.; Kiss, Z. Smart watch accelerometry for analysis and diagnosis of tremor. *J. Neurosci. Methods* **2014**, *230*, 1–4.

18. Wijaya, R.; Setijadi, A.; Mengko, T.; Mengko, R. Heart rate data collecting using smart watch. In Proceedings of the 2014 IEEE 4th International Conference on System Engineering and Technology (ICSET 2014), Bandung, Indonesia, 24–25 November 2014.

19. Militara, A.; Frandes, M.; Lungeanu, D. Smart wristbands as inexpensive and reliable non-dedicated solution for self-managing type 2 diabetes. In Proceedings of the 2015 E-Health and Bioengineering Conference (EHB 2015), Iasi, Romania, 19–21 November 2015.

20. Perry, J.; Burnfield, J.M. *Gait Analysis: Normal and Pathological Function*; Slack: Thorofare, NJ, USA, 1992.

21. Lopez-Meyer, P.; Fulk, G.D.; Sazonov, E.S. Automatic detection of temporal gait parameters in poststroke individuals. *IEEE Trans. Inf. Technol. Biomed.* **2011**, *15*, 594–601.

22. Balasubramanian, C.K.; Neptune, R.R.; Kautz, S.A. Variability in spatiotemporal step characteristics and its relationship to walking performance post-stroke. *Gait Posture* **2009**, *29*, 408–414.

23. Kotiadis, D.; Hermens, H.; Veltink, P. Inertial Gait Phase Detection for control of a drop foot stimulator: Inertial sensing for gait phase detection. *Med. Eng. Phys.* **2010**, *32*, 287–297.

24. Blaya, J.A.; Herr, H. Adaptive control of a variable-impedance ankle-foot orthosis to assist drop-foot gait. *IEEE Trans. Neural Syst. Rehabil. Eng.* **2004**, *12*, 24–31.

25. Lyons, G.M.; Sinkjær, T.; Burridge, J.H.; Wilcox, D.J. A review of portable FES-based neural orthoses for the correction of drop foot. *IEEE Trans. Neural Syst. Rehabil. Eng.* **2002**, *10*, 260–279.

26. Meyer-Heim, A.; Ammann-Reiffer, C.; Schmartz, A.; Schaefer, J.; Sennhauser, F.H.; Heinen, F.; Knecht, B.; Dabrowski, E.; Borggraefe, I. Improvement of walking abilities after robotic-assisted locomotion training in children with cerebral palsy. *Arch. Dis. Child.* **2009**, *94*, 615–620.

27. Thaler-Kall, K.; Peters, A.; Thorand, B.; Grill, E.; Autenrieth, C.S.; Horsch, A.; Meisinger, C. Description of spatio-temporal gait parameters in elderly people and their association with history of falls: results of the population-based cross-sectional KORA-Age study. *BMC Geriatr.* **2015**, *15*, 1.

28. Nguyen, L.V.; La, H.M.; Sanchez, J.; Vu, T. A Smart Shoe for building a real-time 3D map. *Autom. Constr.* **2016**, *71*, 2–12.

29. Foxlin, E. Pedestrian tracking with shoe-mounted inertial sensors. *IEEE Comput. Graph. Appl.* **2005**, *25*, 38–46.

30. Jiménez, A.R.; Seco, F.; Prieto, J.C.; Guevara, J. Indoor pedestrian navigation using an INS/EKF framework for yaw drift reduction and a foot-mounted IMU. In Proceedings of the 2010 7th IEEE Workshop on Positioning Navigation and Communication (WPNC), Dresden, Germany, 11–12 March 2010; pp: 135–143.

31. Van Nguyen, L.; La, H.M. A human foot motion localization algorithm using IMU. In Proceedings of the 2016 American Control Conference (ACC), Boston, MA, USA, 6–8 July 2016; pp. 4379–4384.

32. Rueterbories, J.; Spaich, E.G.; Larsen, B.; Andersen, O.K. Methods for gait event detection and analysis in ambulatory systems. *Med. Eng. Phys.* **2010**, *32*, 545–552.

33. Taborri, J.; Palermo, E.; Rossi, S.; Cappa, P. Gait partitioning methods: A systematic review. *Sensors* **2016**, *16*, 66.

34. Mansfield, A.; Lyons, G.M. The use of accelerometry to detect heel contact events for use as a sensor in FES assisted walking. *Med. Eng. Phys.* **2003**, *25*, 879–885.

35. Aminian, K.; Najafi, B.; Büla, C.; Leyvraz, P.F.; Robert, P. Spatio-temporal parameters of gait measured by an ambulatory system using miniature gyroscopes. *J. Biomech.* **2002**, *35*, 689–699.

36. Williamson, R.; Andrews, B.J. Gait event detection for FES using accelerometers and supervised machine learning. *IEEE Trans. Rehabil. Eng.* **2000**, *8*, 312–319.

37. Mannini, A.; Sabatini, A.M. Gait phase detection and discrimination between walking–jogging activities using hidden Markov models applied to foot motion data from a gyroscope. *Gait Posture* **2012**, *36*, 657–661.

38. Taborri, J.; Rossi, S.; Palermo, E.; Patanè, F.; Cappa, P. A novel HMM distributed classifier for the detection of gait phases by means of a wearable inertial sensor network. *Sensors* **2014**, *14*, 16212–16234.

39. Van Nguyen, L.; La, H.M. Real-Time Human Foot Motion Localization Algorithm With Dynamic Speed. *IEEE Trans. Hum.-Mach. Syst.* **2016**.

40. Smith, B.T.; Coiro, D.J.; Finson, R.; Betz, R.R.; McCarthy, J. Evaluation of force-sensing resistors for gait event detection to trigger electrical stimulation to improve walking in the child with cerebral palsy. *IEEE Trans. Neural Syst. Rehabil. Eng.* **2002**, *10*, 22–29.

41. Edgar, S.R.; Swyka, T.; Fulk, G.; Sazonov, E.S. Wearable shoe-based device for rehabilitation of stroke patients. In Proceedings of the 2010 Annual International Conference of the IEEE Engineering in Medicine and Biology, 2010, Buenos Aires, Argentina, 31 August–4 September 2010; pp. 3772–3775.

42. Kong, K.; Tomizuka, M. Smooth and continuous human gait phase detection based on foot pressure patterns. In Proceedings of the IEEE International Conference on Robotics and Automation (ICRA 2008), Pasadena, CA, USA; 19–23 May 2008; pp. 3678–3683.

43. Pappas, I.P.; Popovic, M.R.; Keller, T.; Dietz, V.; Morari, M. A reliable gait phase detection system. *IEEE Trans. Neural Syst. Rehabil. Eng.* **2001**, *9*, 113–125.

44. Senanayake, C.M.; Senanayake, S.A. Computational intelligent gait-phase detection system to identify pathological gait. *IEEE Trans. Inf. Technol. Biomed.* **2010**, *14*, 1173–1179.

45. Hegde, N.; Sazonov, E. SmartStep: A Fully Integrated, Low-Power Insole Monitor. *Electronics* **2014**, *3*, 381–397.

46. Tareco, J.M.; Miller, N.H.; MacWilliams, B.A.; Michelson, J.D. Defining flatfoot. *Foot Ankle Int.* **1999**, *20*, 456–460.

47. Mueller, M.J.; Hastings, M.; Commean, P.K.; Smith, K.E.; Pilgram, T.K.; Robertson, D.; Johnson, J. Forefoot structural predictors of plantar pressures during walking in people with diabetes and peripheral neuropathy. *J. Biomech.* **2003**, *36*, 1009–1017.

48. Sazonov, E.S.; Bumpus, T.; Zeigler, S.; Marocco, S. Classification of plantar pressure and heel acceleration patterns using neural networks. In Proceedings of the 2005 IEEE International Joint Conference on Neural Networks, Montreal, QC, Canada, 31 July–4 August 2005; Volume 5, pp. 3007–3010.

49. Sazonov, E.S.; Fulk, G.; Hill, J.; Schutz, Y.; Browning, R. Monitoring of posture allocations and activities by a shoe-based wearable sensor. *IEEE Trans. Biomed. Eng.* **2011**, *58*, 983–990.

50. Tang, W.; Sazonov, E.S. Highly accurate recognition of human postures and activities through classification with rejection. *IEEE J. Biomed. Health Inform.* **2014**, *18*, 309–315.

51. Fulk, G.D.; Sazonov, E. Using sensors to measure activity in people with stroke. *Top. Stroke Rehabil.* **2011**, *18*, 746–757.

52. Zhang, T.; Fulk, G.D.; Tang, W.; Sazonov, E.S. Using decision trees to measure activities in people with stroke. In Proceedings of the 2013 35th Annual International Conference of the IEEE Engineering in Medicine and Biology Society (EMBC), Osaka, Japan, 3–7 July 2013; pp. 6337–6340.

53. Zhang, T.; Lu, J.; Uswatte, G.; Taub, E.; Sazonov, E.S. Measuring gait symmetry in children with cerebral palsy using the SmartShoe. In Proceedings of the 2014 IEEE Healthcare Innovation Conference (HIC), Seattle, WA, USA, 8–10 October 2014; pp. 48–51.

54. Bae, J.; Kong, K.; Byl, N.; Tomizuka, M. A mobile gait monitoring system for abnormal gait diagnosis and rehabilitation: A pilot study for Parkinson disease patients. *J. Biomech. Eng.* **2011**, *133*, 041005.

55. F-Scan System. Available online: https://www.tekscan.com/products-solutions/systems/f-scan-system (accessed on 10 August 2016).

56. Pedar System. Available online: http://novel.de/novelcontent/pedar (accessed on 10 August 2016).

57. Lemaire, E.D.; Biswas, A.; Kofman, J. Plantar pressure parameters for dynamic gait stability analysis. In Proceedings of the 28th Annual International Conference of the IEEE Engineering in Medicine and Biology Society, New York, NY, USA, 30 August–3 September 2006; pp. 4465–4468.

58. Catalfamo, P.; Moser, D.; Ghoussayni, S.; Ewins, D. Detection of gait events using an F-Scan in-shoe pressure measurement system. *Gait Posture* **2008**, *28*, 420–426.

59. El Kati, R.; Forrester, S.; Fleming, P. Evaluation of pressure insoles during running. *Procedia Eng.* **2010**, *2*, 3053–3058.

60. FootMoov. Available online: http://www.footmoov.com (accessed on 5 August 2016).

61. Tognetti, A.; Lorussi, F.; Carbonaro, N.; de Rossi, D. Wearable Goniometer and Accelerometer Sensory Fusion for Knee Joint Angle Measurement in Daily Life. *Sensors* **2015**, *15*, 28435.

62. Sciavicco, L.; Siciliano, B. *Modelling and Control of Robot Manipulators*; Springer-Verlag: London, UK, 2000.

63. Karantonis, D.M.; Narayanan, M.R.; Mathie, M.; Lovell, N.H.; Celler, B.G. Implementation of a real-time human movement classifier using a triaxial accelerometer for ambulatory monitoring. *IEEE Trans. Inf. Technol. Biomed.* **2006**, *10*, 156–167.

64. Luinge, H.J.; Veltink, P.H. Inclination measurement of human movement using a 3-D accelerometer with autocalibration. *IEEE Trans. Neural Syst. Rehabil. Eng.* **2004**, *12*, 112–121.

65. Sabatini, A.M.; Martelloni, C.; Scapellato, S.; Cavallo, F. Assessment of walking features from foot inertial sensing. *IEEE Trans. Biomed. Eng.* **2005**, *52*, 486–494.

66. BTS SMART-DX. Available online: www.btsbioengineering.com/products/kinematics/bts-smart-dx (accessed on 10 August 2016).

67. Truong, P.H.; Lee, J.; Kwon, A.R.; Jeong, G.M. Stride Counting in Human Walking and Walking Distance Estimation Using Insole Sensors. *Sensors* **2016**, *16*, 823.

68. Pappas, I.P.I.; Keller, T.; Mangold, S.; Popovic, M.R.; Dietz, V.; Morari, M. A reliable gyroscope-based gait-phase detection sensor embedded in a shoe insole. *IEEE Sens. J.* **2004**, *4*, 268–274.

69. Winter, D.A. *Biomechanics and Motor Control of Human Gait: Normal, Elderly and Pathological*; University of Waterloo Press: Waterloo, ON, Canada, 1991.

electronics

MDPI

Article

A Novel 12-Lead ECG T-Shirt with Active Electrodes

Anna Boehm *, Xinchi Yu, Wilko Neu, Steffen Leonhardt and Daniel Teichmann

Philips Chair for Medical Information Technology, RWTH Aachen University, Pauwelsstr. 20,
52074 Aachen, Germany; yu@hia.rwth-aachen.de (X.Y.); wilko.neu@rwth-aachen.de (W.N.);
leonhardt@hia.rwth-aachen.de (S.L.); teichmann@hia.rwth-aachen.de (D.T.)
* Correspondence: boehm@hia.rwth-aachen.de; Tel.: +49-241-80-23224

Academic Editors: Enzo Pasquale Scilingo and Gaetano Valenza
Received: 13 August 2016; Accepted: 3 November 2016; Published: 8 November 2016

Abstract: We developed an ECG T-shirt with a portable recorder for unobtrusive and long-term multichannel ECG monitoring with active electrodes. A major drawback of conventional 12-lead ECGs is the use of adhesive gel electrodes, which are uncomfortable during long-term application and may even cause skin irritations and allergic reactions. Therefore, we integrated comfortable patches of conductive textile into the ECG T-shirt in order to replace the adhesive gel electrodes. In order to prevent signal deterioration, as reported for other textile ECG systems, we attached active circuits on the outside of the T-shirt to further improve the signal quality of the dry electrodes. Finally, we validated the ECG T-shirt against a commercial Holter ECG with healthy volunteers during phases of lying down, sitting, and walking. The 12-lead ECG was successfully recorded with a resulting mean relative error of the RR intervals of 0.96% and mean coverage of 96.6%. Furthermore, the ECG waves of the 12 leads were analyzed separately and showed high accordance. The P-wave had a correlation of 0.703 for walking subjects, while the T-wave demonstrated lower correlations for all three scenarios (lying: 0.817, sitting: 0.710, walking: 0.403). The other correlations for the P, Q, R, and S-waves were all higher than 0.9. This work demonstrates that our ECG T-shirt is suitable for 12-lead ECG recordings while providing a higher level of comfort compared with a commercial Holter ECG.

Keywords: ECG; unobtrusive measurement; wearables

1. Introduction

Unobtrusive sensing of vital signs, such as cardiac activity and respiration, has been increasingly applied in the past decade. The aging of our society has resulted in an increasing demand on medical staff, which cannot always be met. As a result, an increasing number of technical solutions, the so-called personal healthcare systems, are being developed. They aim at enabling sick and elderly patients to stay at home for a longer period , rather than facing prolonged hospital stays. When staying at home, patients generally benefit from increased comfort, which may accelerate their recovery. In turn, costs for the healthcare system will be reduced by shortening the stay in hospital. This is the main rationale for developing long-term monitoring solutions for the home environment.

One of the established long-term cardiac monitoring devices is the Holter. This is a portable electrocardiography (ECG) device with up to 12 leads for long-time application. These ECG recorders are often used to diagnose cardiac conditions over the duration of several days. For this, patients wear the device while continuing their daily routine.

Commercial Holter devices consist of a portable ECG recorder with adhesive electrodes. However, these electrodes have one major problem: the gel that ensures good conductivity can lead to skin allergies. Moreover, the longer the gel is applied, the greater the possibility that more problems arise. Signal quality is deteriorated if the gel dries up, which is highly probable during long-term monitoring. In addition, in some cases (e.g., if patients are sweating), the electrodes detach themselves, requiring reapplication. If this occurs, the patient may not reattach them in the correct place. In order to address these problems and to improve patient comfort, we developed a 12-lead ECG T-shirt with active electrodes and a portable ECG recorder.

Various textile ECG T-shirts have been investigated in the last decade. Whereas some systems were developed mainly for research purposes only [1–4], some shirts are commercially available: CardioLeaf (Clearbridge VitalSigns, Singapore) and hWear (HealthWatch Technologies, Kfar Saba, Israel) [5,6]. The latter type is a shirt that is compatible with a 12-lead ECG recorder.

Furthermore, textile electrodes have been used for ECG acquisition in research [7–12]. It was discovered that certain dry textile electrodes can achieve comparable results to conventional electrodes in resting electrocardiography but have lessened signal quality during movement [13]. However, most dry electrodes have very high contact impedances to the skin. To improve signal quality, active circuits with buffers are used on the electrodes, forming an active electrode. An active electrode is comprised of a sensing part (conductive textile), front-end electronics and shielding [14]. The input impedance is very high and the output impedance is low. The output can drive a long screened cable, reducing the power line interference and cable movement.

The principle of active electrodes has been widely applied for capacitive ECGs and can be found in multiple systems, such as ECG chairs [15], car seats [16], beds [17–20], toilet seats [21], a bathtub [22], and clothing [8,9].

Because none of the commercially available ECG T-shirts were compatible with our design for active electrodes, we developed a custom T-shirt with 10 electrodes. The electrode material is an electrically conductive fabric sewn into the inside of a sports T-shirt. Our T-shirt does not require electrode gel. Furthermore, it can be worn as an undergarment. The T-shirt is attached to a (relatively small) hand-held ECG recording device.

This article is structured as follows: first, we describe the design of the measurement system and the experimental design. This is followed by the results, the experimental data, a discussion of the results, and, lastly, conclusions are presented.

2. Materials and Methods

The portable 12-lead ECG measurement system consists of a T-shirt, active electrodes and an ECG recorder. The active electrodes of the capacitive measurement system record the potentials on the body's surface. The analogue signals from the active electrodes are digitalized in the ECG recorder, which also calculates the 12 ECG leads. The signals of the leads are then processed by a microcontroller and stored on an SD card in the ECG recorder. In terms of practicability, it would have been an option to send the data wirelessly. The two main reasons why no wireless module was used are medical data protection and power consumption. A wireless module would drastically reduce battery life, especially when sending data of 12 channels at a high sample rate.

2.1. 12-Lead ECG

A 12-lead ECG requires 10 electrodes on the patient's limbs and chest: 10 physical channels are recorded (3 limb leads, 6 thoracic leads, 1 RL lead), as shown in Figure 1.

In Holter ECGs, the electrodes are placed only on the chest. Figure 1 shows that, additional to the 3-lead setup for Einthoven and Goldberger, 6 Wilson leads are included. Einthoven and Goldberger leads are calculated from three electrodes forming a triangle, namely the left leg (LL), left arm (LA) and right arm (RA), also called limb electrodes. The fourth electrode applied is the neutral or right leg electrode. The bipolar Einthoven leads are calculated as follows:

$$I = LA - RA, \tag{1}$$

$$II = LL - RA, \tag{2}$$

$$III = LL - LA, \tag{3}$$

so that $III = II - I$.

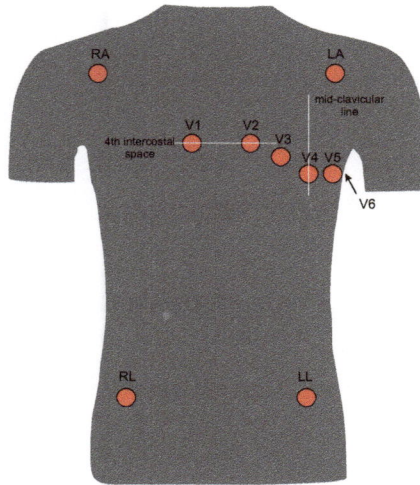

Figure 1. Positions of the 12-lead electrocardiograph (ECG).

The regular shape of the Einthoven II lead is shown in Figure 2. The ECG can be divided into different waves that are referred to as P to T. They reflect the stages of the electrical excitation propagation of the heart. The P-wave is the depolarization of the atria. The QRS complex (QRS waves) reflects the depolarization of the ventricles and the T-wave is their repolarization.

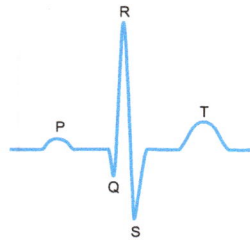

Figure 2. ECG morphology.

The unipolar Goldberger leads are also recorded for the 12-lead ECG. The Goldberger leads, denoted aVR, aVL and aVF, are augmented leads. These leads are formed using an augmented reference electrode, which is a combination of the two other limb electrodes. This is calculated as follows:

$$aVR = RA - \frac{1}{2}(LA + LL), \tag{4}$$

$$aVL = LA - \frac{1}{2}(RA + LL), \tag{5}$$

$$aVF = LL - \frac{1}{2}(RA + LA). \tag{6}$$

The Wilson leads are placed around the left side of the rib cage. Wilson leads are used to detect local irregularities of electric cardiac function, such as infarctions. The leads are labeled V1 to V6.

Their reference is called the Wilson central terminal (WCT). It is a reference potential that is formed by connecting all three limb electrodes to 5 k ohm resistors resulting in their average [23].

2.2. Measurement System

2.2.1. T-Shirt

The T-shirt is a commercially available breathable sports T-shirt (Nike Legend Pro DRI-FIT, Beaverton, OR, USA) . Ten textile patches made of electrically conductive fabric (Shieldex Med-tex P180, Statex, Bremen, Germany) serve as electrodes. The patches (4 cm × 4 cm) are sewn into the interior of the T-shirt (see Figure 3). This fabric is silver plated with 99% silver and has been used as electrodes by two other groups [24,25]. While other conductive textile materials exist, silver coating was selected. It was found that silver electrodes are advantageous even at recording low frequencies [26].

Figure 3. Active electrode. (**a**) active circuit printed circuit board (PCB); and (**b**) textile electrode on the interior of the T-shirt.

The locations are chosen according to the common 12-lead ECG setup (see Figure 1). The driven right leg (DRL) electrode has a larger area to ensure good contact (30 cm × 5 cm). Each electrode has a snap fastener connection, where the amplifier boards (or in the case of the DRL electrode, the cable) that lead to the ECG recorder are fastened.

The T-shirt needs to fit relatively tightly, since signal quality improves with contact pressure of the electrodes. Therefore, we added a few Velcro straps to "tighten" the T-shirts and keep the electronics in place.

2.2.2. Electrodes

The textile electrodes are followed by active circuits. The aim was to improve the signal quality of a dry or an optionally capacitive setup.

The main principle of the active electrodes is shown in Figure 4. As mentioned, the textile patches serve as electrodes and they are connected to the outside of the T-shirt by snap fasteners. An active circuit PCB is placed on the snap fasteners from the exterior.

More precisely, an impedance converter with OPA129U (Texas Instruments, Dallas, TX, USA) on the PCB decouples the ECG signal at the snap fastener from the following electronics. A highly resistive bias resistor (10 G ohm) prevents the build-up of static charge at the input on the impedance converter. Then, the analogue output of the impedance converter is led to the ECG recorder with a shielded six-wire cable. This cable also carries extra wires for the power supplies (see Figure 3).

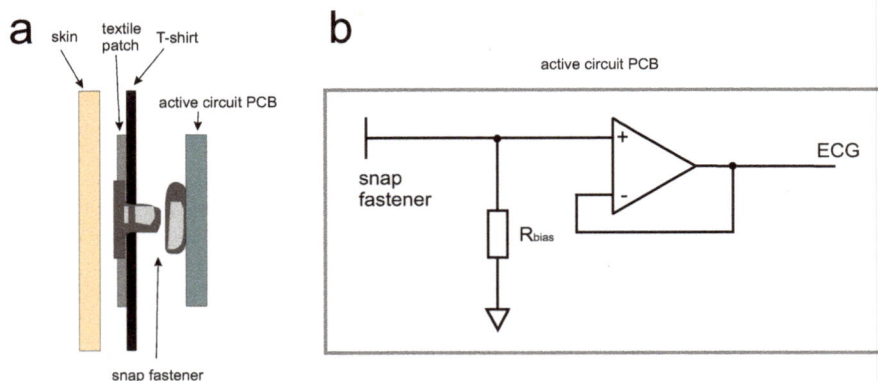

Figure 4. Electrode principle. (**a**) structure of the connection of the skin to the electrodes on the T-shirt; and (**b**) simplified diagram of an active electrode circuit.

The tenth electrode is the so-called DRL electrode, which attenuates common mode signals and produces a better signal-to-noise ratio. The DRL signal is generated in the ECG recorder on an integrated chip and is fed into the electrode patch of the right leg. The T-shirt and the active electrodes are presented in Figure 5.

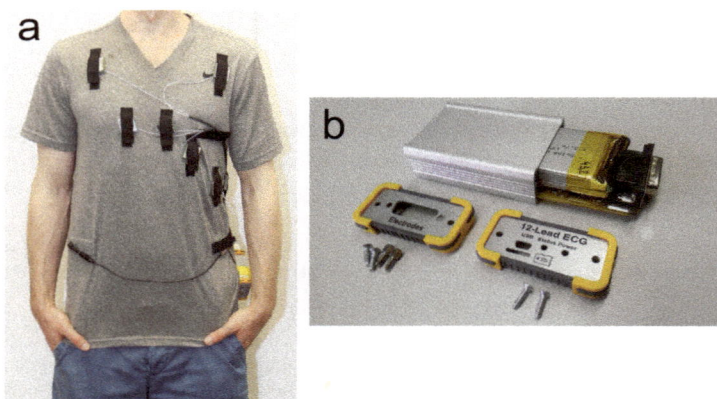

Figure 5. (**a**) ECG T-shirt with electrodes; and (**b**) ECG recorder.

2.2.3. 12-Lead ECG Recorder

The 12-lead ECG recorder is a portable device with the following dimensions: 70 mm × 65 mm × 30 mm (Figure 5) and weighs 180 g.

Figure 6 presents details of the recorder. The design is based on two ADS1298 analogue digital converters (ADC) (Texas Instruments, Dallas, TX, USA). The ADS1298 chip is a designated ECG ADC converter. It has eight differential ADC input channels with 24-bit resolution and a sample rate that can be set from 250 samples per second (SPS) to 32,000. It also has the capability of calculating the 12-lead ECG and a DRL signal.

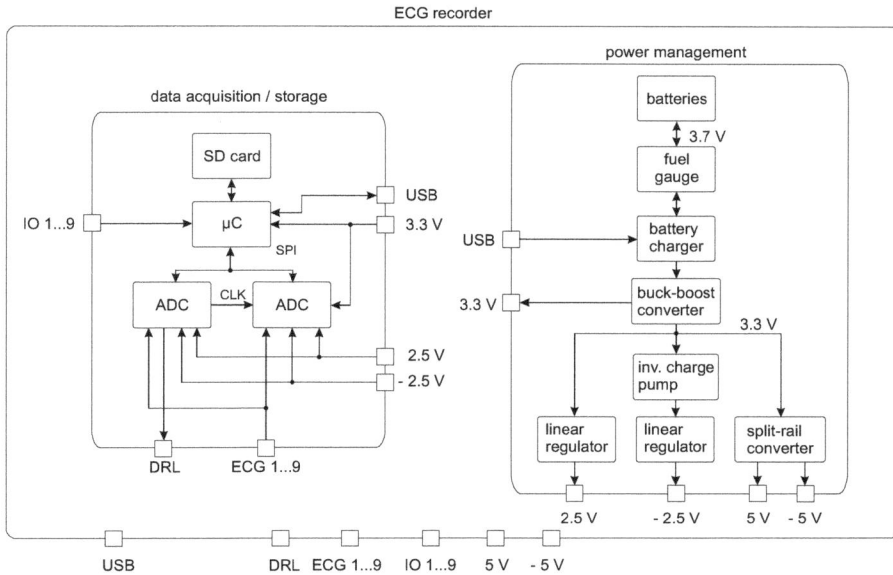

Figure 6. Complete block diagram of the ECG recorder.

As our design consists of two ADS1298, 16 channels are available and a true 12-lead ECG can be recorded. This means that the 12-lead ECG is recorded with 10 electrodes and is not estimated from a smaller number of electrodes. Both chips are connected in parallel to one serial peripheral interface (SPI) port of the microcontroller using separate chip select (CS) lines (see Figure 7). One of the ADS1298 is the master and provides its clock for synchronization and the other is the slave. They are started with a START signal. The ADS1298 signals the microcontroller when data are available and ready to be sent. Since both ADS1298 are synchronized, data from both chips can be retrieved. After retrieving data from the first chip, the CS line is switched and the second one is read out. The sampling rate is set to 500 SPS ; no other sampling rates are used.

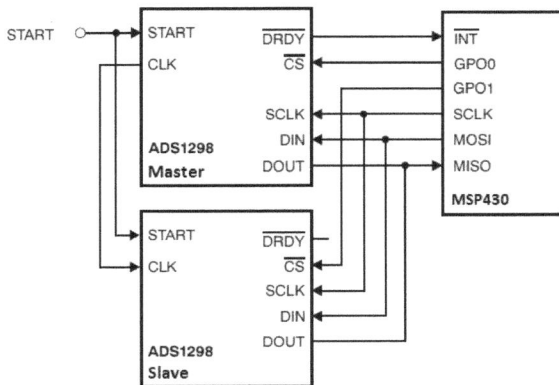

Figure 7. Diagram of the master and slave ADS1298 with the microcontroller.

The ADS1298 has differential inputs that can be enhanced by a gain factor. The bipolar Einthoven leads II and III are recorded on the master ADS1298 by the differences of its respective electrodes (see Equations (1)–(3)). The Wilson Central Terminal (WCT) is generated internally using the three limb leads. In addition, a DRL circuitry is on board the ADS1298 using the WCT signal.

The second ADS1298 is primarily used to generate the Goldberger leads and to record the Einthoven lead I. The augmented leads are created according to Equations (4)–(6). The reference is formed by averaging the other two limb electrode potentials. Because only five of the eight channels are used for Goldberger, the three remaining channels can be used for other sensors.

A microcontroller MSP430F5529 (Texas Instruments, Dallas, TX, USA) was used. This microcontroller has low power consumption, but is fast enough to write on a micro SD with 200,000 bytes/s. Additionally, the MSP430F5529 has an SPI to access the ADS1298 and to store the ECG data onto the microSD, with up to 64 GB storage.

In order to run a portable device, the power is provided by two 3.7 V rechargeable lithium polymer batteries with 1950 mAh and 1400 mAh, respectively. Two different sizes of batteries were chosen to enable a perfect fit into the small casing. Power management had to be devised carefully, since not only the internal components of the ECG recorder require power, but also the active electrodes. Moreover, as the system is intended for long-time recording purposes, battery life is an important issue.

A safety circuit uncouples the battery if the voltage of the cells is below 2.7 V. A fuel gauge BQ27441 (Texas Instruments, Dallas, TX, USA) is used to estimate the remaining battery capacity and the state of charge. The batteries are charged via a mini USB connector on the recorder with an external power source. The loading of the batteries is solved with a BQ24075 (Texas Instruments, Dallas, TX, USA) chip, which can load with up to 1.5 A current and is set to 1.317 A. Charging begins as soon as the USB port is connected.

Several power sources are required. The microcontroller needs a constant voltage supply of 3.3 V, which is provided by a TPS63030 (Texas Instruments, Dallas, TX, USA) buck-boost converter. The two ADS1298 are supplied by a symmetrical power supply of ±2.5 V, which is generated by TPS799 (Texas Instruments, Dallas, TX, USA) and the TPS723 (Texas Instruments, Dallas, TX, USA) negative output linear regulator. It also needs 3.3 V digital supply.

The OPA129U chip of the active electrodes requires a minimum of ±5 V. Other than with conventional wet electrodes, higher baselines can occur; therefore, the higher power supply would be beneficial to avoid saturation of the amplifier through DC offsets. However, only ±5 V was chosen. This improves battery running time and satisfies the concerns about patient safety. This ±5 V power source was implemented by using the TPS65131 (Texas Instruments, Dallas, TX, USA) split-rail converter, which has a positive and negative voltage output. In total, around 260 mW are needed for recording, which leads to an estimated maximum runtime of 47 h.

2.3. Measurement Setup

The design was experimentally tested on three healthy male volunteers. The subjects were asked to wear both the 12-lead ECG T-shirt and the reference Holter ECG BT12 (Corscience, Erlangen, Germany) at the same time. The BT12 has a software that displays the ECG channels and the heart rate in real-time. However, only raw data that can be exported from the software, which does not include the RR (beat-to-beat) or heart rate information. Data collection with the ECG T-shirt hardware is as follows: after a recording is started, the data is stored on the SD card in the recorder. The raw data can then be transferred to a PC for further processing.

Three experimental scenarios were conducted for 5 min each: (i) lying in bed; (ii) sitting; and (iii) walking at 1.49 mph. We hypothesized that whilst lying down, fewer movement artifacts would occur than during the other two scenarios. In addition, heart rate at rest was expected to be lower compared with sitting and walking.

The data were post-processed on a PC with an Intel Core i5-2500 CPU @ 3.30 GHz (Intel Corporation, Santa Clara, CA, USA) and 8 GB RAM. The first step was to filter the obtained

ECG data with a 3rd order Butterworth bandpass filter between 2 and 20 Hz. In order to obtain the RR intervals (beat-to-beat intervals) , a method for beat-to-beat interval identification of multi-channel data based on self-similarity features was applied for the reference and T-shirt data [27]. Using this method, beat-to-beat intervals in any kind of cardiac-related data can be determined as long as a repetitive pattern is detectable. In this way, the RR intervals of the shirt and the reference were calculated.

The data of the two systems had to be synchronized, since the devices could not be started at exactly the same time. The idea was to make use of heart rate variability. Every healthy person has a slight variation in heart rate over time and, thus, a variation in RR intervals. This fact was used to match both systems by aligning intervals of the same length.

For that, a prominent sequence, which appears in the intervals of both the T-shirt ECG and the reference ECG, was manually selected (see the vertical dotted lines in Figure 8).

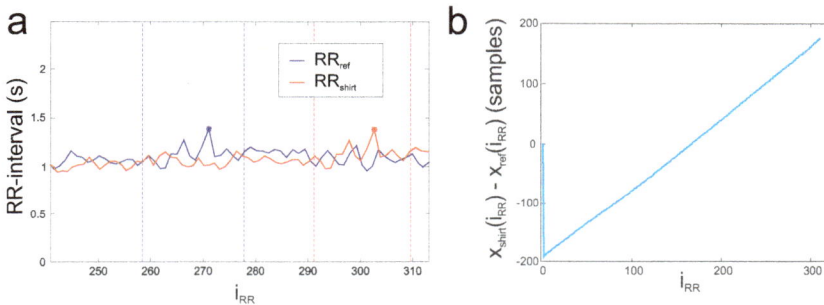

Figure 8. (**a**) synchronization using RR intervals; and (**b**) the difference between corresponding samples $x_{\text{shirt}}(i_{RR}) - x_{\text{ref}}(i_{RR})$ rises linearly.

Afterwards, the respective maximum was automatically detected in each of the two sequences (see markers in Figure 8). Finally, the time shift was automatically computed using the difference between the two maximums, and the timestamps of the T-shirt ECG and the reference ECG were corrected accordingly.

However, the data did not yet correspond because the sampling rates of the devices were slightly off. Despite the sampling rate of $fs = 500$ Hz, the data did not correspond. Therefore, the next step was to determine the samples $x_{\text{ref}}(i_{RR})$ and $x_{\text{shirt}}(i_{RR})$ of the corresponding intervals $i_{RR} = \{1, 2, ..., m\}$ and to calculate their difference $\Delta x = x_{\text{shirt}} - x_{\text{ref}}$. This difference increases linearly over the course of the measurement, as can be seen in Figure 8. This linearity was used to approximate a function $\Delta x = p_1 \cdot x_{\text{shirt}} + p_0$. The mean was $p_0 = 0.0025$ and $p_1 = -296.16$. We adjusted the time stamps of the commercial device by adding $\Delta x / fs$. This was calculated for every measurement separately. The ECG data were matched with the new time base and could then be adequately compared. Mathematical modeling was only done for the purpose of evaluation and for the comparison of our system to the reference system. In future, the reference system is not intended for the use at the same time.

3. Results

The experimental data obtained from the novel device were analyzed. Extracted RR intervals as well as the morphology of the ECG wave were compared with the reference signal.

3.1. RR Interval Analysis

The relative error for the RR intervals was computed for all three scenarios with all subjects. The duration of each experiment was 5 min. The relative error and signal coverage (percentage of the entire measurement time during which the signal was evaluable) were computed (Table 1). In one experiment (S1 walking), the ECGs were exceptionally noisy: the coverage was only 49.5% with a relative error of 75.5%. This particular measurement was disregarded for further analysis. The mean relative error for the RR intervals in every scenario was 0.96% and the mean coverage of the signal was 96.6%.

Table 1. Relative error of RR intervals and coverage for RR interval determination.

Subject	Scenario	Relative Error	Coverage	Duration (s)	Beats
S1	lying	0.0024	0.9942	314.75	305
S1	sitting	0.0022	0.9939	296.31	344
S1	walking *	0.755	0.4952	350.01	468
S2	lying	0.0022	0.9775	295.90	284
S2	sitting	0.0029	0.9898	297.54	323
S2	walking	0.0031	0.9938	290.57	368
S3	lying	0.0216	0.8881	300.82	295
S3	sitting	0.0208	0.949	296.31	305
S3	walking	0.0218	0.9449	296.31	365
Mean		0.0096 *	0.9664 *	304.28	340

* S1 walking was excluded from the mean.

The Bland–Altman diagram for all subjects is presented in Figure 9; the RR intervals extracted from both devices are compared in this diagram. The RR intervals used in this section were extracted from the Einthoven II lead with the algorithm by Brueser [27]. Furthermore, only artifact-free data were selected for the diagram, so the mean coverage was 87.2%. The bias was at −1.9 ms and the 95% limits of agreement were −20.1 ms and −23.9 ms.

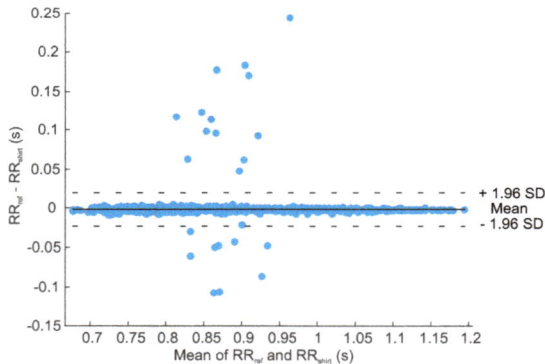

Figure 9. Bland–Altman diagram. The bias is at −1.9 ms, whereas the 95% limits of agreement are at −20.1 ms and −23.9 ms, respectively. The solid line indicates the bias/mean difference and the dashed lines indicate the 95% of limits of agreement. The mean coverage was 87.2%. S1 walking was excluded.

3.2. ECG Wave Analysis

In addition to RR interval analysis, the ECG leads were segmented into their P, Q, R and S-waves. Data with artifacts were selected manually and subsequently discarded. Finally, the resulting ECG waves were compared. For that purpose, a script was written to automatically identify the R, P, Q and S-waves. Since the location of the R-wave is easily found, the other waves could be identified as peaks relative to that position.

Figure 10 shows an excerpt from the Einthoven II lead from subject S1. The data of the T-shirt and the reference were bandpass-filtered (2–20 Hz) beforehand. The waves are highlighted for both ECG curves. The R-peaks of the lead and the other waves correspond.

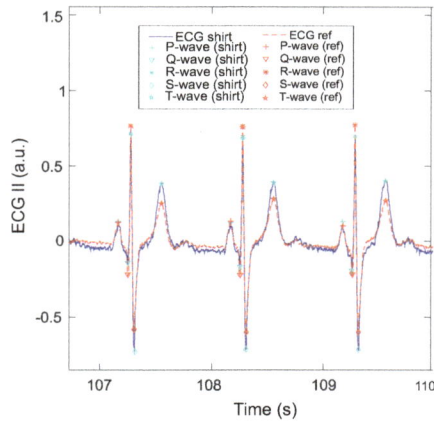

Figure 10. Exemplary comparison of Einthoven lead II from S1 (lying), bandpass filtered (2–20 Hz) from ECG T-shirt and reference.

Next, a closer analysis of the P, Q, R and S-waves was performed. Tables 2–4 present the results for the three scenarios: lying, sitting and walking. The temporal correlation of the data from both devices was calculated for each wave separately using Pearson's linear correlation coefficient. For each wave, the time stamp of the prominent peak was correlated. The mean correlation of each wave per lead is shown in Tables 2–4. The correlation signifies the matching of the waves extracted from both devices. A correlation close to zero indicates poor correspondence of the signals, whereas a correlation close to 1 indicates high correspondence. Additionally, the median of the time differences $t_{shirt}(i_{wave}) - t_{ref}(i_{wave})$ for the occurrence $i_{wave} = \{1, 2, ...n\}$ are determined for each P, Q, R and S wave per lead. These values indicate the time difference of the estimated waves in ms. A difference close to 0 ms is a perfect match. If this value is higher, the match is less precise.

Table 2. Data on statistics of all subjects (lying) with coverage: 86.79% $\pm$ 4.13%. Mean correlation and median of time difference ($t_{shirt} - t_{ref}$) for all subjects for P, Q, R, S, and T waves.

Lead	Correlation					Difference (ms)				
	P	**Q**	**R**	**S**	**T**	**P**	**Q**	**R**	**S**	**T**
I	0.900	0.952	0.957	0.957	0.762	2	10	2	−4	−2
II	0.879	0.957	0.957	0.957	0.957	2	0	2	2	−2
III	0.939	0.957	0.957	0.958	0.882	2	2	2	4	0
aVR	0.947	0.957	0.957	0.957	0.863	2	−2	2	0	8
aVL	0.871	0.953	0.957	0.953	0.732	26	2	2	14	−50
aVF	0.942	0.957	0.957	0.958	0.807	2	2	2	2	4
V1	0.952	0.955	0.957	0.954	0.706	2	4	2	−4	4
V2	0.943	0.957	0.957	0.955	0.826	2	2	2	0	12
V3	0.936	0.956	0.957	0.957	0.875	−2	0	2	−2	22
V4	0.917	0.957	0.957	0.957	0.757	0	4	2	4	6
V5	0.892	0.957	0.957	0.957	0.787	0	4	2	4	8
V6	0.920	0.957	0.957	0.958	0.850	0	4	2	6	8
Mean	0.920	0.956	0.957	0.957	0.817	3	3	2	2	2

For lying down, the data of all subjects were evaluable and the overall coverage was 86.79% $\pm$ 4.13%. Mean correlation was 92.14% $\pm$ 6.05% and the time difference was 2.4 $\pm$ 0.5 ms.

For sitting, the ECG data of all subjects were evaluable and the overall coverage was 94.37% $\pm$ 3.3%. Mean correlation was 91.1% $\pm$ 11.55% and the time difference was 0.6 $\pm$ 10.9 ms.

For walking, the ECG data of only subjects S2 and S3 were acceptable and the overall coverage was 83.47% ± 5.09%. Mean correlation was 76.9% ± 22.39% and the time difference was 6.6 ± 14.7 ms.

Table 3. Data on statistics of all subjects (sitting) with coverage: 94.37% ± 3.30%. Mean correlation and median of time difference ($t_{shirt} - t_{ref}$) for all subjects for P, Q, R, S, and T waves.

Lead	Correlation					Difference (ms)				
	P	**Q**	**R**	**S**	**T**	**P**	**Q**	**R**	**S**	**T**
I	0.941	0.975	0.978	0.976	0.739	−4	4	−4	−6	130
II	0.909	0.977	0.978	0.977	0.928	−4	−6	−4	−4	−22
III	0.914	0.978	0.978	0.977	0.685	−4	−2	−4	−2	−22
aVR	0.944	0.977	0.978	0.977	0.633	−4	−10	−4	−6	156
aVL	0.874	0.975	0.978	0.968	0.808	−8	−2	−4	−2	−22
aVF	0.931	0.978	0.978	0.977	0.648	−4	−2	−4	−2	−2
V1	0.946	0.973	0.978	0.970	0.641	−6	−13	−4	−8	−6
V2	0.934	0.977	0.978	0.977	0.742	−4	−4	−4	−14	8
V3	0.928	0.974	0.978	0.976	0.827	−8	−6	−4	−4	24
V4	0.884	0.977	0.978	0.978	0.624	−4	−2	−4	0	−6
V5	0.872	0.976	0.978	0.977	0.632	−4	0	−4	0	4
V6	0.903	0.977	0.978	0.977	0.608	−8	0	−4	2	2
Mean	0.915	0.976	0.978	0.976	0.710	−5	−4	−4	−4	20

Table 4. Data on statistics of subjects S2 and S3 (walking) with coverage: 83.47% ± 5.09%. Mean correlation and median of time difference ($t_{shirt} - t_{ref}$) for all subjects for P, Q, R, S, and T waves.

Lead	Correlation					Difference (ms)				
	P	**Q**	**R**	**S**	**T**	**P**	**Q**	**R**	**S**	**T**
I	0.739	0.901	0.920	0.913	0.580	8	8	2	−4	150
II	0.608	0.921	0.920	0.918	0.600	0	0	2	0	−10
III	0.678	0.921	0.920	0.918	0.439	2	0	2	0	0
aVR	0.739	0.920	0.920	0.918	0.397	2	−2	2	0	140
aVL	0.713	0.910	0.920	0.909	0.384	−20	12	2	14	−20
aVF	0.724	0.920	0.920	0.918	0.440	−2	2	2	2	8
V1	0.701	0.903	0.920	0.895	0.240	−16	2	2	-2	8
V2	0.731	0.893	0.920	0.890	0.274	−12	−2	2	0	30
V3	0.748	0.904	0.920	0.904	0.477	6	2	2	2	24
V4	0.663	0.913	0.920	0.910	0.296	−12	4	2	4	10
V5	0.691	0.912	0.920	0.909	0.346	−10	4	2	4	20
V6	0.704	0.908	0.920	0.907	0.368	−14	2	2	4	20
Mean	0.703	0.910	0.920	0.909	0.403	−6	3	2	2	32

4. Discussion

We have presented a 12-lead textile ECG T-shirt and portable ECG recorder. This device was evaluated using a standard 12-lead Holter as a reference and by conducting a study with healthy volunteers. Three experimental scenarios were devised (lying down, sitting and walking). Evaluation was performed by extracting the RR intervals and comparing the results with the gold standard. Finally, the ECG waves were examined in detail.

The first step was to synchronize the data of both devices. As the devices were not started at exactly the same time, matching was necessary. Moreover, because the sampling rate of 500 Hz produced a different amount of samples for the two devices, this also had to be taken into account. It is most likely that the clocks in the two devices run at a slightly different rate.

By comparing the RR intervals, a very low mean relative error of 0.96% was produced with a coverage of 96.6%. Furthermore, the Bland–Altman diagram (see Figure 9) shows accurate agreement of both systems with a bias of −1.9 ms and 95% limits of agreement of −20.1 ms and −23.9 ms. The bias represents a systematic error between the devices, which equals the possible temporal resolution for 500 Hz (2 ms). Although this is only a small error, it was relevant for the comparison of the ECG waves and the reason why the data were synchronized.

The mean RR interval length was 1.012 ± 0.071 s for lying, 0.898 ± 0.079 s for sitting, and 0.800 ± 0.047 s for walking. Although we had hypothesized that walking would produce more artifacts, the coverage was comparable for all three scenarios. We managed to avoid artifacts by applying sufficiently high contact pressure to the electrodes by putting adjustable Velcro on the back of the T-shirt. At the same time, the pressure of the T-shirt was not so high as to obstruct the subjects' breathing. However, if the T-shirt is planned to be worn by a particular group of patients, other safety measures might be necessary to consider.

One measurement of S1 walking was unsuccessful and consisted mostly of noise. An explanation for this could be the buildup of static charge or triboelectric effects, as explained by Wartzek et al. [28]. According to Wartzek, local triboelectric effects are the source of severe artifacts. They are mainly caused by unmatched electrodes and a bad connection to the DRL. Unmatched electrodes decrease the common mode rejection ratio (CMRR), while the CMRR should be as high as possible for good measurements.

Static effects can be minimized with some arrangements. As little electrode movement as possible is desirable to prevent charges. This is partially accomplished by maintaining sufficient electrode contact. Furthermore, if excess charges occur, a bias resistor in the active electrode allows discharge. Finally, the materials of the contact areas could be selected carefully [14]. An insulated electrode would cause larger electrostatic charges than a metal electrode [28].

After inspection of the RR intervals, the ECG waves were evaluated. The highest correlation was found for the R-waves, which were used to synchronize the data. Furthermore, in every scenario, the Q and S-waves had correlations of over 0.9; in contrast, the T-wave had significantly lower correlations. For lying, the correlation of the P-wave was 0.9198 and the correlation of the T-wave was 0.8171. For sitting, the P-wave still had a correlation of 0.915, but the T-wave only 0.7095. In walking, the P-wave deteriorated to 0.7032 and the T-wave to only 0.4034. This can also be observed in the time differences, which were high for the T-waves. A possible explanation for this is an inaccurate algorithmic determination of the waves. Another reason could be a subpar electrode contact leading to signal deformation, which is determined by the electrode-body interface [14,29]. Sufficient and constant contact pressure is known to be important for textile ECG measurements [1,11,30]. Assuming that the electrodes are the reason for the deformations, the results from Tables 2 and 3 were examined. Lead I, aVR, aVL, V2 and V3 clearly have high time differences. Lead I, aVR and aVL all depend on electrode RA and/or LA. Moreover, V2 and V3 seem to be affected. It is possible that these electrodes did not have sufficient contact pressure.

The coupling of the electrode-skin interface is most likely responsible for these aberrations. The electrode fabric is a silver plated knitted fabric with 99% pure silver. Typically, gel establishes conductive contact and decreases the impedance between the skin and the electrode. For dry contact, this impedance is rather high. That is why active electrodes with voltage followers are used. The input impedance is increased, while the output impedance is decreased [14].

In our experiments, we experienced deteriorated signal quality if the shirt was too loose. In that case, there was not sufficient contact between the skin and the electrode so that the electrode-skin interface became both resistive and capacitive. This combination possibly created a high-pass filter and caused signal deformation. Further challenges with dry and non-contact electrodes have been discussed by e.g., Chi et al. [9].

The system has not been tested with moisture so far. According to Chi et al., the signal of dry electrodes improves with moisture because it establishes a conductive contact between the electrode and the skin [9]. Additionally, tests with more subjects need to be conducted to ensure long-term functionality and stability. Moreover, adding extra sensors to observe the skin–electrode impedance is recommended, as this will allow to examine morphological differences between the T-shirt ECG.

5. Conclusions

This paper introduces a novel ECG T-shirt for 12-lead measurements with fully active and dry electrodes. A portable 12-lead ECG recorder was developed, which is compatible with the T-shirt. The system is portable and has a battery life of two days. To our knowledge, a 12-lead ECG T-shirt specifically with active electrodes has not been developed before. In a study with three volunteers, the functionality of the device was successfully compared with a commercial device in everyday scenarios. The relative error of the RR intervals was 0.96% with a mean coverage of 96.6%. The P-wave had a correlation of 0.703 for walking subjects, while the T-wave demonstrated lower correlations for all three scenarios (lying: 0.817, sitting: 0.710, walking: 0.403). The other correlations for the P, Q, R, and S-waves were all higher than 0.9. This work shows that a comfortable ECG T-shirt with active electrodes is suitable for 12-lead ECG recordings.

Acknowledgments: This work was funded by the Excellence Initiative of the German federal and state governments (OPBF074).

Author Contributions: W.N., X.Y. and A.B. developed the device. X.Y., W.N. and D.T. conceived and designed the experiments. A.B., X.Y. and D.T. performed the data analysis. A.B., D.T. and S.L. supervised the manuscript writing phase.

Conflicts of Interest: The authors declare no conflict of interest.

Abbreviations

The following abbreviations are used in this manuscript:

ADC	analogue digital converter
aVF	Goldberger lead
aVL	Goldberger lead
aVR	Goldberger lead
CMRR	common mode rejection ratio
CS	chip select line
DRL	driven right leg
ECG	electrocardiography
I	Einthoven I lead
II	Einthoven II lead
III	Einthoven III lead
LA	left arm
LL	left leg
PCB	printed circuit board
RA	right arm
SPI	serial peripheral interface
SPS	samples per second
V1 - V6	Wilson leads
WCT	Wilson central terminal

References

1. Kim, S.; Leonhardt, S.; Zimmermann, N.; Kranen, P.; Kensche, D.; Müller, E.; Quix, C. Influence of contact pressure and moisture on the signal quality of a newly developed textile ECG sensor shirt. In Proceedings of the IEEE 5th International Summer School and Symposium on Medical Devices and Biosensors (ISSS-MDBS 2008), Hong Kong, China, 1–3 June 2008; pp. 256–259.
2. Linz, T.; Kallmayer, C.; Aschenbrenner, R.; Reichl, H. Fully integrated EKG shirt based on embroidered electrical interconnections with conductive yarn and miniaturized flexible electronics. In Proceedings of the BSN 2006: International Workshop on Wearable and Implantable Body Sensor Networks, Cambridge, MA, USA, 3–5 April 2006; pp. 23–26.

3. Ottenbacher, J.; Romer, S.; Kunze, C.; Grosmann, U.; Stork, W. Integration of a Bluetooth Based ECG System into Clothing. In Proceedings of the Eighth International Symposium on Wearable Computers, Arlington, VA, USA, 31 October–3 November 2004; pp. 186–187.

4. Karlsson, J.; Wiklund, U. Wireless Monitoring of Heart Rate and Electromyographic Signals using a Smart T-shirt. In Proceedings of the International Workshop on Wearable Micro and Nanosystems for Personalised Health, Valencia, Spain, 21–23 May 2008.

5. Cardio Leaf. Available online: http://www.clearbridgevitalsigns.com/shirt.html (accessed on 19 May 2016).

6. hWear. Available online: http://www.personal-healthwatch.com/ (accessed on 19 May 2016).

7. Meziane, N.; Webster, J.G.; Attari, M.; Nimunkar, A.J. Dry electrodes for electrocardiography. *Physiol. Meas.* **2013**, *34*, 47–69.

8. Silva, M.; Catarino, A.; Carvalho, H.; Rocha, A.; Monteiro, J.; Montagna, G. Study of vital sign monitoring with textile sensors in swimming pool environment. In Proceedings of the Industrial Electronics Conference (IECON), Porto, Portugal, 3–5 November 2009; pp. 4426–4431.

9. Chi, Y.M.; Jung, T.P.; Cauwenberghs, G. Dry-contact and noncontact biopotential electrodes: Methodological review. *IEEE Rev. Biomed. Eng.* **2010**, *3*, 106–119.

10. Puurtinen, M.M.; Komulainen, S.M.; Kauppinen, P.K.; Malmivuo, J.A.V.; Hyttinen, J.A.K. Measurement of noise and impedance of dry and wet textile electrodes, and textile electrodes with hydrogel. In Proceedings of the Annual International Conference of the IEEE Engineering in Medicine and Biology, New York, NY, USA, 31 August–3 September 2006; pp. 6012–6015.

11. Cömert, A.; Hyttinen, J. Investigating the possible effect of electrode support structure on motion artifact in wearable bioelectric signal monitoring. *Biomed. Eng. Online* **2015**, *14*, 1–18.

12. Meziane, N.; Yang, S.; Shokoueinejad, M.; Webster, J.G.; Attari, M.; Eren, H. Simultaneous comparison of 1 gel with 4 dry electrode types for electrocardiography. *Physiol. Meas.* **2015**, *36*, 513–529.

13. Pani, D.; Dessì, A.; Saenz-Congolo.; J. F., Barabino, G.; Fraboni, B.; Bonfilgio, A. Fully Textile, PEDOT:PSS Based Electrodes for Wearable ECG Monitoring Systems. *IEEE Trans. Biomed. Eng.* **2016**, *63*, 540–549.

14. Sun, Y.; Yu, X. Capacitive Biopotential Measurement for Electrophysiological Signal Acquisition: A Review. *IEEE Sens. J.* **2016**, *16*, 2832–2853.

15. Lim, Y.G.; Kim, K.K.; Park, K.S. ECG measurement on a chair without conductive contact. *IEEE Trans. Biomed. Eng.* **2006**, *53*, 956–959.

16. Leonhardt, S.; Aleksandrowicz, A. Non-contact ECG monitoring for automotive application. In Proceedings of the 5th International Workshop on Wearable and Implantable Body Sensor Networks, Hong Kong, China, 1–3 June 2008; pp. 183–185.

17. Ishijima, M. Monitoring of Electrocardiograms in Bed Without Utilizing Body Surface Electrodes. *IEEE Trans. Biomed. Eng.* **1993**, *40*, 593–594.

18. Lim, Y.G.; Kim, K.K.; Park, K.S. ECG recording on a bed during sleep without direct skin-contact. *IEEE Trans. Biomed. Eng.* **2007**, *54*, 718–725.

19. Wu, K.F.; Zhang, Y.T. Contactless and continuous monitoring of heart electric activities through clothes on a sleeping bed. In Proceedings of the 2008 International Conference on Information Technology and Applications in Biomedicine, Shenzhen, China, 30–31 May 2008; pp. 282–285.

20. Ueno, A.; Yama, Y. Unconstrained monitoring of ECG and respiratory variation in infants with underwear during sleep using a bed-sheet electrode unit. In Proceedings of the 30th Annual International Conference of the IEEE Engineering in Medicine and Biology Society, Vancouver, BC, Canada, 20–24 August 2008; pp. 2329–2332.

21. Kim, K.K.; Lim, Y.K.; Park, K.S. The electrically non-contacting ECG measurement on the toilet seat using the capacitively-coupled insulated electrodes. In Proceedings of the Annual International Conference of the IEEE Engineering in Medicine and Biology Society, Buenos Aires, Argentina, 1–5 September 2004; pp. 2375–2378.

22. Lim, Y.K.; Kim, K.K.; Park, K.S. The ECG measurement in the bathtub using the insulated electrodes. In Proceedings of the Annual International Conference of the IEEE Engineering in Medicine and Biology Society, Buenos Aires, Argentina, 1–5 September 2004.

23. Malmivuo, J.; Plonsey, R. *Bioelectromagnetism: Principles and Applications of Bioelectric and Biomagnetic Fields*; Oxford University Press: Oxford, MI, USA, 2012; pp. 1–506.

24. Sanchez, B.; Praveen, A.; Bartolome, E.; Soundarapandian, K.; Bragos, R. Minimal implementation of an AFE4300-based spectrometer for electrical impedance spectroscopy measurements. *J. Phys.* **2013**, *434*, 012014.

25. Cömert, A.; Hyttinen, J. Impedance spectroscopy of changes in skin–electrode impedance induced by motion. *Biomed. Eng. Online* **2014**, *13*, 149.

26. Tallgren, P.; Vanhatalo, S.; Kaila, K.; Voipio, J. Evaluation of commercially available electrodes and gels for recording of slow EEG potentials. *Clin. Neurophysiol.* **2005**, *116*, 799–806.

27. Bruser, C.; Kortelainen, J.M.; Winter, S.; Tenhunen, M.; Parkka, J.; Leonhardt, S. Improvement of Force-Sensor-Based Heart Rate Estimation Using Multichannel Data Fusion. *IEEE J. Biomed. Health Inform.* **2015**, *19*, 227–235.

28. Wartzek, T.; Lammersen, T.; Eilebrecht, B.; Walter, M.; Leonhardt, S. Triboelectricity in capacitive biopotential measurements. *IEEE Trans. Biomed. Eng.* **2011**, *58*, 1268–1277.

29. Eilebrecht, B.; Czaplik, M.; Wartzek, T.; Schauerte, P.; Leonhardt, S. Analysis of influences on capacitive ECG measurements based on a closed loop model. In Proceedings of the 6th Meeting of the European Study Group on Cardiovascular Oscillations (ESGCO 2010), Berlin, Germany, 12–14 April 2010; pp. 12–15.

30. Reyes, B.A.; Posada-Quintero, H.F.; Bales, J.R.; Clement, A.L.; Pins, G.D.; Swiston, A.; Riistama, J.; Florian, J.P.; Shykoff, B.; Qin, M.; et al. Novel electrodes for underwater ECG monitoring. *IEEE Trans. Biomed. Eng.* **2014**, *61*, 1863–1876.

electronics

MDPI

Article

A Pulsed Coding Technique Based on Optical UWB Modulation for High Data Rate Low Power Wireless Implantable Biotelemetry

Andrea De Marcellis [1,*], Elia Palange [1], Luca Nubile [1], Marco Faccio [1], Guido Di Patrizio Stanchieri [1] and Timothy G. Constandinou [2]

[1] Department of Industrial and Information Engineering and Economics, University of L'Aquila, L'Aquila 67100, Italy; elia.palange@univaq.it (E.P.); luca.nubile@graduate.univaq.it (L.N.); marco.faccio@univaq.it (M.F.); guido.dipatriziostanchieri@student.univaq.it (G.D.P.S.)
[2] Centre for Bio-Inspired Technology, Imperial College London, London SW7 2AZ, UK; t.constandinou@imperial.ac.uk
*   Correspondence: andrea.demarcellis@univaq.it; Tel.: +39-086-243-4424

Academic Editors: Enzo Pasquale Scilingo and Gaetano Valenza
Received: 14 June 2016; Accepted: 10 October 2016; Published: 17 October 2016

Abstract: This paper reports on a pulsed coding technique based on optical Ultra-wideband (UWB) modulation for wireless implantable biotelemetry systems allowing for high data rate link whilst enabling significant power reduction compared to the state-of-the-art. This optical data coding approach is suitable for emerging biomedical applications like transcutaneous neural wireless communication systems. The overall architecture implementing this optical modulation technique employs sub-nanosecond pulsed laser as the data transmitter and small sensitive area photodiode as the data receiver. Moreover, it includes coding and decoding digital systems, biasing and driving analogue circuits for laser pulse generation and photodiode signal conditioning. The complete system has been implemented on Field-Programmable Gate Array (FPGA) and prototype Printed Circuit Board (PCB) with discrete off-the-shelf components. By inserting a diffuser between the transmitter and the receiver to emulate skin/tissue, the system is capable to achieve a 128 Mbps data rate with a bit error rate less than 10^{-9} and an estimated total power consumption of about 5 mW corresponding to a power efficiency of 35.9 pJ/bit. These results could allow, for example, the transmission of an 800-channel neural recording interface sampled at 16 kHz with 10-bit resolution.

Keywords: high data rate low power link; optical ultra-wideband modulation; pulsed coding technique; wireless implantable biotelemetry

1. Introduction

Emerging implantable biomedical systems need to transmit large amounts of data through skin/tissue to achieve high accuracy measurements, high dimensionality and real-time control of complex prosthetic devices like brain machine interfaces [1–3]. These systems require wireless biotelemetry links with high data rate, reduced power consumption, small Bit Error Rate (BER) and good electromagnetic compliance [3–7]. Solutions that make use of carrier-based narrow-band and Ultra-wideband (UWB) Radio Frequency (RF) links, employing Impulse Radio (IR) signal modulation, pose significant challenges when requiring high data rates due to their low power efficiency and electromagnetic compatibility [8–13]. Optical biotelemetry links, employing semiconductor modulated/pulsed lasers as data transmitters and photodiodes as data receivers, allow the performances of the RF-based systems to be enhanced [4,14–17]. In these regards, further improvements have been obtained by increasing the laser power and by using On-Off Keying (OOK) based modulations and large sensitive area photodiodes. However, this solution increases the laser response time and BER so limiting system bandwidth and data rate up to 100 Mbps with 21 pJ/bit power efficiency [4,15].

This Paper presents a data coding technique based on optical UWB modulation employing sub-nanosecond laser pulses with duration much smaller than the half bit period as typically employed in OOK modulation-based solutions. In particular, the proposed approach, already investigated in IR-UWB systems [13], implements an optical synchronised-OOK modulation allowing for the clock recovery and a proper synchronisation between the transmitter and the receiver. In general, optical UWB communication systems using modulated/pulsed signals have the advantages of low power spectral density, null RF interference, immunity to multipath fading and high power efficiency, so suitable for short-range wireless links [17–20]. Based also on these characteristics, the proposed approach allows us to strongly reduce the system overall power consumption and the laser response time reaching data rates up to 128 Mbps with 800 ps laser pulses, a bit period of 7.8 ns and a power efficiency of 35.9 pJ/bit. These results have been obtained by adjusting the amplitude and duration of the laser pulses to achieve the optimal trade-off between power consumption, data rate and BER. Respect to the state-of-the-art, the implemented pulsed coding technique is capable to increase the data rate (i.e., the transmission bandwidth) and, at the same time, to strongly reduce the power consumption (i.e., better power efficiency) of the optical communication system that is mainly related to the generation of the sub-nanosecond laser pulses. The presented architecture has been implemented on Field-Programmable Gate Array (FPGA) and prototype Printed Circuit Board (PCB) developed with discrete off-the-shelf components. The achieved results enable, for example, the transmission of an 800-channel neural recording interface (i.e., extracellular electrodes) sampled at 16 kHz with 10-bit resolution [21,22].

2. The Pulsed Coding Technique: Theory, Modelling and Methods of the Optical Approach

Figure 1 illustrates the top-level complete architecture of an optical wireless transcutaneous biotelemetry system. The dashed black square box highlights the scheme implementing the proposed optical pulsed UWB modulation technique performing the data coding process. The overall system typically includes two units, implantable and external, each one consisting of a main sub-block: the transmitter and the receiver.

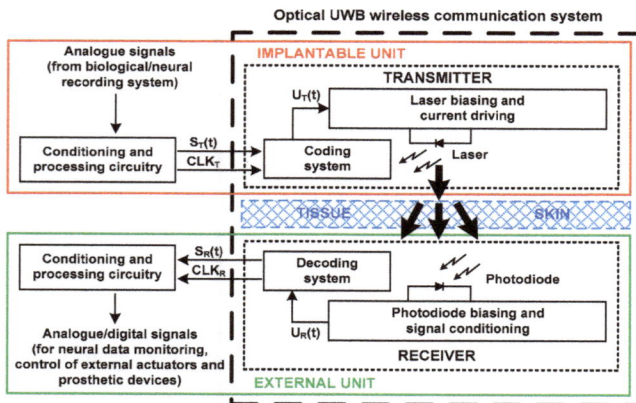

Figure 1. Complete scheme of an optical wireless transcutaneous biotelemetry system: in the dashed box, the architecture implementing the proposed data coding approach based on optical pulsed Ultra-wideband (UWB) modulation technique.

The conditioning and processing circuitry block in the implantable unit handles analogue signals related to neural signals (e.g., coming from neural recording systems) [21,22]. The analogue signals are converted into a serial digital data stream $S_T(t)$ to be transmitted, after its coding in pulsed signal $U_T(t)$, from the implantable to the external unit. The receiver in the external unit provides the decoded/regenerated serial digital data stream $S_R(t)$ to the related conditioning and processing circuitry block. The latter, in this case, generates analogue and/or digital signals associated to the transmitted neural analogue signals for the control of external actuators as well as for neural data monitoring.

More in detail, the transmitter takes, as input signals, the main clock CLK_T and the data stream $S_T(t)$ to be coded and transferred, while the receiver provides the recovered clock signal CLK_R and the decoded data stream $S_R(t)$. In the transmitter, the coding system generates the analogue pulsed voltage signal $U_T(t)$ converted by the bias and drive circuits into current pulses that generate the laser pulses (i.e., gain switching operation of the laser). In the receiver, the photodiode provides photocurrent pulses converted by the biasing and signal conditioning circuitry into analogue voltage pulses $U_R(t)$ processed by the decoding system that recovers the clock CLK_R and the data stream $S_R(t)$. The transmitter uses a high speed Vertical Cavity Surface Emitting Laser (VCSEL) generating sub-nanosecond optical pulses and the receiver employs a wide bandwidth Si photodiode.

Figure 2 illustrates an example of a timing diagram, with the corresponding laser operating conditions, describing the proposed data coding process that allows the optical pulsed UWB modulation performed by the transmitter to be achieved. During each bit period T, the VCSEL driving current is maintained at a value I_{min} just above the VCSEL threshold current I_{th} (corresponding to a minimum laser power P_{min}) to provide high output power stability, few picoseconds rise times and negligible jitter respect to the driving current pulses. In order to simultaneously transmit the CLK_T and $S_T(t)$, short current pulses with a maximum value I_{max} drive the VCSEL that generates optical pulses with a power P_{max}. Therefore, only during the laser pulse generation the transmitter achieves the maximum power consumption. Depending on the receiver performances, the current pulse amplitude and duration can be adjusted to obtain the optimal trade-off between power consumption, data rate and BER. More in detail, referring to Figure 2, at the start of each bit period T a short laser pulse is generated, independently from the symbol being transmitted, so allowing for the clock signal to be simultaneously transmitted with the symbol. This is needed to synchronise the transmitter and the receiver (i.e., the clock recovery procedure). Then, at half bit period $T/2$, if the symbol 1 has to be transmitted, a second laser pulse is generated, while for the transmission of the symbol 0 the VCSEL driving current is maintained at the value I_{min}. In this way, the transmitter power consumption is mainly due to the current I_{min} that is constant during each bit period T, while the contribution of the current pulses, with an amplitude equal to I_{max}, is minimal due to the very small duty cycle. These operating conditions allow us to strongly reduce the system overall power consumption and the VCSEL response time.

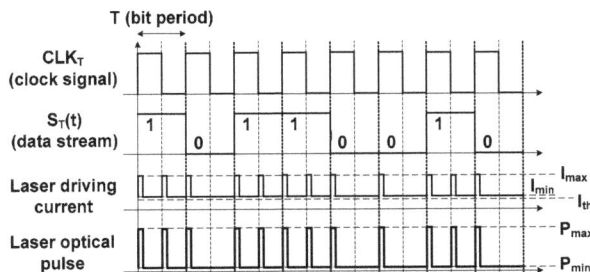

Figure 2. Example of a timing diagram describing the proposed data coding process and the corresponding laser operating conditions.

Figure 3 shows the logic block level implementation of the coding and decoding systems inside the transmitter and the receiver, respectively. The coding system receives in input CLK_T and $S_T(t)$ considering the bit period T equal to the clock period (i.e., one bit transmitted at each bit period T), as shown in Figure 2 and detailed in Figure 4a.

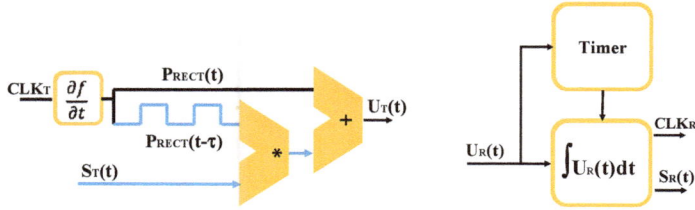

Figure 3. Description, at logic block level, of the implementation of the coding (on the **left**); and decoding (on the **right**) systems.

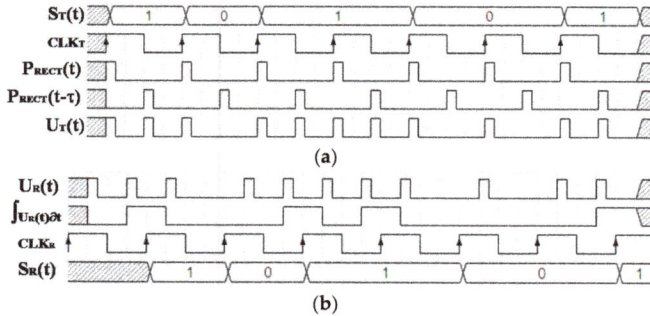

Figure 4. Correlations between the signals $CLK_{T,R}$, $S_{T,R}(t)$ and the corresponding coded data $U_{T,R}(t)$ related to the coding (**a**); and decoding (**b**) systems.

Thus, the coding system generates a train of "rect" pulses $P_{RECT}(t) = \sum_{n=-\infty}^{+\infty} \left[\text{rect}_{T_0} (t - nT) \right]$ having a period T with a very small duty cycle T_0 and the same signal, delayed by a time period $\tau = T/2$, defined as $P_{RECT}(t - \tau) = \sum_{n=-\infty}^{+\infty} \left[\text{rect}_{T_0} (t - nT - \tau) \right]$. The transmitted pulsed signal $U_T(t)$ is obtained as the combination of $P_{RECT}(t)$ and $P_{RECT}(t - \tau)$ and can be expressed as follows:

$$U_T(t) = \sum_{n=-\infty}^{+\infty} \left[\text{rect}_{T_0} (t - nT) + S_T(t) * \text{rect}_{T_0} (t - nT - \tau) \right] \tag{1}$$

On the other hand, $U_T(t)$ contains the constant pulses $P_{RECT}(t)$ used for the clock recovery and the pulses $P_{RECT}(t - \tau)$ related to the transmitted data $S_T(t)$ (i.e., the bit symbols 1 or 0). Inversely, $U_R(t)$ is the input of the decoding system that operates clock and data recovery so providing the synchronisation signal CLK_R and the decoded data stream $S_R(t)$. In particular, the decoding system integrates the signal $U_R(t)$ during a bit period T, starting from $T/2$ after the first synchronisation pulse through the "Timer" block (see Figure 3), according to the following relation:

$$S_R(t) = \begin{cases} 1 \text{ if } \int_{T/2}^{T} U_R(t) \, dt \geq t_H \\[2mm] 0 \text{ if } \int_{T/2}^{T} U_R(t) \, dt \leq t_L \end{cases} ; \text{ Err } \forall \tag{2}$$

where t_H and t_L are the high and low threshold levels defined by the decoding system. Consequently, the generated output signals $S_R(t)$ and CLK_R can be described as detailed in Figure 4b. More in detail, since the clock recovery as well as the synchronisation between the transmitter and the receiver is a critical issue in communication systems, the "Timer" block includes a control unit. Starting from the pulsed signal $U_R(t)$, the control unit operates the clock recovery by managing the regeneration of the new clock signal CLK_R and its synchronisation with the received pulses $U_R(t)$. Simultaneously, the overall "Timer" block also controls the synchronisation between the recovered clock CLK_R and the integrator block that, in turn, performs the recovery and regeneration of the data stream $S_R(t)$. In particular, the control unit accomplishes these operations by regulating the relative time delays (i.e., the timing alignments) between the received pulse train $U_R(t)$ and the recovered clock CLK_R together with the integration window, from $T/2$ to T, of the integrator block to correctly recover the data stream $S_R(t)$.

3. Implementation of the Wireless Communication System Based on the Pulsed Coding Technique

The optical pulsed UWB wireless communication system, performing the proposed data coding process, has been characterised by implementing the coding and decoding digital systems (see Figure 3) on an FPGA board (Xilinx VIRTEX-6 ML605, Xilinx Inc., San Jose, CA, USA) by a hardware description (VHDL, that is VHSIC Hardware Description Language where VHSIC means Very High Speed Integrated Circuits) of the functional/logic blocks previously described. The FPGA was used to: (i) supply the CLK_T and $S_T(t)$ signals at the input of the coding system that generates $U_T(t)$; (ii) recover the CLK_R and $S_R(t)$ from the received $U_R(t)$ through the decoding system. The resulting digital architecture is composed by a 2:1 multiplexer, a 1:2 demultiplexer, three D-type flip-flops and 22 logic gates (AND, OR, NOT). The corresponding circuitry can be implemented in standard Complementary Metal-Oxide Semiconductor (CMOS) technology as a full-custom solution with less than 150 transistors in a Si area smaller than 1 mm^2.

The employed laser is a high-speed VCSEL-850 (Thorlabs Inc., Newton, NJ, USA) with 2.2 mA threshold current emitting at λ = 845 nm. This wavelength has been chosen considering the best light "penetration window" of human skin that ranges between 800 nm and 1300 nm [23].

The photodiode is the Si high-speed FDS025 (Thorlabs Inc., Newton, NJ, USA) with 47 ps rise time and 49,000 μm^2 sensitive area. All analogue circuits have been implemented on prototype PCB using high-bandwidth discrete off-the-shelf components. In particular, the VCSEL biasing and driving current circuitry has been implemented by employing the driver ADN2871 (Analog Devices Inc., Norwood, MA, USA) operating a voltage-to-current conversion. The output DC bias current was set to 2.25 mA (i.e., I_{min}, see Figure 2) just about 2.5% above the VCSEL threshold current I_{th}, while the peak value of the current pulses was regulated up to a maximum value of 4.5 mA (i.e., $I_{max} = 2I_{min}$). The photodiode biasing and signal conditioning circuitry has been implemented by using the ultra-low noise voltage-feedback operational amplifier OPA847 (Texas Instruments Inc., Dallas, TX, USA) configured as a transimpedance amplifier to convert the generated pulsed photocurrent into voltage pulses (i.e., a current-to-voltage conversion). This stage is combined with ADN2915 (Analog Devices Inc., Norwood, MA, USA) that includes a limiting amplifier and a signal level detector.

4. Experimental Results and Discussion

Experimental measurements were performed with the implemented overall system operating at different data rates up to 128 Mbps employing a Pseudo-Random Bit Sequence (PRBS) of length $2^{31}-1$ (i.e., the data stream) generated by the FPGA board with an average number of pulses transmitted for each bit equal to 1.5 pulses/bit. Moreover, a 1.5 mm thick diffuser ED1-C20-MD (Thorlabs Inc., Newton, NJ, USA) was inserted between the VCSEL and the photodiode to emulate the presence of skin or tissue [11,23]. For the considered distances between the VCSEL and the photodiode of only few millimetres, the diffuser reduces the laser power of about a factor 10. Furthermore, the transmission channel could be considered as a "noisy channel" and the BER has been directly evaluated through experimental measurements since it depends on the channel condition mainly related to the employed diffuser. More in detail, the BER has been evaluated through the FPGA board considering the system under different working conditions: variations of the laser pulse amplitude and duration as well as changes of the distance between the VCSEL and the photodiode. In particular, in the operating conditions at 128 Mbps, the bit period T is 7.8 ns and the laser pulse duration was reduced to about 800 ps corresponding to a duty-cycle T_0 of 10%. This system set-up allowed us to achieve the optimum trade-off between data rate, BER and power consumption. In principle, for a fixed data rate, the duty-cycle T_0 of the pulses (i.e., the laser pulse duration) can be reduced providing lower power consumptions at the expenses of an increase of the BER value. On the contrary, for achieving a minimum fixed BER value, it is possible to decrease the duty-cycle of the laser pulses providing a lower power consumption against a reduction of the data rate that, in some case, could imply a worst system power efficiency. In our case, at 128 Mbps transmission data rate and T_0 = 10% laser pulse duty-cycle, the highest measured BER is less than 10^{-9}, achieved for a total maximum distance lower than 4 mm including the presence of the diffuser, with an estimated maximum total power consumption of the overall system of about 5 mW, the 10% of which is related to the coding and decoding digital systems. In this regard, any clock synchronisation mismatch affecting a correct data communication has been also taken into account in the evaluation of the BER since timing misalignments of the clock signal (i.e., clock synchronisation errors) provide bit errors in the decoding system. In addition, it is important to consider that in the proposed architecture, the coding system in the transmitter block does not include any error-correction process that further would improve the BER. In this sense, for example, Hamming and Golay error-correction codes could be considered as suitable solutions to be implemented by an additional low power circuitry enhancing the BER of few orders of magnitude with an additional energy consumption of only 3–4 pJ/bit [19,20].

Figure 5 reports a picture of the overall implemented system together with the related block scheme of the experimental set-up showing its main components. In particular, the transmitter and the receiver blocks are connected to the FPGA board and the two XYZ translation stages allow us both to perform the optical alignment between the VCSEL and the photodiode and change their relative distance. Figure 6 shows an example of the main signals related to the transmitter and the receiver blocks operating at 128 Mbps. From top to bottom: the green trace is CLK_T, the yellow $S_T(t)$, the purple $U_T(t)$ and the blue $S_R(t)$. In the insets are reported: (A) multiple acquisitions of $U_T(t)$ for the evaluation of the amplitude and phase jitters resulting lower than 150 mV and 100 ps, respectively; (B) the $U_R(t)$ and the related $S_R(t)$ signals; (C) the $S_R(t)$ eye diagram showing amplitude and phase jitters lower than 100 mV and 150 ps, respectively. All the measurements have been performed by using WAVEMASTER 8600A digital oscilloscope (LeCroy). Figure 6 demonstrates that the received and decoded data stream $S_R(t)$ correctly matches with the transmitted one ($S_T(t)$) showing a time delay of about 18 ns mainly due to clock recovery and data decoding processes. Moreover, referring to the inset B) of Figure 6, it has been experimentally evaluated also the overall impulse response of the transmission channel by means of the time-domain analysis of the received pulsed signal $U_R(t)$.

Furthermore, it is important to highlight that the proposed system is not based on IR-UWB modulation and does not employs RF electrical pulses and/or antennas to transmit the modulated/pulsed signals. On the other hand, the duration of the generated optical pulses is lower than 1 ns. Thus, in principle, the proposed architecture implementing the optical pulsed coding technique complies with the standards UWB specifications on the generation of pulsed signals. In this regards, Figure 7 reports the overall power spectrum of the transmitted pulsed signal $U_T(t)$ generated by the coding system showing high spikes that, in principle, could affect the overall power efficiency of the system. Nevertheless, an effective trade-off between transmission data rate, power efficiency and BER has been achieved. In particular, the experimental data demonstrate that the pulse characteristics and specifications comply and satisfy the standard mask defined by the Federal Communications Commission (FCC) on the power spectrum of modulated/pulsed signals (i.e., the power emission limits of the signals for communications) [24]. Moreover, Figure 7 shows a primary lobe with a bandwidth of about 1.25 GHz corresponding to the inverse of the laser pulse duration (i.e., 1/800 ps) so validating and confirming the theoretical expectations.

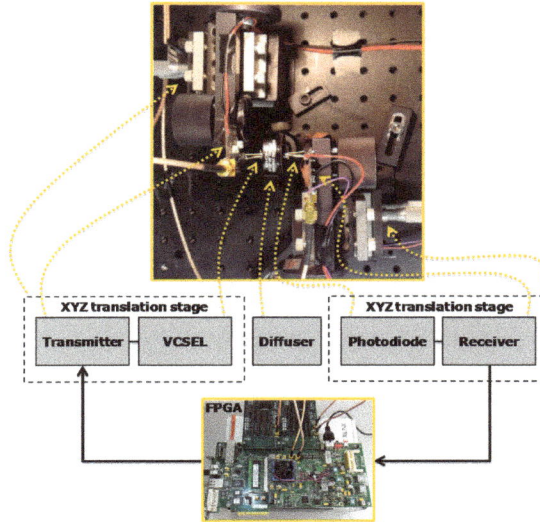

Figure 5. Experimental set-up: picture of the overall implemented system and the related block scheme showing its main components.

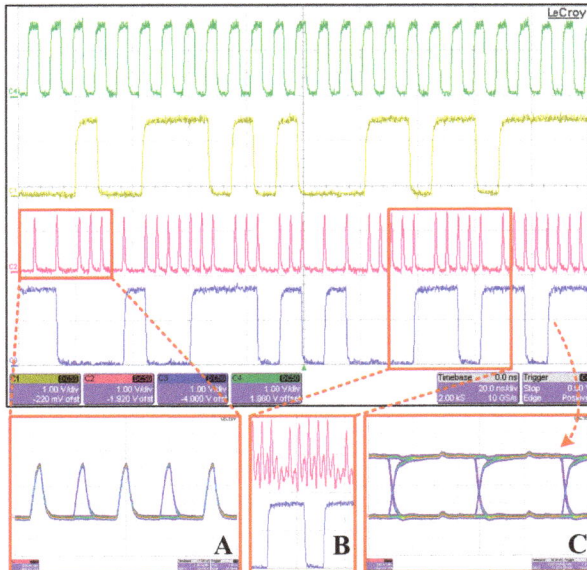

Figure 6. Main signals related to the transmitter and the receiver blocks. From top to bottom: CLK_T (green), $S_T(t)$ (yellow), $U_T(t)$ (purple), $S_R(t)$ (blue). In the insets: (**A**) multiple acquisitions of $U_T(t)$ for the evaluation of the amplitude and phase jitters; (**B**) $U_R(t)$ and the related $S_R(t)$; (**C**) the eye diagram of $S_R(t)$.

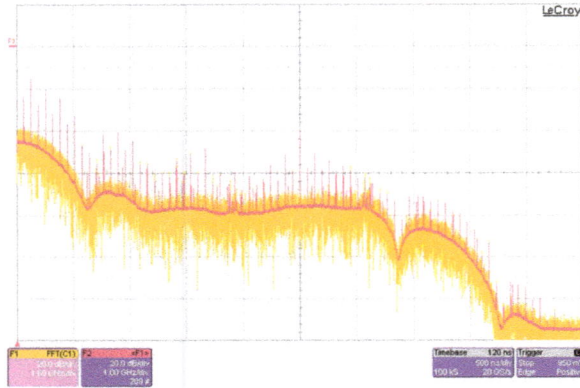

Figure 7. Power spectrum of the transmitted pulsed signal $U_T(t)$ (orange trace) and its average value (purple trace).

Finally, in order to perform a comparative analysis with the state-of-the-art on transcutaneous biotelemetry, Table 1 summarises the main performances of different solutions reported in literature and those ones presented in this Paper. More in detail, in RF-based solutions, high data rates are achieved at low power consumptions and good power efficiencies with high BER values [9,10,12]. A very low BER with a satisfactory data rate are obtained in [11,13] even if at the expenses of a strong increase of the RF system power consumption and power efficiency. Considering now the solutions based on optical wireless links, 100 Mbps is the highest achieved data rate with reduced power consumptions and good power efficiencies and BER values [4,14,15]. This represents the best trade-off among the main parameters of interest. On the other hand, the solution proposed in this paper shows the highest data rate and the best BER value among the optical link solutions. The mean value of the transmitter power consumption is 4.6 mW corresponding to a 35.9 pJ/bit power efficiency of the overall implantable unit. It is worth noting that these values have been obtained by implementing the complete architecture on prototype PCB with discrete off-the-shelf commercial components and using a VCSEL with a relatively high threshold current. Nevertheless, the proposed optical pulsed UWB modulation technique implemented on-chip in a standard CMOS Si technology using VCSEL with lower threshold currents (e.g., Philips-ULM Photonics VCSEL-ULM850-10-TTN0101U with I_{th} = 0.5 mA [15]) could allow the reduction of the transmitter power consumption down to about 1 mW corresponding to a power efficiency of 7.8 pJ/bit. In this case, the estimated maximum total power consumption of the overall system is about 1.5 mW.

Table 1. Main performances of different biotelemetry links.

Reference (Year)	Data Rate (Mbps)	Transmitter Power (mW)	Power Efficiency (pJ/bit)	BER (Bit Error Rate)
[9] (2015)	67	2	30	$<10^{-7}$
[10] (2010)	136	3	22	$<10^{-3}$
[11] (2004)	80	45	562.5	$<10^{-14}$
[12] (2013)	135	1.4	10	-
[13] (2011)	10	1.84	97.5	-
[14] (2007)	40	4.3	107.5	$<10^{-5}$
[15] (2014)	100	2.1	21	$<10^{-7}$
[4] (2015)	100	3.2	32	-
This work	128	4.6 (measured with a VCSEL having I_{th} = 2.2 mA)	35.9	$<10^{-9}$
		1 (estimated with a VCSEL having I_{th} = 0.5 mA)	7.8	

5. Conclusions

The Paper reported on a data coding technique for high data rate low power optical wireless biotelemetry using sub-nanosecond pulsed laser as data transmitter and small sensitive area photodiode as data receiver. The implementation of this approach, based on an optical pulsed UWB synchronised-OOK modulation, through FPGA board and prototype PCB with discrete off-the-shelf components allowed us to achieve a 128 Mbps data rate, a BER less than 10^{-9} and a power efficiency of 35.9 pJ/bit. These values have been obtained by inserting a diffuser between the transmitter and the receiver to emulate skin/tissue barrier. The achieved results demonstrate that the proposed approach enables, for example, the transmission of an 800-channel neural recording interface sampled at 16 kHz with 10-bit resolution for high performances (i.e., high data rate and good power efficiency) optical wireless implantable biotelemetry. The overall architecture that implements the developed technique has been designed to be integrated on-chip in a compact Si footprint lower than 1 mm^2 with about 180 transistors. In this regard, a suitable design of the complete circuitry in a proper standard technology (e.g., Bi-CMOS) and the use of fast pulsed lasers and high-speed photodiodes, will allow us to achieve high performances wireless optical links with data rates up to few Gbps. Finally, the presented optical pulsed coding technique can be also applied to bidirectional transcutaneous biomedical platforms for very high data rate and ultra low power optical wireless biotelemetry that would integrate a transmitter-receiver in the implantable unit and the corresponding receiver-transmitter in the external unit.

Author Contributions: The Authors contributed equally to this work. In particular: Andrea De Marcellis developed the new technique, implemented the analogue circuitry and performed experimental measurements. He contributed to writing and editing the manuscript. He coordinated the manuscript elaboration. Elia Palange optimised the use of the optoelectronic/photonic components, implemented the experimental set-up and analysed the data. He contributed to writing and editing the manuscript. Luca Nubile developed, simulated and implemented the digital architectures on FPGA. He contributed to writing and editing the manuscript. Marco Faccio developed the digital architectures and analysed the data. He contributed to writing and editing the manuscript. Guido Di Patrizio Stanchieri simulated and implemented the digital architectures on FPGA and performed tests and experimental measurements. He contributed to writing and editing the manuscript. Timothy G. Constandinou supplied the overall system specifications and constrains, contributed in theoretical discussions and analysed the data. He contributed to writing and editing the manuscript.

Conflicts of Interest: The authors declare no conflict of interest.

References

1. Park, J.H.; Jeong, J.; Moon, H.; Kim, C.; Kim, S.J. Feasibility of LCP as an encapsulating material for photodiode-based retinal implants. *IEEE Photonics Technol. Lett.* **2016**, *28*, 1018–1021. [CrossRef]
2. Keith-Hynes, P.; Mize, B.; Robert, A.; Place, J. The diabetes assistant: A smartphone-based system for real-time control of blood glucose. *Electronics* **2014**, *3*, 609–623. [CrossRef]
3. Yuan, H.; He, B. Brain–computer interfaces using sensorimotor rhythms: Current state and future perspectives. *IEEE Trans. Biomed. Eng.* **2014**, *61*, 1425–1435. [CrossRef] [PubMed]
4. Liu, T.; Anders, J.; Ortmanns, M. Bidirectional optical transcutaneous telemetric link for brain machine interface. *IET Electron. Lett.* **2015**, *51*, 1969–1971. [CrossRef]
5. Duncan, K.; Etienne-Cummings, R. Selecting a safe power level for an indoor implanted UWB wireless biotelemetry link. In Proceedings of the 2013 IEEE Biomedical Circuits and Systems Conference, Rotterdam, The Netherlands, 31 October–2 November 2013; pp. 230–233.
6. Rush, A.D.; Troyk, P.R. A power and data link for a wireless-implanted neural recording system. *IEEE Trans. Biomed. Eng.* **2012**, *59*, 3255–3262. [CrossRef] [PubMed]
7. Vračar, L.; Prijić, A.; Nešić, D.; Dević, S.; Prijić, Z. Photovoltaic energy harvesting wireless sensor node for telemetry applications optimized for low illumination levels. *Electronics* **2016**, *5*. [CrossRef]
8. Miranda, H.; Gilja, V.; Chestek, C.A.; Shenoy, K.V.; Meng, T.H. A high-rate long-range wireless transmission system for simultaneous multichannel neural recording applications. *IEEE Trans. Biomed. Circuits Syst.* **2010**, *4*, 181–191. [CrossRef] [PubMed]

9.   Ebrazeh, A.; Mohseni, P. 30 pJ/b, 67 Mbps, centimeter-to-meter range data telemetry with an IR-UWB wireless link. *IEEE Trans. Biomed. Circuits Syst.* **2015**, *9*, 362–369. [CrossRef] [PubMed]

10. Jung, J.; Zhu, S.; Liu, P.; Emery-Chen, Y.J.; Heo, D. 22-pj/bit energy-efficient 2.4-GHz implantable OOK transmitter for wireless biotelemetry systems: In vitro experiments using rat skin-mimic. *IEEE Trans. Microw. Theory Tech.* **2010**, *58*, 4102–4111. [CrossRef]

11. Guillory, K.S.; Misener, A.K.; Pungor, A. Hybrid RF/IR transcutaneous telemetry for power and high-bandwidth data. In Proceedings of the 2004 IEEE Engineering in Medicine and Biology Societ, San Francisco, CA, USA, 1–5 September 2004; pp. 4338–4340.

12. Elzeftawi, M.; Theogarajan, L. A 10 pJ/bit 135 Mbps IR-UWB transmitter using Pulse Position Modulation and with on-chip LDO regulator in 0.13 μm CMOS for biomedical implants. In Proceedings of the 2013 IEEE Topical Conference on Biomedical Wireless Technologies, Networks, and Sensing Systems, Austin, TX, USA, 20–23 January 2013; pp. 37–39.

13. Crepaldi, M.; Li, C.; Fernandes, J.; Kinget, P. An ultra-wideband impulse-radio transceiver chipset using synchronized-OOK modulation. *IEEE J. Solid-State Circuits* **2011**, *46*, 2284–2299. [CrossRef]

14. Ackermann, D. High Speed Transcutaneous Optical Telemetry Link. Master's Thesis, Case Western Reserve University, Cleveland, OH, USA, May 2007.

15. Liu, T.; Bihr, U.; Becker, J.; Anders, J.; Ortmanns, M. In vivo verification of a 100 Mbps transcutaneous optical telemetric link. In Proceedings of the 2014 IEEE Biomedical Circuits and Systems Conference, Lausanne, Switzerland, 22–24 October 2014; pp. 580–583.

16. Kuchta, D.M.; Rylyakov, A.V.; Doany, F.E.; Schow, C.L.; Proesel, J.E.; Baks, C.W.; Westbergh, P.; Gustavsson, J.S.; Larsson, A. A 71-Gb/s NRZ modulated 850-nm VCSEL-based optical link. *IEEE Photonics Technol. Lett.* **2015**, *27*, 577–580. [CrossRef]

17. Wen, Y.H.; Feng, K.M. A simple NRZ-OOK to PDM RZ-QPSK optical modulation format conversion by bidirectional XPM. *IEEE Photonics Technol. Lett.* **2015**, *27*, 935–938. [CrossRef]

18. Chenhui, Y.; Kun, Z.; Hongyan, F.; Sailing, H. An all-optical transformer from differential NRZ data to Ultra-Wideband pulse stream. *IEEE Photonics Technol. Lett.* **2011**, *23*, 579–581.

19. Desset, C.; Fort, A. Selection of channel coding for low-power wireless systems. In Proceedings of the 2003 IEEE Vehicular Technology Conference, Orlando, FL, USA, 22–25 April 2003; pp. 1920–1924.

20. Hannan, M.A.; Abbas, S.M.; Samad, S.A.; Hussain, A. Modulation Techniques for Biomedical Implanted Devices and Their Challenges. *Sensors* **2012**, *12*, 297–319. [CrossRef] [PubMed]

21. Barsakcioglu, Y.D.; Liu, Y.; Bhunjun, P.; Navajas, J.; Eftekhar, A.; Jackson, A.; Quiroga, R.Q.; Constandinou, T.G. An analogue front-end model for developing neural spike sorting systems. *IEEE Trans. Biomed. Circuits Syst.* **2014**, *8*, 216–227. [CrossRef] [PubMed]

22. Montoye, A.H.; Dong, B.; Biswas, S.; Pfeiffer, K.A. Use of a Wireless Network of Accelerometers for Improved Measurement of Human Energy Expenditure. *Electronics* **2014**, *3*, 205–220. [CrossRef] [PubMed]

23. Bashkatov, A.N.; Genina, E.A.; Kochubey, V.I.; Tuchin, V.V. Optical properties of human skin, subcutaneous and mucous tissues in the wavelength range from 400 to 2000 nm. *J. Phys. D Appl. Phys.* **2005**, *38*, 2543–2555. [CrossRef]

24. Federal Communications Commission (FCC). *Revision of Part 15 of the Commission's Rules Regarding Ultra-Wideband Transmission Systems, First Report and Order*; Federal Communications Commission: Washington, DC, USA, February 2002; pp. 1–118.

![electronics logo] electronics

MDPI

Article

A Wearable System for the Evaluation of the Human-Horse Interaction: A Preliminary Study

Andrea Guidi [1], Antonio Lanata [1,*,†], Paolo Baragli [2], Gaetano Valenza [1] and Enzo Pasquale Scilingo [1]

[1] Research Center "E.Piaggio", School of Engineering, University of Pisa, Largo Lucio Lazzarino 1, Pisa 56122, Italy; andrea.guidi@for.unipi.it (A.G.); g.valenza@iet.unipi.it (G.V.); e.scilingo@centropiaggio.unipi.it (E.P.S.)

[2] Department of Veterinary Sciences, Laboratory of Equine Behavior and Physiology, University of Pisa, Viale delle Piagge 2, Pisa 56124, Italy; paolo.baragli@unipi.it

* Correspondence: antonio.lanata@unipi.it; Tel.: +39-050-2217604

† Current address: Department of Information Engineering, School of Engineering, University of Pisa, Via Caruso 16, Pisa 56122, Italy.

Academic Editor: Mostafa Bassiouni
Received: 26 June 2016; Accepted: 18 September 2016; Published: 26 September 2016

Abstract: This study reports on a preliminary estimation of the human-horse interaction through the analysis of the heart rate variability (HRV) in both human and animal by using the dynamic time warping (DTW) algorithm. Here, we present a wearable system for HRV monitoring in horses. Specifically, we first present a validation of a wearable electrocardiographic (ECG) monitoring system for horses in terms of comfort and robustness, then we introduce a preliminary objective estimation of the human-horse interaction. The performance of the proposed wearable system for horses was compared with a standard system in terms of movement artifact (MA) percentage. Seven healthy horses were monitored without any movement constraints. As a result, the lower amount of MA% of the wearable system suggests that it could be profitably used for reliable measurement of physiological parameters related to the autonomic nervous system (ANS) activity in horses, such as the HRV. Human-horse interaction estimation was achieved through the analysis of their HRV time series. Specifically, DTW was applied to estimate dynamic coupling between human and horse in a group of fourteen human subjects and one horse. Moreover, a support vector machine (SVM) classifier was able to recognize the three classes of interaction with an accuracy greater than 78%. Preliminary significant results showed the discrimination of three distinct real human-animal interaction levels. These results open the measurement and characterization of the already empirically-proven relationship between human and horse.

Keywords: wearable systems; e-textile; human interaction; biomedical signal processing; non-stationary signal

1. Introduction

In the last few decades, the interest in decoding the human-horse relationship and interaction (HHRI) has increased dramatically. This was guided by the strong empirical evidence of the positive outcomes in equine assisted therapy (EAT) and horseback riding in therapeutic programs [1], as well as the positive impact of animal companionship on human quality of life [2], where the equine is an important element of these therapeutic practices, with its feelings and behavior. For this purpose, the investigation of the modality in which both human and horse can communicate might be crucial. Measuring and evaluating the impact of the interaction experience might be relevant [3]. Some studies investigated the equine perception of humans in terms of positive, negative or neutral valence [4]. A study on how human psychological and physiological state can be perceived by horses was performed in [5] via the study of the heart rate. A more relaxed equine behavior was observed when humans showed positive attitudes toward them [6,7], while an equine increased heart rate was observed when humans were engaged in negative thinking [8]. A nervous mood can be transmitted from humans to horses under handling and riding conditions [9]. Voice [10–12], posture [13,14], facial expression [15,16], autonomic signals [17–19], hormones [20–23] and pheromones [24] might be used to fruitfully describe and characterize the emotional content [25]. Non-verbal communication between human and horse was also investigated in [26,27]. Heart rate and behavior resulted in being sensitive and reliable indicators of fear or anxiety in horses [28,29]. Horses that are in discomfort were observed to be more aggressive toward humans [30] or to be characterized by an increased heart rate, motor activity and vocalizations [31]. The effect of the gender of the person interacting with the horse was discussed in [32]. Although a recent review described a parallel behavior between the human multi-sensorial perception and the demonstrated equine cross-modal recognition [33,34], an interdisciplinary approach is mandatory to reach a deeper knowledge of human-horse interaction. In fact, these achieved experimental findings pointed out complex and multidimensional aspects of the interaction, which involve medical, bioengineering, physics and veterinary science [2,35,36]. In [9], the heart rate of both human and horse were monitored simultaneously under different experimental handling or riding conditions. In this study, Keeling et al. asserted that the analysis of heart rate is an important tool to investigate horse-human interactions. Again, hormones, heart rate and some standard heart rate variability-related indices were investigated in [20] during both training and performance. Different feature trends were observed between human and horse when they were obtained from ECG (electrocardiographic) recordings related to public or private sessions. A body sensors network technology was used to real-time monitor the horse-rider dyad in [37]. The aim of such a monitoring was evaluating the human-horse interaction. Specifically, based on a study concerning the equine emotional response during physical activity [38], the evaluation was based on the measurements of heart rate and physical activities via a mathematical model. Such a model was proposed to decompose the equine heart rate into two different components. The first one was concerning the physical component, while the second one contained information about equine emotional state [38]. Finally, in [39], the link between horse and human was also investigated by studying their electroencephalograph signals (EEG), revealing a higher synchronicity in EEG waves at increasing interactions. A correlation analysis between human and equine hormone concentrations was performed in [20,21].

Therefore, the human-horse relationship appears to be a complex interaction affected by several psychological factors [40]. The perception of humans by horses seems to be based on experience and repeated interactions, with horses that form a memory of humans that impacts their reactions in subsequent interactions [5]. Hence, previous negative experiences with human contact could lead horses toward a negative emotional reaction [7] or, vice versa, previous positive experience could lead them toward positive feelings with humans [41].

In our hypothesis, human and horse are considered as complex systems that interact through a coupling process. In this frame, we hypothesize that coupling can be modulated by the type and time duration of the contact itself. Specifically, we analyzed the level of coupling by studying their heart rate variability (HRV) time series [42] through dynamic time warping (DTW). As is well known from the literature, HRV can be considered as a non-linear time series, in which complex oscillations are present [43]. Therefore, we aimed at measuring this biological coupling over time [44]. DTW is, generally, used for studying time series dynamics of non-stationary systems. It calculates the best possible warp alignment between two time series, by selecting the one with the minimum distortion. Specifically, DTW is also defined as a measure of similarity between two time series, and it is calculated as the minimum mapping distance between them. DTW was widely used in many contexts, including data mining [45], speech processing [46] and medicine [47].

In this study, DTW was adopted to evaluate how heart activities evolve in a similar or dissimilar way. For example, if during an experiment with increasing exciting levels we detect an increasing DTW between human and equine HRVs, it indicates that the distance of the two HRV dynamics is increasing, and therefore, the heart activities are following different patterns. In order to perform a continuous, comfortable and non-invasive monitoring of the interaction in a natural environment, we developed, and here present, a wearable monitoring system for horses. The amount of advantages that wearable systems have brought to physiological signal monitoring for humans is well known, occupying an even larger space in the research. Moreover, the continuous technological development and the increasing demand of smart systems push wearables as the most used and suitable systems for ubiquitous and pervasive investigations. In addition, their flexibility allows researchers and clinicians to face the large variability of biomedical signals and tasks in monitoring subjects during their daily activities [48–52]. However, the biggest limitation in using wearable systems with humans and animals is due to motion artifacts (MA) [53–58], which are the major source of noise in biomedical signal acquisition, inducing the loss and alteration of informative content. For instance, the electrocardiographic (ECG) signal acquired in a naturalistic environment without movement constraints can be severely affected by important artifacts, and a great amount of data might be lost in contrast to the signal quality easily achievable in controlled environments and protocols [59]. As a matter of fact, cardiac stress tests or simply respiration can generate a big amount of MAs that can alter the signal [59]. Moreover, it is important to highlight that restraining horse is usually discouraged since it is unnatural and stressful and induces an increase of the sympathetic contribution of the heart control [60] that leads to misleading ECG interpretations [61]. In this work, we present a textile-based system for the ECG monitoring in horses, where the electrodes are completely made of fabric (electro-textile or e-textile). Normally, textile materials are insulators, but for this application, conductive yarns are integrated into the fabric during the manufacturing process [62]. In the human biomedical field, e-textiles are considered as higher value-added textiles and are prominently developed for being used in smart clothing. Smart clothing refers to a new garment that is able to acquire and process information, as well as actuate responses [63]. The potentialities deriving from these kinds of textile sensors enable the application of wearable systems in a great variety of experimental settings. As a matter of fact, many human studies showed reliable recordings of biomedical signals [64–68]: for example, ECG in [69–76], respiration in [69,70,73,74,77–79], electrodermal response (EDR) in [51,80] and, finally, PhotoPlethysmoGraphy (PPG)and blood pressure in [81,82]. In this kind of application, physiological signals are monitored and recorded in order to evaluate or follow the health status of the person who is wearing the wearable device [48,83].

Similarly in the veterinary research field, some authors [84–87] proposed to use Holter devices in equine applications and subsequently systems with radiotelemetry. Here, we propose the use of wearable systems in both animals and humans simultaneously, in order to acquire their cardiac activity in a reliable and artifact-free way, as was generally demonstrated in [72,88–92]. The aim is to infer autonomic nervous system responses enabling the detection of uncontrolled responses of animals when elicited with emotional stimuli coming from humans and vice versa. The manuscript is organized as follows: Section 2 deals with the materials' and methods' description; specifically, it describes the wearable systems, the experimental protocol of the interaction and the signal processing chain. Section 3 reports the achieved results, and Section 4 is focused on the discussion and conclusion of the study highlighting the future perspectives of this pioneering work.

2. Materials and Methods

In this study, two wearable monitoring systems, i.e., one for humans and the other for horses, were employed. Each system was comprised of a smart garment and portable electronics. The human system (Figure 1) was designed as a sensorized t-shirt (Smartex, Pisa, IT, Italy), and it is exhaustively described in [48,79,93–95]. Differently, the equine system (Figure 2) was comprised of an elastic smart belt (Smartex, Pisa, IT, Italy) [89,91] fastened around the chest behind the shoulder area. In both systems, two textiles electrodes (Smartex, Pisa, IT, Italy) and a strain gauge sensor (Smartex, Pisa, IT, Italy) were integrated to acquire the ECG (with a sampling frequency of 250 Hz) and respiration activity (with a sampling frequency of 25 Hz). The strain gauge is carried out by textile sensors that monitor the cross-sectional variations of the rib cage. The respiration sensor along with electrodes are integrated in an elastic band through a one-step process in the fabric by means of a circular knitting machine [96]. They are developed by Smartex s.r.l.; many details can be found in [92,96]. Moreover, equine ECG was recorded by placing the electrodes according to the modified base-apex configuration [97]. It is worthwhile to note that the use of a dry textile-based electrode provides several advantages. Firstly, the system is easy to use through an automatic placement of the sensors and allows high comfort. Secondly, electrodes are made of a special multilayer structure of textile material that increases the amount of sweat and reduces the rate of evaporation reaching very rapidly an electrochemical equilibrium between skin and electrode. This means that the signal quality [72] is remarkably improved and kept as constant as possible. These materials are knitted together and are fully integrated in the garment without any mechanical and physical discontinuity, creating areas with different functionalities (see Figure 1). For each system, the two ECG e-textile electrodes and the strain gauge sensor are finally connected to portable electronics through a simple plug that can be easily unplugged when necessary.

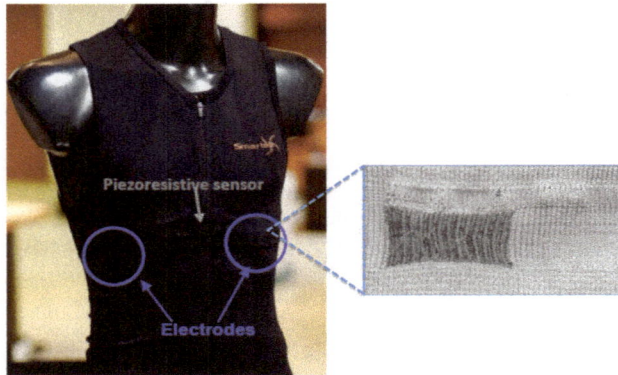

Figure 1. Wearable monitoring system for humans [48]. As is possible to note, from the box on the right, the e-textile electrodes for ECG (electrocardiographic) acquisition are knitted and completely integrated into the garment.

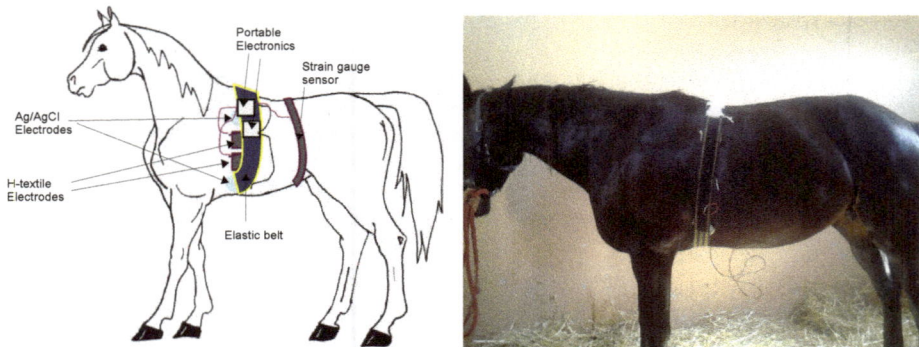

Figure 2. Systems placement on the horse. Elastic smart belt and Ag/AgCl electrodes placement. A scheme is presented on the left side and a real picture is on the right side.

Moreover, an inertial platform (triaxial accelerometer) (Smartex, Pisa, IT, Italy) integrated into the portable electronics that was positioned on the back of the horse allowed the monitoring of the horse's physical activity at a sampling frequency of 25 Hz. Finally, the wearable system was wirelessly connected to a smartphone where a dedicated app enabled checking the status of the sensors and remotely controlling the storing process of the physiological information in a secure digital (SD) card.

2.1. Equine Textile-Based ECG: Test and Validation

This section is focused on testing and validating the reliability of the equine ECG traces coming from textile electrodes. For this study, seven healthy standard bred mares in anestrus (mean age 8.4 ± 1.3 years) were enrolled. Equine subjects were socially housed in a paddock (75×75 m) and were provided ad libitum access to both hay and water. Horses were used as receivers in the embryo transfer program of the Department of Veterinary Sciences (University of Pisa, Italy) where this study was performed. Mares were in healthy condition and not pregnant at the time of this protocol.

The signal quality test of e-textile electrodes was performed by means of a comparative study in terms of the motion artifact (MA) between e-textiles and conventional Ag/AgCl electrodes. Each horse was simultaneously monitored by the wearable system, i.e., equine elastic smart belt and Ag/AgCl electrodes (see Figure 2). Two identical electronics were employed for both couples of electrodes Ag/AgCl and the textile ones. All data were acquired in a stall (4×4 m), where horses were left free to spontaneously move for 60 min. At the end of the recording session, all of the acquired ECGs were visually examined by one expert, and all of the ECG segments, corrupted by MAs, were marked. The goal was the estimation of the percentage of corrupted signal. Such a percentage was calculated as the sum of the time intervals in which an MA was observed over the whole time length of the recording.

A nonparametric Wilcoxon signed-rank test for paired samples was used to compare the percentage of corrupted signals between the signals acquired by the two kinds of electrodes, i.e., e-textile and Ag/AgCl. Specifically, the test was designed to compare the performance of the two kinds of electrodes while simultaneously recording the heart activity of each horse. A significant result of such a test would indicate a coherent different percentage of corrupted signal between the traces acquired by means of the two kinds of electrodes.

2.2. Protocol of Interaction

Here, the design of the interaction protocol is reported. Fourteen subjects (25 ± 3 years old, 4 males) were enrolled. The participants did not show any past or current experience of mental or personality disorder. In addition, one standard bred mare out of the seven previously described was enrolled (age: 8 years, weight: 560 kg, height: 160 cm). Informed consent was signed by the participants according to this specific protocol approved by the Ethical Committee of the University of Pisa.

During the whole experimental protocol, the autonomic nervous system (ANS) response of both human and horse were monitored. Specifically, two different systems were used to acquire the ECG signal. The fabric-based monitoring system [48,79,93–95] was used to record the human ECG, while the elastic smart belt [89,91] was used to record the equine one.

The experimental protocol consisted of three phases, each one lasting 4 min. During the first phase, P_1, the human and the mare were in different stalls (4×4 m). In this phase, considered as the resting phase, the subject sat on a chair, while the horse was free to move. Successively, the horse was moved from her stall to the human's. In this phase, P_2, the subject was asked to keep himself/herself still on the chair, while the horse was free to move and explore the environment. This phase implies a visual and olfactory interaction. Finally, in the third phase, P_3, all of the participants were asked to stand up and groom the horse. It is worthwhile to note that in order to keep the visual conditions of the horse as constant as possible, all of the subjects were asked to wear an azure plain t-shirt.

2.3. Signal Processing Chain for Human-Horse Interaction

This section deals with the description of the signal processing chain, which can be summarized in three steps. First, the HRV of both human and horse were estimated. Then, the DTW coupling index is computed. Finally, in the third step, the statistical analysis of DTW in the three sessions and the pattern recognition of the three phases is performed.

2.3.1. Heart Rate Variability Estimation

HRV is considered to be one of the most important ANS-related series. HRV describes the variations in the beat-to-beat intervals or in the instantaneous heart rate, since it reflects the regulatory mechanism of the cardiac activity by the ANS [98]. ANS is divided into sympathetic and parasympathetic branches. According to a simple model, sympathetic activity is responsible for the increasing of the heart rate (HR) and of the decreasing of HRV, while parasympathetic activity is usually considered to be in charge of decreasing HR and increasing HRV [99]. Many studies have focused on the significant relationship between ANS and HRV. Specially, frequency domain indexes, e.g., the LF/HF ratio (i.e., it is the ratio between the power in the low frequency bandwidth and the power in the high frequency bandwidth), have been deeply investigated [100], as well as temporal [19], spectral [101] and non-linear [102–104] indices. Several tools have been developed for its analysis [105], since its estimation is quite simple. In fact, HRV can be estimated as the series obtained from the interpolation of the beat-to-beat distances, i.e., RR distances (i.e. it means the distance between consecutive R-peaks of QRS complex). For this purpose, ECG signals were first of all digitally filtered with a zero-phase Butterworth infinite impulse response band-pass filter with cut-off frequencies between 0.5 and 40 Hz. Then, different R-peak detection algorithms were used to process human and equine ECG signals with the aim of estimating the HRV. More precisely, R-peaks related to the human ECG signals were detected by means of the well-known Pan–Tompkins method [106], while the method proposed in [91] was used to detect the R-peaks in the equine ECG signals. The Pan–Tompkins method [106] is an algorithm based on a pre-processing phase, including band-pass filtering, squaring of the data samples and moving average filtering, and on a decision rule phase, which includes an amplitude threshold to detect R-peaks. Differently, the algorithm proposed in [91] to detect R-peaks in equine ECG was based on the estimation of the energy of the second derivative of the ECG signal. R-peaks were detected by performing a thresholding of the obtained energy-signal. Finally, a cubic spline interpolation was applied to the irregular RR sampled series with a new sampling frequency equal to 10 Hz [105]. To do that, the cubic spline interpolation was used to create an evenly-distributed sampled series. This procedure is necessary since RRseries are irregularly sampled, due to the physiological variability of the heart activity, and an evenly-distributed sampled series is a mandatory condition in several analyses. Then, since the RR interval sampling frequency is usually chosen among 2, 4, 6, 8 and 10 Hz [107], we decided to obtain a sampling frequency equal to 10 Hz as performed in [108,109]. In fact, according to [107], it is important to highlight that the bandwidth within which the autonomic nervous system has a significant response is 0–1 Hz.

2.3.2. Feature Estimation

This section reports on the computation of the dynamic time warping [110]. DTW is usually used for studying time series dynamics of non-stationary systems. It computes the best possible warp alignment between two time series, by selecting the one with the minimum distortion. DTW was widely used in many contexts including data mining [45], speech processing [46] and medicine [47]. Specifically, DTW is a measure of the similarity between two temporal series. To estimate it, a temporal non-linear warping is performed to find the optimal match between the two sequences. In this frame, given two sequences, x and y of length N and M, respectively, it is possible to define a (N,M)-warping path as a sequence $p = (p_1, ..., p_L)$, with $p_l = (n_l, m_l) \in [1 : N] \times [1 : M]$ for $l \in [1 : L]$, able to align the elements of x and y. Such an alignment assigns the element x_{n_l} of x to the element y_{m_l} of y. Hence, the total cost $c_p(x, y)$ of a warping-path p between x and y can be defined, with respect to the local cost measure or local distance measure c, as reported in Equation (1).

$$c_p(x, y) = \sum_{l=1}^{L} c(x_{n_l}, y_{m_l}) \tag{1}$$

Finally, the optimal warping path between x and y is the warping path p^* corresponding to the minimal total cost. DTW distance is defined as the cost of p^*, according to (Equation (2)):

$$\text{DTW}(x, y) = c_{p^*}(x, y) = \min(c_p(x, y)) \tag{2}$$

where p is a (N,M)-warping path. This means that a low DTW distance implies a short warping path, which results in a higher similarity between the two time series.

In this study, DTW is used to estimate the degree of similarity/dissimilarity between the two subjects' heart activities (i.e., horse and human), as well as the degree of similarity/dissimilarity among all of the phases within each single subject (i.e., only human or horse). To this aim, the DTW method is applied to both human and horse HRV series.

2.3.3. Statistical Analysis and Pattern Recognition

A Wilcoxon signed rank test for paired data was used to compare the features estimated during the different phases of the experimental protocol. All of the possible combinations of the 2-class problems were taken into account. Specifically, the DTW, estimated between human and horse in each phase, was compared with the one obtained from the other phases, i.e., P_1 vs. P_2, P_1 vs. P_3 and P_2 vs. P_3, to evaluate possible statistically-significant differences between the human-horse interactions occurring during the different phases. Moreover, DTW was also estimated among the HRV series of each single enrolled subject between the different phases, i.e., only horse and only human. In this case, we aimed at comparing the degree of similarity/dissimilarity passing through the different phases within the same subject. A correction factor for multiple comparisons was applied according to the Benjamini and Hochberg method [111]. Such a correction was necessary to cope with the rate of Type I errors in null hypothesis testing when conducting multiple comparisons. The Benjamini and Hochberg method, which is a false discovery rate (FDR)-controlling procedure, takes into account the expected proportion of rejected null hypotheses that were incorrect rejections ("false discoveries") [111]. A significance level equal to 0.05 was used.

The classification process was performed by means of a supervised learning method, which aimed at performing the recognition of the experimental phases, i.e., the human-horse interaction, and evaluating the discriminant power of the DTW feature, as estimated between human and horse. Specifically, we implemented a support vector machine (SVM) classifier, with a radial-based kernel function (see Equation (3)):

$$K(x, x') = \exp(-\gamma|x - x'|^2) \tag{3}$$

where γ is equal to 1 and x and x' stand for the two samples. Moreover, a Leave-One-Subject-Out (LOSO) architecture was developed to apply our pattern recognition approach on the estimated DTW. Of note, the LOSO architecture is appositely designed to test the developed classifier on untreated and unknown data. Specifically, if N is the number of enrolled subjects (in this case, $N = 14$), the classifier is trained $N - 1$ times on the $N - 1$ subjects and, therefore, tested on the remaining 1 subject. To this aim, first of all, the data related to each subject were Z-transformed and then given as input to the LOSO architecture.

3. Results

3.1. Results of the E-Textiles' Validation and Testing

The percentages of artifact signal recorded both by using Ag/AgCl electrodes and e-textile electrodes were estimated. Such an estimation was performed by manually placing the labels. A person with experience in the field of equine cardiology was in charge of this issue. The obtained results are displayed in Table 1.

Table 1. Percentage of corrupted signal obtained by means of hand labeling of the two categories of signals: signals recorded by using Ag/AgCl electrodes and signals recorded by using textile electrodes.

	Ag/AgCl	E-Textile
h1	77, 48	32, 37
h2	54, 50	35, 00
h3	51, 49	39, 74
h4	30, 44	27, 51
h5	54, 44	32, 66
h6	47, 02	41, 96
h7	47, 89	35, 03

In addition, a non-parametric Wilcoxon signed-rank test for paired samples was implemented to analyze the statistical significance of the differences observed between the two kinds of electrodes in terms of the percentage of corrupted signal. A significant *p*-value equal to 0.0156 was achieved (Figure 3). This indicated a statistically-significant difference between the performance of the two kinds of electrodes. In fact, as is easily detectable in Figure 3 and in Table 1, the percentage of corrupted signal was significantly lower when ECG traces were recorded via e-textile electrodes.

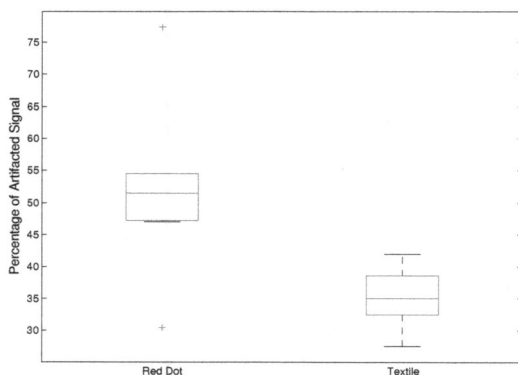

Figure 3. Boxplot of the percentage of artifact signals observed in the two categories of data acquired. The central box represents the central 50% of the data. In fact, its lower and upper boundary lines are respectively at the 25% and 75% quantile of the data. A central red line indicates the median of the data. Finally, two vertical lines indicate the remaining data that are outside the central box and not considered as outliers.

3.2. Results of the Human-Horse Interaction

In Figure 4, an example of two HRVs estimated from both equine and human ECG traces is shown. There, it is possible to observe, within a smaller signal portion, how warping-paths are defined.

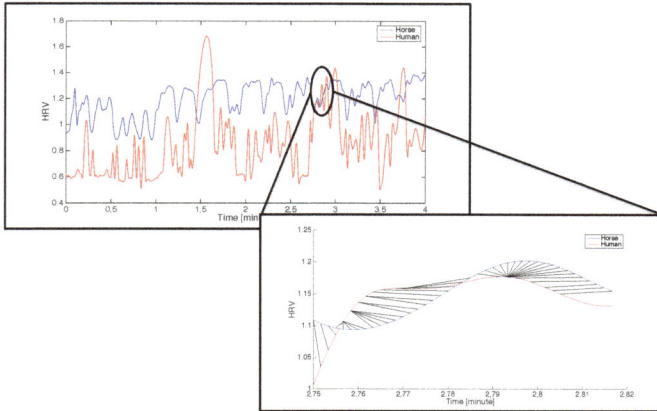

Figure 4. Example of both human and equine heart rate variability (HRV) signals. Within the smaller portion of signal, it is shown how the warping-paths are defined.

The statistical analysis performed on the DTW estimated between the human and equine HRV traces of the three different phases is reported in Table 2 in terms of p-values. All of the pairwise comparisons between different phases were statistically significantly different after performing the p-value correction via the Benjamini and Hochberg method. Specifically, as highlighted in Table 2 by means of the rising arrows, a loss of similarity, i.e., an increased DTW distance value, can be observed as the human-horse interaction becomes progressively stronger. These trends can be also observed in Figure 5. In fact, in 12 out of 14 subjects, the DTW value can be observed to be lower or equal in P_1 compared to in P_2, i.e., human and equine HRV are more similar in the P_1 phase than in P_2. In 12 out of 14 subjects, the DTW value can be detected to be higher in the P_2 phase than in P_1, but lower than in P_3, while 12 out of 14 subjects show a higher DTW value in the P_3 phase with respect to the other ones.

Table 2. p-Values related to pairwise statistical analysis of the DTW (dynamic time warping) values, as estimated between human and horse, during the different phases (P_1, P_2, P_3).

Feature	P_1 vs. P_2	P_1 vs. P_3	P_2 vs. P_3
DTW	0.0353 (↑)	0.0026 (↑)	0.0018 (↑)

P_1: resting state, P_2: visual and olfactory interaction, P_3: grooming.

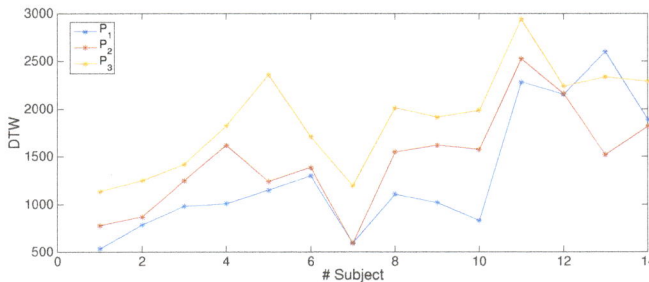

Figure 5. DTW (dynamic time warping) values, as estimated between human and horse, obtained for each subject in each phase.

In Table 3, the statistical results obtained by comparing only the human DTW and only horse DTW, separately, between the different phases are reported. Observing the reported *p*-values and arrows, a significant difference is reported for horse DTW between P_1–P_2 and P_2–P_3 with a higher value for the P_2–P_3. For humans, a significant difference was found in the comparison of DTW between P_1–P_2 and P_1–P_3, showing a higher value for the P_1–P_3, and the comparison of DTW between P_1–P_3 and P_2–P_3, showing a higher value for the P_1–P_3. These results seem to show that the P_3 phase is the more responsive for both humans and horses.

Table 3. *p*-Values related to pairwise statistical analysis of DTW values, as estimated within each enrolled subject, during the different phases (P_1, P_2, P_3).

	$DTW_{P_1-P_2}$ vs. $DTW_{P_1-P_3}$	$DTW_{P_1-P_2}$ vs. $DTW_{P_2-P_3}$	$DTW_{P_1-P_3}$ vs. $DTW_{P_2-P_3}$
human	**7.32E-04** (↑)	4.63E-01	**3.66E-03** (↓)
horse	1.38E-01	**4.4E-02** (↑)	8.21E-01

P_1: resting state, P_2: visual and olfactory interaction, P_3: grooming.

The results of the pattern recognition are returned as a confusion matrix (Table 4). In the main diagonal, each element represents the correct recognition percentage. Upper and lower triangular matrices report the percentage of incorrect recognition. The overall accuracy is about 85.71%.

Table 4. Confusion matrix obtained by applying the Support Vector Machine (SVM) classifier to the leave-one-subject-out architecture.

	P_1	P_2	P_3
P_1	**85.7143**	7.1429	7.1429
P_2	21.4286	**78.5714**	0.0000
P_3	0.0000	7.1429	**92.8571**

P_1: resting state, P_2: visual and olfactory interaction, P_3: grooming.
Note: the percentage of the right recognised classes are marked in bold.

4. Discussion and Conclusions

The objective of the study here presented was two-fold: the first one was to investigate the reliability of using wearable systems for monitoring physiological signal in horses; the second objective was to estimate and quantify the human horse interaction in a specific experimental protocol. As a matter of fact, concerning the first objective, the obtained results showed that the e-textile-based system outperformed the standard monitoring one, in terms of comfortability and reduction of the amount of movement artifacts. This latter one was demonstrated by the comparison of the blind visual inspection performed by an expert applied to both signals. Specifically, an amount of 7 h of equine ECG recording was analyzed, and the statistical analysis showed that ECG signals acquired with the textile electrodes had a lower percentage of corruption, revealing a greater robustness against movement artifacts. It is worthwhile to note that textile-based wearable monitoring systems offer interesting advantages in horse applications. In fact, it resulted in being very easy to use and very comfortable, and it could be automatically placed without any adhesive, glue and adhesive bandages. Furthermore, the multilayer structure of the textile electrode increased the amount of sweat and reduced the rate of evaporation, reaching an optimal electrochemical equilibrium between skin and electrode, rapidly. As a matter of fact, the signal quality was notably improved and kept as constant as possible. Moreover, wearable systems allow us to reliably monitor heart activity and parameters related to the involvement of ANS activity also without any restriction of movement. Indeed, it is possible to record an artifact-free ECG that enables non-linear signal processing techniques' application to horse's HRV signal. This allowed us to perform an objective measurement of the complex ANS responses for the investigation of human-horse interaction over a temporal dynamic evolution. Specifically, by analyzing the dynamic of the HRV of both human and horse, a similarity estimator, i.e., DTW, has been studied to describe different interaction phases. DTW comparisons for horse showed that the transition from P_2 to P_3 is more significant with respect to the transition from P_1 to P_2, whilst for humans, we found that the difference between P_1–P_3 is higher than both P_1–P_2 and P_2–P_3. These results seem to show that the P_3 phase is the more responsive for both humans and horses. Moreover, the achieved results in the horse-human DTW estimation showed that it was able to significantly discriminate all of the three phases of the experimental protocol. Furthermore, a significantly continuous increment from P_1 to P_2 and then to P_3 was observed in human-horse DTW. This may indicate a monotonic variation of the coupling between the human and horse cardiovascular systems. Specifically, as highlighted in Table 2, a loss of similarity, i.e., an increasing DTW distance value, can be observed as the human-horse interaction changes. These trends can be also observed in Figure 5. In fact, almost all of the involved subjects showed a DTW value lower in P_1 than in P_2 and than in P_3. A possible justification of this phenomenon could be found in the nature of the interaction. Probably, when human and horse are in physical contact, human and horse may suspiciously behave with a decrease of similarity and an increase of the distance between the two heart activities. The ability of the classification procedure of recognizing all of the three phases (i.e., accuracy greater than 85%) means that there exists a hyperplane able to differentiate these phases, whose distribution is not explained only by statistics, but it could be due to a non-linear dataset configuration in the feature space [112]. As a matter of fact, the achieved results are not sufficient, at the moment, to generalize and confirm the described phenomena. Hence, an increased number of measures, including further variables, is needed to clarify the phenomena involved in the human-horse interaction. In conclusion, this work poses the basis for the application of novel high level signal processing techniques for stationary and non-stationary signals [113,114], already used in investigating human-to-human interaction [115–117], to animals, and particularly to horses, in order to objectively reveal interesting responses of both the central and autonomic nervous system (ANS) for a particular uncommon stimulation. Moreover, the achieved results lead us to conclude that a quantitative measure of the human-horse interaction is viable, and it could be very effective in many fields of application, for example in therapy assisted by equine (EAT) [2,118] or for controlling the effect of therapeutic horseback riding [119]. In this field, our paradigm could permit analyzing the emotional interaction of a human patient with a "standardized horse" (a horse specifically trained and managed to create a controlled positive emotional background with a human). From a research point of view, these results can open a new scenario in which the emotional interaction [120] between human [121] and horse may be detectable and measurable, thus extending the current knowledge on comparative neuroscience.

5. Ethical Statement

This study was carried out in accordance with the recommendations in the Italian Animal Care Act (Decree Law 116/92). The experimental protocols were approved by the Ethics Committee on Animal Experimentation of the University of Pisa. Consent to participation in the test was signed by each horse owner.

Author Contributions: All authors contributed equally to this work.

Conflicts of Interest: The authors declare no conflict of interest.

References

1. Anderson, M.K.; Friend, T.H.; Evans, J.W.; Bushong, D.M. Behavioral assessment of horses in therapeutic riding programs. *Appl. Anim. Behav. Sci.* **1999**, *63*, 11–24.
2. Schultz, P.N.; Remick-Barlow, G.; Robbins, L. Equine-assisted psychotherapy: A mental health promotion/intervention modality for children who have experienced intra-family violence. *Health Soc. Care Community* **2007**, *15*, 265–271.

3. Hausberger, M.; Roche, H.; Henry, S.; Visser, E.K. A review of the human–horse relationship. *Appl. Anim. Behav. Sci.* **2008**, *109*, 1–24.
4. Waiblinger, S.; Boivin, X.; Pedersen, V.; Tosi, M.V.; Janczak, A.M.; Visser, E.K.; Jones, R.B. Assessing the human–animal relationship in farmed species: A critical review. *Appl. Anim. Behav. Sci.* **2006**, *101*, 185–242.
5. Fureix, C.; Jego, P.; Sankey, C.; Hausberger, M. How horses (*Equus caballus*) see the world: Humans as significant objects? *Anim. Cogn.* **2009**, *12*, 643–654.
6. Chamove, A.S.; Crawley-Hartrick, O.J.; Stafford, K.J. Horse reactions to human attitudes and behavior. *Anthrozoös* **2002**, *15*, 323–331.
7. Lesimple, C.; Fureix, C.; Menguy, H.; Hausberger, M. Human direct actions may alter animal welfare, a study on horses (*Equus caballus*). *PLoS ONE* **2010**, *5*, doi:10.1371/journal.pone.0010257.
8. Hama, H.; Yogo, M.; Matsuyama, Y. Effects of stroking horses on both humans' and horses' heart rate responses1. *Jpn. Psychol. Res.* **1996**, *38*, 66–73.
9. Keeling, L.J.; Jonare, L.; Lanneborn, L. Investigating horse–human interactions: The effect of a nervous human. *Vet. J.* **2009**, *181*, 70–71.
10. Koolagudi, S.G.; Rao, K.S. Emotion recognition from speech: A review. *Int. J. Speech Technol.* **2012**, *15*, 99–117.
11. Guidi, A.; Vanello, N.; Bertschy, G.; Gentili, C.; Landini, L.; Scilingo, E.P. Automatic analysis of speech F0 contour for the characterization of mood changes in bipolar patients. *Biomed. Signal Process. Control* **2015**, *17*, 29–37.
12. Guidi, A.; Scilingo, E.; Gentili, C.; Bertschy, G.; Landini, L.; Vanello, N. Analysis of running speech for the characterization of mood state in bipolar patients. In Proceedings of the 2015 AEIT International Annual Conference (AEIT), Naples, Italy, 14–16 October 2015.
13. Bianchi-Berthouze, N.; Cairns, P.; Cox, A.; Jennett, C.; Kim, W.W. On posture as a modality for expressing and recognizing emotions. In Proceedings of the Emotion and HCI Workshop at BCS HCI London, London, UK, 11–15 September 2006.
14. Baragli, P.; Gazzano, A.; Martelli, F.; Sighieri, C. How Do Horses Appraise Humans' Actions? A Brief Note over a Practical Way to Assess Stimulus Perception. *J. Equine Vet. Sci.* **2009**, *29*, 739–742.
15. Ekman, P. Facial expression and emotion. *Am. Psychol.* **1993**, *48*, 384–392.
16. Keltner, D.; Ekman, P.; Gonzaga, G.C.; Beer, J. Facial expression of emotion. *Am. Psychol.* **1993**, *48*, 173–183.
17. Ekman, P.; Levenson, R.W.; Friesen, W.V. Autonomic nervous system activity distinguishes among emotions. *Science* **1983**, *221*, 1208–1210.
18. Pluta, M.; Osiński, Z. Variability of heart rate in primitive horses and their relatives as an indicator of stress level, behavioral conduct towards humans and adaptation to living in wild. *Bull. Vet. Inst. Pulawy* **2014**, *58*, 495–501.
19. Nardelli, M.; Valenza, G.; Greco, A.; Lanata, A.; Scilingo, E.P. Recognizing emotions induced by affective sounds through heart rate variability. *IEEE Trans. Affect. Comput.* **2015**, *6*, 385–394.
20. von Lewinski, M.; Biau, S.; Erber, R.; Ille, N.; Aurich, J.; Faure, J.M.; Möstl, E.; Aurich, C. Cortisol release, heart rate and heart rate variability in the horse and its rider: Different responses to training and performance. *Vet. J.* **2013**, *197*, 229–232.
21. Strzelec, K.; Kędzierski, W.; Bereznowski, A.; Janczarek, I.; Bocian, K.; Radosz, M. Salivary cortisol Levels in horses and their riders during three-day-events. *Bull. Vet. Inst. Pulawy* **2013**, *57*, 237–241.
22. Kang, O.D.; Lee, W.S. Changes in Salivary Cortisol Concentration in Horses during Different Types of Exercise. *Asian-Australas. J. Anim. Sci.* **2016**, *29*, doi:10.5713/ajas.16.0009.
23. Fazio, E.; Medica, P.; Aveni, F.; Ferlazzo, A. The potential role of training sessions on the temporal and spatial physiological patterns in young Friesian horses. *J. Equine Vet. Sci.* **2016**, *47*, 84–91.
24. Van Dyke Parunak, H.; Bisson, R.; Brueckner, S.; Matthews, R.; Sauter, J. A model of emotions for situated agents. In Proceedings of the Fifth International Joint Conference on Autonomous Agents and Multiagent Systems, Hakodate, Japan, 8–12 May 2006; ACM: New York, NY, USA, 2006; pp. 993–995.
25. Smith, A.V.; Proops, L.; Grounds, K.; Wathan, J.; McComb, K. Functionally relevant responses to human facial expressions of emotion in the domestic horse (*Equus caballus*). *Biol. Lett.* **2016**, *12*, doi:10.1098/rsbl.2015.0907.
26. Brandt, K. A language of their own: An interactionist approach to human-horse communication. *Soc. Anim.* **2004**, *12*, 299–316.

27. Rochais, C.; Henry, S.; Sankey, C.; Nassur, F.; Gorecka-Bruzda, A.; Hausberger, M. Visual attention, an indicator of human-animal relationships? A study of domestic horses (*Equus caballus*). *Front. Psychol.* **2014**, *5*, doi:10.3389/fpsyg.2014.00108.

28. Gehrke, E.K.; Baldwin, A.; Schiltz, P.M. Heart rate variability in horses engaged in equine-assisted activities. *J. Equine Vet. Sci.* **2011**, *31*, 78–84.

29. Visser, E.; Van Reenen, C.; Van der Werf, J.; Schilder, M.; Knaap, J.; Barneveld, A.; Blokhuis, H. Heart rate and heart rate variability during a novel object test and a handling test in young horses. *Physiol. Behav.* **2002**, *76*, 289–296.

30. Popescu, S.; Diugan, E.A. The relationship between behavioral and other welfare indicators of working horses. *J. Equine Vet. Sci.* **2013**, *33*, 1–12.

31. Forkman, B.; Boissy, A.; Meunier-Salaün, M.C.; Canali, E.; Jones, R. A critical review of fear tests used on cattle, pigs, sheep, poultry and horses. *Physiol. Behav.* **2007**, *92*, 340–374.

32. Birke, L.; Brandt, K. Mutual corporeality: Gender and human/horse relationships. In *Women's Studies International Forum*; Elsevier: Amsterdam, The Netherlands, 2009; Volume 32, pp. 189–197.

33. Proops, L.; McComb, K.; Reby, D. Cross-modal individual recognition in domestic horses (Equus caballus). *Proc. Natil. Acad. Sci. USA* **2009**, *106*, 947–951.

34. Proops, L.; McComb, K. Cross-modal individual recognition in domestic horses (Equus caballus) extends to familiar humans. *Proc. R. Soc. Lond. B Biol. Sci.* **2012**, *279*, 3131–3138.

35. MacLean, B. Equine-assisted therapy. *J. Rehabil. Res. Dev.* **2011**, *48*, ix–xii.

36. Christian, J.E. All creatures great and small utilizing equine-assisted therapy to treat eating disorders. *J. Psychol. Christianity* **2005**, *24*, 65–67.

37. Piette, D.; Norton, T.; Exadaktylos, V.; Berckmans, D. Real-time monitoring of the horse-rider dyad using body sensor network technology. In Proceedings of the 2016 IEEE 13th International Conference on Wearable and Implantable Body Sensor Networks (BSN), San Francisco, CA, USA, 14–17 June 2016; IEEE: New York, NY, USA, 2016; pp. 287–291.

38. Jansen, F.; Van der Krogt, J.; Van Loon, K.; Avezzu, V.; Guarino, M.; Quanten, S.; Berckmans, D. Online detection of an emotional response of a horse during physical activity. *Vet. J.* **2009**, *181*, 38–42.

39. Crews, D. The Bond Between a Horse and a Human. Arizona State University: Phoenix, Arizona, July 2009.

40. Sankey, C.; Henry, S.; André, N.; Richard-Yris, M.A.; Hausberger, M. Do horses have a concept of person? *PLoS ONE* **2011**, *6*, doi:10.1371/journal.pone.0018331.

41. Sankey, C.; Richard-Yris, M.A.; Leroy, H.; Henry, S.; Hausberger, M. Positive interactions lead to lasting positive memories in horses, *Equus caballus*. *Anim. Behav.* **2010**, *79*, 869–875.

42. Electrophysiology, Task Force of the European Society of Cardiology the North American Society of Pacing. Heart rate variability standards of measurement, physiological interpretation, and clinical use. *Eur. Heart J.* **1996**, *17*, 354–381.

43. Narayanan, K.; Govindan, R.; Gopinathan, M. Unstable periodic orbits in human cardiac rhythms. *Phys. Rev. E* **1998**, *57*, 4594–4602.

44. Poon, C.S.; Merrill, C.K. Decrease of cardiac chaos in congestive heart failure. *Nature* **1997**, *389*, 492–495.

45. Keogh, E.J.; Pazzani, M.J. Scaling up dynamic time warping for datamining applications. In Proceedings of the Sixth ACM SIGKDD International Conference on Knowledge Discovery and Data Mining, Boston, MA, USA, 20–23 August 2000; ACM: New York, NY, USA, 2000; pp. 285–289.

46. Rabiner, L.; Juang, B.H. *Fundamentals of Speech Recognition*; Signal Processing Series; Prentice Hall: Englewood Cliffs, NJ, USA, 1993.

47. Caiani, E.; Porta, A.; Baselli, G.; Turie, M.; Muzzupappa, S.; Pieruzzi, F.; Crema, C.; Malliani, A.; Cerutti, S. Warped-average template technique to track on a cycle-by-cycle basis the cardiac filling phases on left ventricular volume. In Proceedings of the 1998 Computers in Cardiology, Cleveland, OH, USA, 13–16 September 1998; IEEE: Cleveland, OH, USA, 1998; pp. 73–76.

48. Valenza, G.; Nardelli, M.; Lanata, A.; Gentili, C.; Bertschy, G.; Kosel, M.; Scilingo, E.P. Predicting Mood Changes in Bipolar Disorder through Heartbeat Nonlinear Dynamics. *IEEE J. Biomed. Health Inform.* **2016**, *20*, 1034–1043.

49. Guidi, A.; Salvi, S.; Ottaviano, M.; Gentili, C.; Bertschy, G.; de Rossi, D.; Scilingo, E.P.; Vanello, N. Smartphone Application for the Analysis of Prosodic Features in Running Speech with a Focus on Bipolar Disorders: System Performance Evaluation and Case Study. *Sensors* **2015**, *15*, 28070–28087.

50. Lorussi, F.; Rocchia, W.; Scilingo, E.P.; Tognetti, A.; De Rossi, D. Wearable, redundant fabric-based sensor arrays for reconstruction of body segment posture. *IEEE Sens. J.* **2004**, *4*, 807–818.

51. Lanata, A.; Valenza, G.; Scilingo, E.P. A novel EDA glove based on textile-integrated electrodes for affective computing. *Med. Biol. Eng. Comput.* **2012**, *50*, 1163–1172.

52. Greco, A.; Valenza, G.; Nardelli, M.; Bianchi, M.; Citi, L.; Scilingo, E.P. Force-Velocity Assessment of Caress-like Stimuli through the Electrodermal Activity Processing: Advantages of a Convex Optimization Approach. *IEEE Trans. Hum.-Mach. Syst.* **2016**, 1–10, doi:10.1109/THMS.2016.2586478.

53. Ödman, S.; Öberg, P.A. Movement-induced potentials in surface electrodes. *Med. Biol. Eng. Comput.* **1982**, *20*, 159–166.

54. Webster, J. *Medical Instrumentation: Application and Design*; John Wiley & Sons: Hoboken, NJ, USA, 2009.

55. De Talhouet, H.; Webster, J.G. The origin of skin-stretch-caused motion artifacts under electrodes. *Physiol. Meas.* **1996**, *17*, 81–93.

56. Thakor, N.; Webster, J. The origin of skin potential and its variations. In Proceedings of the 31st Annual Conference on Engineering in Medicine and Biology, Atlanta, GA, USA, 1978; Volume 20, p. 212.

57. Milanesi, M.; Martini, N.; Vanello, N.; Positano, V.; Santarelli, M.; Paradiso, R.; De Rossi, D.; Landini, L. Multichannel techniques for motion artifacts removal from electrocardiographic signals. In Proceedings of the 28th Annual International Conference of the IEEE Engineering in Medicine and Biology Society, EMBS'06, New York, NY, USA, 31 August –3 September 2006; IEEE: New York, NY, USA, 2006; pp. 3391–3394.

58. Martini, N.; Milanesi, M.; Vanello, N.; Positano, V.; Santarelli, M.; Landini, L. A real-time adaptive filtering approach to motion artefacts removal from ECG signals. *Int. J. Biomed. Eng. Technol.* **2010**, *3*, 233–245.

59. Scheffer, C.; van Oldruitenborgh-Oosterbaan, M.S. Computerized ECG recording in horses during a standardized exercise test. *Vet. Q.* **1996**, *18*, 2–7.

60. Vitale, V.; Balocchi, R.; Varanini, M.; Sgorbini, M.; Macerata, A.; Sighieri, C.; Baragli, P. The effects of restriction of movement on the reliability of heart rate variability measurements in the horse (*Equus caballus*). *J. Vet. Behav. Clin. Appl. Res.* **2013**, *8*, 400–403.

61. Young, L.; Van Loon, G. Diseases of the heart and vessels. In *Equine Sports Medicine and Surgery: Basic and Clinical Sciences of Equine Athlete*; Elsevier: Amsterdam, The Netherlands, 2013; pp. 695–744.

62. Stoppa, M.; Chiolerio, A. Wearable electronics and smart textiles: A critical review. *Sensors* **2014**, *14*, 11957–11992.

63. Suh, M. E-textiles for wearability: Review of integration technologies. Available online: http://www.textileworld.com/textile-world/features/2010/04/e-textiles-for-wearability-review-of-integration-technologies/ (accessed on 20 April 2010).

64. Gargiulo, G.; Bifulco, P.; Calvo, R.A.; Cesarelli, M.; Jin, C.; Van Schaik, A. Mobile biomedical sensing with dry electrodes. In Proceedings of the 2008 International Conference on Intelligent Sensors, Sensor Networks and Information Processing, ISSNIP 2008, Sydney, Australia, 15–18 December 2008; IEEE: New York, NY, USA, 2008; pp. 261–266.

65. Seoane, F.; Ferreira, J.; Alvarez, L.; Buendia, R.; Ayllón, D.; Llerena, C.; Gil-Pita, R. Sensorized garments and textrode-enabled measurement instrumentation for ambulatory assessment of the autonomic nervous system response in the atrec project. *Sensors* **2013**, *13*, 8997–9015.

66. Carvalho, H.; Catarino, A.P.; Rocha, A.; Postolache, O. Health monitoring using textile sensors and electrodes: An overview and integration of technologies. In Proceedings of the 2014 IEEE International Symposium on Medical Measurements and Applications (MeMeA), ISCTE, University of Lisbon, Lisbon, Portugal, 11–12 June 2014; IEEE: New York, NY, USA, 2014; pp. 1–6.

67. Lanatà, A.; Valenza, G.; Greco, A.; Gentili, C.; Bartolozzi, R.; Bucchi, F.; Frendo, F.; Scilingo, E.P. How the Autonomic Nervous System and Driving Style Change With Incremental Stressing Conditions during Simulated Driving. *IEEE Trans. Intell. Transp. Syst.* **2015**, *16*, 1505–1517.

68. Zito, D.; Pepe, D.; Neri, B.; De Rossi, D.; Lanata, A.; Tognetti, A.; Scilingo, E.P. Wearable system-on-a-chip UWB radar for health care and its application to the safety improvement of emergency operators. In Proceedings of the 2007 29th Annual International Conference of the IEEE Engineering in Medicine and Biology Society, Lyon, France, 23–26 August 2007; IEEE: New York, NY, USA, 2007; pp. 2651–2654.

69. Ishijima, M. Cardiopulmonary monitoring by textile electrodes without subject-awareness of being monitored. *Med. Biol. Eng. Comput.* **1997**, *35*, 685–690.

70. Paradiso, R.; Gemignani, A.; Scilingo, E.; De Rossi, D. Knitted bioclothes for cardiopulmonary monitoring. In Proceedings of the 25th Annual International Conference of the IEEE Engineering in Medicine and Biology Society, Cancun, Mexico, 17–21 September 2003; IEEE: New York, NY, USA, 2003; Volume 4, pp. 3720–3723.

71. Milanesi, M.; Vanello, N.; Positano, V.; Santarelli, M.; Paradiso, R.; Rossi, D.D.; Landini, L. Frequency domain approach to blind source separation in ECG monitoring by wearable system. In Proceedings of the 2005 Computers in Cardiology, Lyon, France, 25–28 September 2005; IEEE: New York, NY, USA, 2005; pp. 767–770.

72. Scilingo, E.P.; Gemignani, A.; Paradiso, R.; Taccini, N.; Ghelarducci, B.; De Rossi, D. Performance evaluation of sensing fabrics for monitoring physiological and biomechanical variables. *IEEE Trans. Inf. Technol. Biomed.* **2005**, *9*, 345–352.

73. Watanabe, K.; Watanabe, T.; Watanabe, H.; Ando, H.; Ishikawa, T.; Kobayashi, K. Noninvasive measurement of heartbeat, respiration, snoring and body movements of a subject in bed via a pneumatic method. *IEEE Trans. Biomed. Eng.* **2005**, *52*, 2100–2107.

74. Paradiso, R.; Bianchi, A.; Lau, K.; Scilingo, E. Psyche: Personalised monitoring systems for care in mental health. In Proceedings of the 2010 Annual International Conference of the IEEE Engineering in Medicine and Biology Society (EMBC), Buenos Aires, Argentina, 1–4 September 2010; IEEE: New York, NY, USA, 2010; pp. 3602–3605.

75. Peltokangas, M.; Verho, J.; Vehkaoja, A. Night-time EKG and HRV monitoring with bed sheet integrated textile electrodes. *IEEE Trans. Inf. Technol. Biomed.* **2012**, *16*, 935–942.

76. Lanata, A.; Valenza, G.; Mancuso, C.; Scilingo, E.P. Robust multiple cardiac arrhythmia detection through bispectrum analysis. *Expert Syst. Appl.* **2011**, *38*, 6798–6804.

77. Zito, D.; Pepe, D.; Neri, B.; Zito, F.; De Ross, D.; Lanatà, A. Feasibility study and design of a wearable system-on-a-chip pulse radar for contactless cardiopulmonary monitoring. *Int. J. Telemed. Appl.* **2008**, *2008*, 6.

78. Zito, D.; Pepe, D.; Mincica, M.; Zito, F.; De Rossi, D.; Lanata, A.; Scilingo, E.; Tognetti, A. Wearable system-on-a-chip UWB radar for contact-less cardiopulmonary monitoring: Present status. In Proceedings of the 2008 30th Annual International Conference of the IEEE Engineering in Medicine and Biology Society, Vancouver, BC, Canada, 21–24 August 2008; IEEE: New York, NY, USA, 2008; pp. 5274–5277.

79. Lanata, A.; Scilingo, E.P.; De Rossi, D. A multimodal transducer for cardiopulmonary activity monitoring in emergency. *IEEE Trans. Inf. Technol. Biomed.* **2010**, *14*, 817–825.

80. Betella, A.; Zucca, R.; Cetnarski, R.; Greco, A.; Lanatà, A.; Mazzei, D.; Tognetti, A.; Arsiwalla, X.D.; Omedas, P.; De Rossi, D.; et al. Inference of human affective states from psychophysiological measurements extracted under ecologically valid conditions. In *Using Neurophysiological Signals that Reflect Cognitive or Affective State, Frontiers in Neuroscience*; Frontiers Media SA: Lausanne, Switzerland, 2014; p. 66.

81. Zhang, Y.T.; Poon, C.C.; Chan, C.H.; Tsang, M.W.; Wu, K.F. A health-shirt using e-textile materials for the continuous and cuffless monitoring of arterial blood pressure. In Proceedings of the 2006 3rd IEEE/EMBS International Summer School on Medical Devices and Biosensors, Cambridge, MA, USA, 4–6 September 2006; IEEE: New York, NY, USA, 2006; pp. 86–89.

82. Chan, C.; Zhang, Y. Continuous and long-term arterial blood pressure monitoring by using h-Shirt. In Proceedings of the 2008 International Conference on Information Technology and Applications in Biomedicine (ITAB 2008), Shenzhen, China, 30–31 May 2008; IEEE: New York, NY, USA, 2008; pp. 267–269.

83. Arshad, A. A Study on Health Monitoring System: Recent Advancements. *IIUM Eng. J.* **2014**, *15*, 87–99.

84. Baha, S. T wave shape as fitness indicator in racehorse. A study by the Holter method. *Rev. Méd. Vét.* **1991**, *142*, 125–129.

85. Raekallio, M. Long term ECG recording with Holter monitoring in clinically healthy horses. *Acta Vet. Scand.* **1991**, *33*, 71–75.

86. Reef, V.; Marr, C.; Hammett, B. Holter monitoring in the management of atrial fibrillation following conversion. In Proceedings of the 11th American College of Veterinary Internal Medicine Forum, Washington, DC, USA, 20 May 1993; pp. 610–613.

87. Scheffer, C.; Robben, J.; Sloet, O.O.M. Continuous monitoring of ECG in horses at rest and during exercise. *Vet. Rec.* **1995**, *137*, 371–374.

88. Pantelopoulos, A.; Bourbakis, N.G. A survey on wearable sensor-based systems for health monitoring and prognosis. *IEEE Trans. Syst. Man Cybern. C (Appl. Rev.)* **2010**, *40*, 1–12.

89. Lanata, A.; Guidi, A.; Baragli, P.; Paradiso, R.; Valenza, G.; Scilingo, E.P. Removing movement artifacts from equine ECG recordings acquired with textile electrodes. In Proceedings of the 2015 37th Annual International Conference of the IEEE Engineering in Medicine and Biology Society (EMBC), Milan, Italy, 25–29 August 2015; IEEE: New York, NY, USA, 2015; pp. 1955–1958.

90. Pandian, P.; Mohanavelu, K.; Safeer, K.; Kotresh, T.; Shakunthala, D.; Gopal, P.; Padaki, V. Smart Vest: Wearable multi-parameter remote physiological monitoring system. *Med. Eng. Phys.* **2008**, *30*, 466–477.

91. Lanata, A.; Guidi, A.; Baragli, P.; Valenza, G.; Scilingo, E.P. A Novel Algorithm for Movement Artifact Removal in ECG Signals Acquired from Wearable Systems Applied to Horses. *PLoS ONE* **2015**, *10*, e0140783.

92. Lanatà, A.; Scilingo, E.P.; Nardini, E.; Loriga, G.; Paradiso, R.; De-Rossi, D. Comparative evaluation of susceptibility to motion artifact in different wearable systems for monitoring respiratory rate. *IEEE Trans. Inf. Technol. Biomed.* **2010**, *14*, 378–386.

93. Valenza, G.; Citi, L.; Gentili, C.; Lanata, A.; Scilingo, E.P.; Barbieri, R. Characterization of depressive states in bipolar patients using wearable textile technology and instantaneous heart rate variability assessment. *IEEE J. Biomed. Health Inform.* **2015**, *19*, 263–274.

94. Valenza, G.; Lanatà, A.; Scilingo, E.P.; Rossi, D.D. Towards a smart glove: Arousal recognition based on textile electrodermal response. In Proceedings of the 32nd Annual International Conference of the IEEE Engineering in Medicine and Biology Society, Buenos Aires, Argentina, 1–4 September 2010; IEEE: New York, NY, USA, 2010; pp. 3598–3601.

95. Valenza, G.; Gentili, C.; Lanatà, A.; Scilingo, E.P. Mood recognition in bipolar patients through the PSYCHE platform: Preliminary evaluations and perspectives. *Artif. Intell. Med.* **2013**, *57*, 49–58.

96. Pacelli, M.; Loriga, G.; Taccini, N.; Paradiso, R. Sensing fabrics for monitoring physiological and biomechanical variables: E-textile solutions. In Proceedings of the 2006 3rd IEEE/EMBS International Summer School on Medical Devices and Biosensors, Cambridge, MA, USA, 4–6 September 2006; pp. 1–4.

97. Verheyen, T.; Decloedt, A.; De Clercq, D.; Deprez, P.; Sys, S.; van Loon, G. Electrocardiography in horses, part 1: How to make a good recording. *Vlaams Diergeneeskd. Tijdschr.* **2010**, *79*, 331–336.

98. Saul, J.P. Beat-to-beat variations of heart rate reflect modulation of cardiac autonomic outflow. *Physiology* **1990**, *5*, 32–37.

99. Berntson, G.G. Heart rate variability: Origins, methods, and interpretive caveats. *Psychophysiology* **1997**, *34*, 623–648.

100. Acharya, U.R.; Joseph, K.P.; Kannathal, N.; Lim, C.M.; Suri, J.S. Heart rate variability: A review. *Med. Biol. Eng. Comput.* **2006**, *44*, 1031–1051.

101. Rodríguez-Liñares, L.; Méndez, A.J.; Lado, M.J.; Olivieri, D.N.; Vila, X.A.; Gómez-Conde, I. An open source tool for heart rate variability spectral analysis. *Comput. Methods Progr. Biomed.* **2011**, *103*, 39–50.

102. Guzzetti, S.; Signorini, M.G.; Cogliati, C.; Mezzetti, S.; Porta, A.; Cerutti, S.; Malliani, A. Non-linear dynamics and chaotic indices in heart rate variability of normal subjects and heart-transplanted patients. *Cardiovasc. Res.* **1996**, *31*, 441–446.

103. Valenza, G.; Nardelli, M.; Bertschy, G.; Lanata, A.; Scilingo, E. Mood states modulate complexity in heartbeat dynamics: A multiscale entropy analysis. *Europhys. Lett.* **2014**, *107*, doi:10.1209/0295-5075/107/18003/meta.

104. Valenza, G.; Greco, A.; Gentili, C.; Lanata, A.; Sebastiani, L.; Menicucci, D.; Gemignani, A.; Scilingo, E.P. Combining electroencephalographic activity and instantaneous heart rate for assessing brain–heart dynamics during visual emotional elicitation in healthy subjects. *Phil. Trans. R. Soc. A.* **2016**, *374*, 441–446.

105. Tarvainen, M.P.; Niskanen, J.P.; Lipponen, J.A.; Ranta-Aho, P.O.; Karjalainen, P.A. Kubios HRV–heart rate variability analysis software. *Comput. Methods Progr. Biomed.* **2014**, *113*, 210–220.

106. Pan, J.; Tompkins, W.J. A real-time QRS detection algorithm. *IEEE Trans. Biomed. Eng.* **1985**, *3*, 230–236.

107. Singh, D.; Vinod, K.; Saxena, S. Sampling frequency of the RR interval time series for spectral analysis of heart rate variability. *J. Med. Eng. Technol.* **2004**, *28*, 263–272.

108. Anosov, O.; Patzak, A.; Kononovich, Y.; Persson, P.B. High-frequency oscillations of the heart rate during ramp load reflect the human anaerobic threshold. *Eur. J. Appl. Physiol.* **2000**, *83*, 388–394.

109. Allen, J.J. Calculating metrics of cardiac chronotropy: A pragmatic overview. *Psychophysiology* **2002**, *39*, S18.

110. Müller, M. Dynamic time warping. In *Information Retrieval for Music and Motion*; Springer-Verlag: Berlin, Heidelberg, 2007; pp. 69–84.

111. Benjamini, Y.; Hochberg, Y. Controlling the false discovery rate: A practical and powerful approach to multiple testing. *J. R. Stat. Soc. Ser. B (Methodol.)* **1995**, *57*, 289–300.

112. Lanata, A.; Greco, A.; Valenza, G.; Scilingo, E.P. A pattern recognition approach based on electrodermal response for pathological mood identification in bipolar disorders. In Proceedings of the 2014 IEEE International Conference on Acoustics, Speech and Signal Processing (ICASSP), Florence, Italy, 4–9 May 2014; IEEE: New York, NY, USA, 2014; pp. 3601–3605.

113. Nardelli, M.; Valenza, G.; Cristea, I.A.; Gentili, C.; Cotet, C.; David, D.; Lanata, A.; Scilingo, E.P. Characterizing psychological dimensions in non-pathological subjects through autonomic nervous system dynamics. *Front. Comput. Neurosci.* **2015**, *9*, doi:10.3389/fncom.2015.00037.

114. Lanata, A.; Valenza, G.; Nardelli, M.; Gentili, C.; Scilingo, E.P. Complexity index from a personalized wearable monitoring system for assessing remission in mental health. *IEEE J. Biomed. Health Inform.* **2015**, *19*, 132–139.

115. Quer, G.; Daftari, J.; Rao, R.R. Heart rate wavelet coherence analysis to investigate group entrainment. *Pervasive Mob. Comput.* **2016**, *28*, 21–34.

116. Lanata, A.; Valenza, G.; Scilingo, E.P. Eye gaze patterns in emotional pictures. *J. Ambient Intell. Humaniz. Comput.* **2013**, *4*, 705–715.

117. Lazzeri, N.; Mazzei, D.; Greco, A.; Rotesi, A.; Lanatà, A.; De Rossi, D.E. Can a humanoid face be expressive? A psychophysiological investigation. *Front. Bioeng. Biotechnol.* **2015**, *3*, doi:10.3389/fbioe.2015.00064.

118. Alfonso, S.V.; Alfonso, L.A.; Llabre, M.M.; Fernandez, M.I. Project Stride: An Equine-Assisted Intervention to Reduce Symptoms of Social Anxiety in Young Women. *Explor. J. Sci. Heal.* **2015**, *11*, 461–467.

119. Bass, M.M.; Duchowny, C.A.; Llabre, M.M. The effect of therapeutic horseback riding on social functioning in children with autism. *J. Autism Dev. Disord.* **2009**, *39*, 1261–1267.

120. Inderbitzin, M.P.; Betella, A.; Lanatá, A.; Scilingo, E.P.; Bernardet, U.; Verschure, P.F. The social perceptual salience effect. *J. Exp. Psychol. Hum. Percept. Perform.* **2013**, *39*, 62–74.

121. Lanatà, A.; Armato, A.; Valenza, G.; Scilingo, E.P. Eye tracking and pupil size variation as response to affective stimuli: A preliminary study. In Proceedings of the 2011 5th International Conference on Pervasive Computing Technologies for Healthcare (PervasiveHealth), Dublin, Ireland, 23–26 May 2011; IEEE: New York, NY, USA, 2011; pp. 78–84.

electronics

MDPI

Article

Automatic Measurement of Chew Count and Chewing Rate during Food Intake

Muhammad Farooq and Edward Sazonov *

Department of Electrical Engineering, The University of Alabama, Tuscaloosa, AL 35487, USA;
mfarooq@crimson.ua.edu
* Correspondance: esazonov@eng.ua.edu; Tel.: +1-205-348-1981

Academic Editors: Enzo Pasquale Scilingo and Gaetano Valenza
Received: 19 July 2016; Accepted: 8 September 2016; Published: 23 September 2016

Abstract: Research suggests that there might be a relationship between chew count as well as chewing rate and energy intake. Chewing has been used in wearable sensor systems for the automatic detection of food intake, but little work has been reported on the automatic measurement of chew count or chewing rate. This work presents a method for the automatic quantification of chewing episodes captured by a piezoelectric sensor system. The proposed method was tested on 120 meals from 30 participants using two approaches. In a semi-automatic approach, histogram-based peak detection was used to count the number of chews in manually annotated chewing segments, resulting in a mean absolute error of $10.40\% \pm 7.03\%$. In a fully automatic approach, automatic food intake recognition preceded the application of the chew counting algorithm. The sensor signal was divided into 5-s non-overlapping epochs. Leave-one-out cross-validation was used to train a artificial neural network (ANN) to classify epochs as "food intake" or "no intake" with an average $F1$ score of 91.09%. Chews were counted in epochs classified as food intake with a mean absolute error of $15.01\% \pm 11.06\%$. The proposed methods were compared with manual chew counts using an analysis of variance (ANOVA), which showed no statistically significant difference between the two methods. Results suggest that the proposed method can provide objective and automatic quantification of eating behavior in terms of chew counts and chewing rates.

Keywords: chewing rate; food intake detection; piezoelectric sensor; artificial neural network; feature computation; chew counting; peak detection

1. Introduction

Excessive eating can cause a drastic increase in weight; therefore, it is important to study the eating behavior of individuals to understand eating patterns contributing towards obesity. Similarly, people suffering from eating disorders experience changes in their normal eating patterns, causing individuals to either eat excessive or insufficient food compared to their body energy requirements. Obesity and eating disorders present a significant public health problem. Statistics from the World Health Organization (WHO) shows that worldwide, obesity is the fifth largest cause of preventable deaths [1]. People with anorexia nervosa have a shorter life expectancy (about 18 times less) in comparison to the people who are not suffering from this condition [2]. People suffering from binge-eating are at higher risk of cardiovascular diseases and high blood pressure [2]. Therefore, there is a need to study the food intake patterns and eating behavior of individuals to better understand patterns and factors contributing to the obesity and eating disorders.

Traditional methods for monitoring food intake patterns such as food frequency questionnaires, food records, and 24-h food recall rely on self-report of the participants [3–5]. Research suggests that, during self-report, participants tend to underestimate their intake where the underestimation varies between 10% and 50% of the food consumed [6]. Apart from being inaccurate, self-reporting also puts the burden on the patients because of the need for their active involvement [6]. Therefore, recent research efforts have focused on the development of methods and techniques that are accurate, objective, and automatic and reduce the patient's burden of self-reporting their intake.

A number of wearable sensor systems and related pattern recognition algorithms have been proposed for monitoring ingestive behavior. The systems relying on chewing as an indicator of food intake use such sensor modalities as in-ear miniature microphones [7,8], accelerometers [9], surveillance video cameras [10], and piezoelectric sensors on the throat [11] or on the jaw [12–14]. In [7], chewing sounds recorded with a miniature microphone in the outer-ear were used to train hidden Markov models (HMMs) to differentiate sequences of food intake from sequences of no intake with an accuracy of 83%. In [15], an earpiece consisting of a 3D gyroscope and three proximity sensors that detected chewing by measuring ear canal deformations during food intake was proposed. Participant-dependent HMMs achieved an average classification accuracy of 93% for food intake detection. In [9], the use of a single axis accelerometer placed on the temporalis was proposed to monitor chewing during eating episodes in laboratory experiments. Several different classification techniques (decision tree (DT), nearest neighbor (NN), multi-layer perceptron (MLP), support vector machine (SVM) and weighted SVM (WSVM)) were compared, and WSVM achieved the highest accuracy of about 96% for detection of food intake. Our research group has been developing systems for monitoring ingestive behavior via chewing [12–14,16]. In [12], features computed from piezoelectric film sensors placed below the ear were used to train SVM and artificial neural network (ANN) models to differentiate between epochs of food intake and no intake with average accuracies of 81% and 86%, respectively. In [13], this system was tested by 12 participants in free living conditions for 24 h each. Sensor signals were divided into 30-s epochs, and for each epoch a feature vector consisting of 68 features was computed. Participant-independent ANN models differentiated between epochs of food intake and no intake with an average accuracy of 89% using leave-one-out cross-validation.

In recent years, researchers have focused on the automatic detection of chewing, but little work has been done on the quantification of chewing behavior, which may be an important factor in studying energy intake. Although no direct relationship has been established between obesity and chewing patterns, several studies have shown that increased mastication before swallowing of the food may reduce the total energy intake [17–21]. Results in [18] showed that obese participants had higher intake rates with lower numbers of chews per 1 g of food (pork pie) compared with the normal weight group. They also showed that an increase in the number of chews per bite decreased final food intake in both obese and normal weight participants. In [19], 45 participants were asked to eat pizza over four lunch sessions. Participants were asked to have 100%, 150%, and 200% of the number of chews of their baseline number (first visit) of chews per bite before swallowing. According to the authors, food intake (total mass intake) reduction of 9.5% and 14.8% was observed for chewing rates of 150% and 200%, respectively, compared with the 100% session. Our research demonstrated that the number of chews per meal may be used in estimating the energy intake if the energy density of the food is given. In [22], individually calibrated models were presented to estimate the energy intake from the counts of chews and swallows.

Most of these studies relied on the manual counting of chews either by the participants or by the investigators, either from videotapes or by watching participants in real time. Thus, there is a need to develop methods to provide objective and automatic estimation of chew counts and chewing rates. In recent studies, semi-automatic chew counting systems utilizing piezoelectric strain sensors have been proposed [23–25]. In [23,24], a modified form of the sensor system proposed in [13] was used to quantify the sucking count of 10 infants by using zero crossing. In [25], an algorithm was proposed for counting chews from a piezoelectric strain sensor and printed sensor signals. A group of five adult participants counted the number of chews taken while eating three different food items and marked each chewing episode with push button signals. A peak detection algorithm was used to count chews from known chewing segments (based on push button signals) with a mean absolute error of 8% (for both sensors). An example of chewing sound use, the chewing rate and bite weight for three different food types was estimated from the sounds captured by a miniature microphone in [8]. A possible limitation of the acoustic-based approach is its sensitivity to the environmental noise, which might require a reference microphone for noise cancellation [7]. For these systems to be useful in free living conditions, fully automated solutions are needed which can not only automatically recognize chewing sequences but can also quantify chewing behavior in terms of chew counts and chewing rates.

The goal of this paper is to present a method for automatic detection and quantification of the chew counts and chewing rates from piezoelectric film sensor signals. Main contributions of this work include the design of the proposed system as means of automatic and objective quantification of chewing behavior (chew counts and chewing rates), which could be used in studying and understanding the chewing behavior of individuals and its relation to the energy intake without relying on manual chew counts. Our previous research has shown that the piezoelectric strain sensor can be used for objective monitoring of eating in unrestricted free living conditions [13]. The approach presented in this work could be extended [21–26] to free living environments, studying the relation of chewing patterns, energy intake, and obesity in community-dwelling individuals. The system implements a fully automatic approach that first detects the intake of foods requiring chewing and then characterizes the chewing in terms of chew counts and chewing rates. Another contribution of this work is the testing of the piezoelectric sensor and related signal processing and pattern recognition algorithms in a relatively larger population in multi-day experiments with a wide variety of foods, which are representative of the daily diet.

2. Methods

2.1. Data Collection Protocol

For this study, 30 participants were recruited. The population consisted of 15 male and 15 female participants with an average age of 29.03 ± 12.20 years and a range of 19–58 years. The average body mass index (BMI) of the population (in kg/m^2) was 27.87 ± 5.51 with a range of 20.5 to 41.7. Each participant came for four visits (a total of 120 experiments). Data from 16 experiments was discarded because of equipment failure. The remaining dataset consisted of a total of 104 visits/experiments. Recruited participants did not show any medical conditions which would hinder their normal eating or chewing. An Institutional Review Board approval for this study was obtained from Clarkson University, Potsdam, NY, and all participants signed a consent form before participation.

Participants were divided into three groups based on meal type, i.e., breakfast, lunch, and dinner, and were asked to make two different meal selections from the food items available at one of the cafeterias at Clarkson University to ensure that there was intra-subject variability in food selection. Overall, 110 distinct food items were selected by the participants, on average each participant consumed 1 to 3 food types and 1 or 2 different beverages. Representative food groups selected by the participants can be found in [22]. The wide spectrum of included food items ensures that the proposed algorithms behave well in the foods with varying physical properties eaten by the general population in their daily routine.

The meals were consumed during a visit to a laboratory instrumented for the monitoring of food intake. An accurate and objective reference (gold standard) was needed for the quantification of chewing sequences. At present, obtaining an accurate reference in free living conditions is virtually impossible. Therefore, the study was conducted in a laboratory environment where close observation of the ingestion process was performed with the sensors and a video recording. During each visit, participants were initially instrumented with the sensors [27]. As the first step of the protocol, the participants were asked to remain in a relaxed seated position for 5 min. Second, they were given unlimited time to eat self-selected foods. Participants were allowed to talk, pause food intake, and move around (within the limitations imposed by the sensor system) during the experiment to replicate normal eating behavior. As the final step of the protocol, participants were asked to remain in a relaxed seated position for another 5 min.

2.2. Sensor System and Annotation

A multimodal sensor system was used to monitor participants [26]. A commercially available piezoelectric film sensor (LDT0-028K, from Measurement Specialties Inc., Hampton, VA, USA) was placed below the ear using a medical adhesive for capturing motions of the lower jaw during chewing/mastication of the food. The selected sensor is comprised of a piezoelectric PVDF polymer film (28-µm thickness) and screen-printed Ag-ink electrodes encapsulated in a polyester substrate (0.125-mm thickness). Vibration of the surface to which the sensor is attached creates strain within the piezo-polymer which in turn generates voltage. The selected sensor has a sensitivity of 10 mV per micro-strains, which has been shown to be enough to detect vibrations at the skin surface caused by chewing [26]. A custom-designed amplifier with an input impedance of about 10 MΩ was used to buffer sensor signals. Sensor signals were in the range of 0–2 V, were sampled at f_s = 44,100 Hz with a data acquisition device (USB-160HS-2AO from the Measurement Computing) with a resolution of 16 bits, and were stored in computer memory. This sampling frequency ensures that the sensor will be able to pick speech signals. The total duration of the sensor signal data was around 60 h, and about 26 h of data belonged to food intake. Figure 1a,b shows an example of the piezoelectric film sensor and its attachment to a participant. Figure 2 shows an example of the sensor signal captured during the experiment.

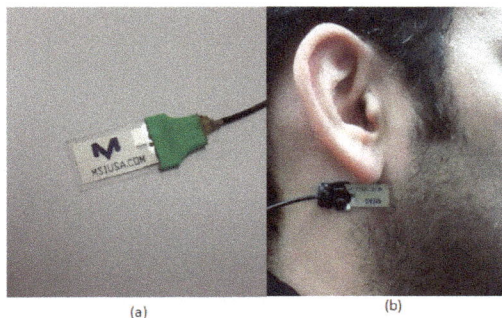

(a)   (b)

Figure 1. (**a**) Piezoelectric film sensor used in the study; (**b**) Sensor attached to a participant.

Experiments were videotaped using a PS3Eye camera (Sony Corp., New York, NY, USA), and videos were time-synchronized with the sensor signals. Custom-built LabVIEW software was used to annotate videos and sensor signals [27]. During the annotation process, videos were reviewed by trained human raters. The custom-built software has the ability to play videos at different speeds and provided a timeline where human raters could mark the start and end of each eating event (bite, chewing, and swallows). The annotation software also allows the user to play the video frame by frame or any specific intervals. This enables the rater to watch the marked chewing segments multiple times for counting chews. Annotated start and end timestamps of chewing along with the corresponding chew counts were stored and used for algorithm development. Further details of the annotation procedure are described in [27]. Inter-rater reliability of the annotation procedure adopted here was established in a previous study, where three raters achieved an intra-class correlation coefficient of 0.988 for chew counting for a sample size of 5 participants [27]. Bites and chewing sequences were marked as food intake, whereas the remaining parts of the sensor signal were marked as non-intake.

For the k-th annotated chewing sequence, annotated chew counts were represented by $CNT(k)$, and the corresponding chewing rate $CR(k)$ was computed as

$$CR(k) = \frac{CNT(k)}{D(k)}, \tag{1}$$

where $D(k)$ is the duration (in seconds) of k-th chewing sequence. The cumulative/total number of annotated chews for each experiment/visit and average chewing rate were represented by $A_{CNT}(n) = \sum_{k=1}^{N} CNT(k)$ and $A_{CR}(n) = \frac{1}{N} \sum_{k=1}^{N} CR(k)$, respectively, where N is the number of chewing sequences in the n-th experiment/visit. The resultant annotated data was used for the development of chew counting algorithms as well as for training and validation of the classification methods.

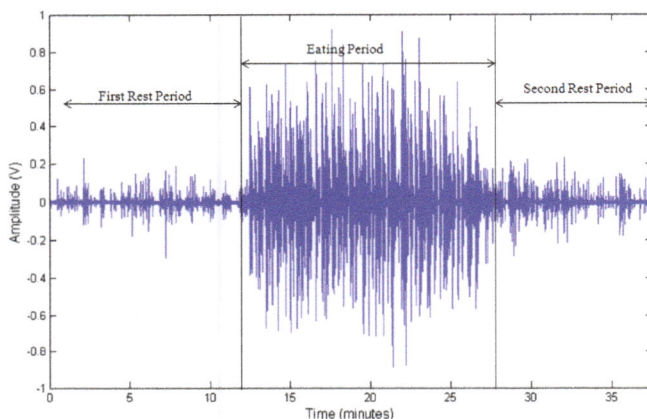

Figure 2. Piezoelectric sensor (raw) signal shows three parts of the experiment. First is a rest period, followed by an eating episode which is followed by a second rest period. Sampling frequency used was 44,100 Hz.

2.3. The Chew Counting Algorithm

To compute the chew counts, the sensor signals were processed in the following manner. All sensor signals were demeaned, i.e., mean amplitude were subtracted from each signal to account for offset drift in the sensor signals. Chewing frequency is in the range of 0.94 to 2 Hz; therefore, a low-pass filter with a cutoff of 3 Hz was used to remove unwanted noise [28]. Mastication/chewing of the food causes bending movements at the jaw which results in variations (peaks and valleys) in the piezoelectric sensor signal. A simple peak detection algorithm could be used to detect the presence of the peaks, and the number of peaks can be used for estimation of the number of chews. To avoid peaks caused by motion artifact such as head movement, a threshold-based peak detection approach was used where peaks were only considered if they were above a given threshold T. For selection of T, a histogram-based approach was adopted, where signal amplitudes were considered as candidates for peaks if they were in the upper α^{th} percentile (details in Section 2.4). Figure 3 shows an example of histogram-based peak detection algorithm where the red line indicates the selected T-value based on the αth percentile. In order for a given amplitude value to be considered as a peak, the value needs to be higher than the selected T-value in the histogram (Figure 3). This example histogram was generated using a single chewing sequence (note sampling frequency is 44,100 Hz). Next, a moving average filter of 100 samples was used to smooth the resultant signal to account for small amplitude variations. The number of peaks in the resultant signal gave an estimate of the number of chews in a given segment. Figure 4 shows an example of the chew counting algorithm. For the nth experiment/visit, the cumulative estimated chew count is given by $E_{CNT}(n)$. For performance evaluation of the proposed chew counting algorithm, the estimated chew counts were compared to the manually annotated chew counts, and errors were computed for each participant (all visits). Both mean errors and mean absolute errors were reported as

$$Error = \frac{1}{M} \sum_{n=1}^{M} \left[\frac{(A_{CNT}(n) - E_{CNT}(n)) * 100}{A_{CNT}(n)} \right], \text{ and} \qquad (2)$$

$$|Error| = \frac{1}{M} \sum_{n=1}^{M} \left| \frac{(A_{CNT}(n) - E_{CNT}(n)) * 100}{A_{CNT}(n)} \right|, \qquad (3)$$

where M was the total number of experiments (visits) in this case. This algorithm was used in two different approaches, i.e., a semi-automatic approach and a fully automatic approach. In the semi-automatic approach, manually annotated chewing segments from the sensor signal were considered for chew counting. In the fully automatic approach, the automatic recognition of food intake (chewing segments) preceded the application of the chew counting algorithm.

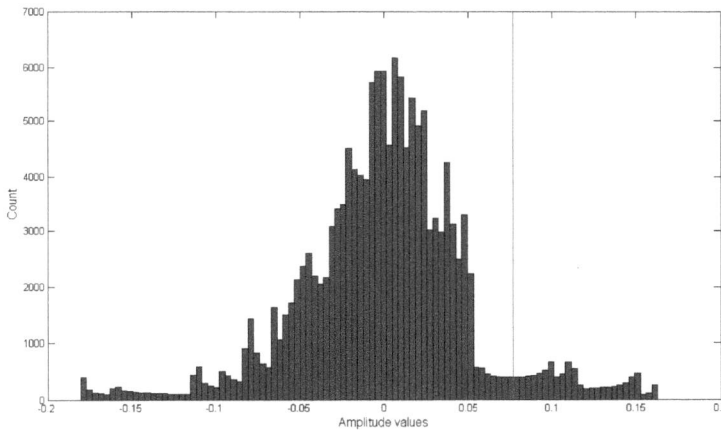

Figure 3. Histogram of a chewing sequence used for the selection of the threshold (T) for peak detection. Leave-one-out cross-validation was performed for the selection of the threshold based on the α-th percentile. The red line shows the selected threshold where only values above the threshold are considered for peak detection.

2.4. Semi-Automatic Approach: Parameter Determination and Validation

In the semi-automatic approach, manually annotated chewing segments were used with the chew counting algorithm. Large amplitude variations were observed in the sensor signal due to different levels of adiposity in the study participants, variations in the sensor placements, and variations in the physical properties of food items requiring different chewing strengths. The peak detection threshold (T) was adapted to the signal amplitude as $T = PERCENTILE(x(k), \alpha)$, where $x(k)$ is the kth manually annotated chewing sequence in the sensor signal and $\alpha \in [0.80, 0.97]$. To avoid the over-fitting of the threshold, a subset of 20 randomly selected visits was used with leave-one-out cross-validation to find the value of α, which gave the minimum mean absolute error (Equation (3)). Data from one visit were withheld, and data from the remaining visits were used to find the value of α, which gave the least mean absolute error by performing a grid search on α in the given range. The selected value of α was used for computing the threshold T for the withheld visit. This process was repeated 20 times such that each visit was used for validation once, resulting in 20 different values of α. An average of these 20 different α values was calculated, and the resultant α was used for chew counting in both semi-automatic and fully automatic approaches. In the semi-automatic approach, for the kth chewing sequence in a visit, the estimated chew counts $CNT\,(k)$ were used for computing corresponding chewing rate $CR\,(k)$ using Equation (1). For the nth visit, for known chewing sequences, cumulative chew counts and average chewing rate were given by $E_{CNT}(n) = \sum_{k=1}^{N} CNT\,(k)$ and $E_{CR}(n) = \frac{1}{N}\sum_{k=1}^{N} CR\,(k)$, respectively. For the semi-automatic approach, the resultant mean error (signed) and mean absolute error (unsigned) were computed using Equations (2) and (3).

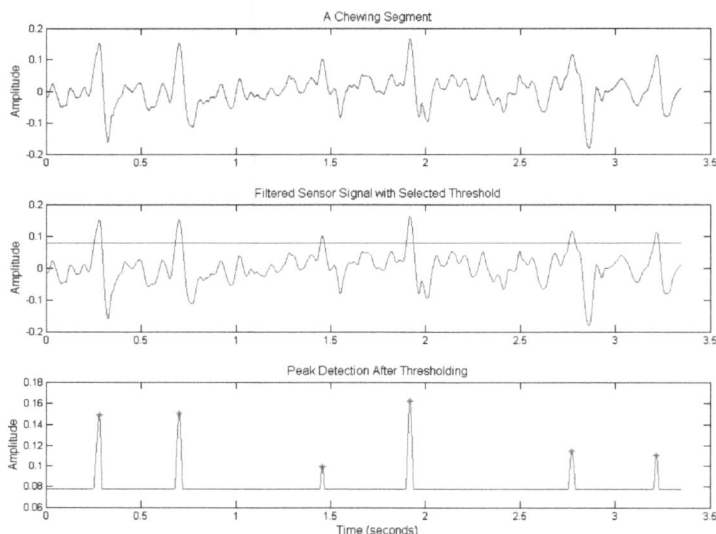

Figure 4. Different stages of signal processing for peak detection in chew counting algorithm. The first row shows the raw sensor signal. The second row shows filtered signal with selected threshold value (horizontal red line). The third row shows signal after thresholding and smoothing. Detected peaks are indicated by a red '*'.

2.5. Fully Automatic Approach: Feature Computation and Classification

The fully automatic approach first recognizes periods of food intake by splitting the sensor signals into short segments called epochs and then applying the chew counting algorithm to the epochs labeled as food intake. The duration of the epochs was estimated to be 5 s based on the following considerations. The average length of a manually annotated chewing sequence was 7.35 ± 5.16 s; therefore, 5-s epochs were able to capture most of the chewing episodes. The frequency of chewing was in the range of 0.94 to 2.0 Hz; therefore, a 5-s epoch ensures that, even for the lower bound of chewing frequency, the epoch will contain multiple chewing events.

This approach results in epochs where some samples may belong to chewing and the rest may not belong to chewing, and vice versa. Such a situation will most likely occur at the start or end of chewing sequences. For this, a 50% determination rule was used to assign a class label for a given epoch, i.e., $C_i \in \{'-1', '+1'\}$. An epoch was labeled as food intake ($C_i = +1$) if at least half of the samples in the epoch belonged to food intake (based on human annotation); otherwise, it was marked as a non-intake epoch ($C_i = -1$). For the ith epoch $x(i)$, 38 time and frequency domain features were computed to find the corresponding feature vector f_i. The set of features used here is the same as the feature extracted from the piezoelectric strain sensor presented in [13]. Each feature vector f_i was associated with its corresponding class label C_i.

ANN is a supervised machine learning technique, which has been shown to perform well in a wide range of classification problems. Advantages of ANN include robustness, flexibility, and the ability to create complex decision boundaries and handling of noisy data [13]. For classification of the epochs, a leave-one-out cross-validation scheme was used to train ANN models [13] that differentiate chewing and non-chewing. Non-chewing can be anything, e.g., absence of chewing, rest, speech, motion artifacts, etc. The classification approach using neural networks has been demonstrated to be robust in free living conditions in its ability to differentiate between chewing and non-chewing in the unrestricted environment [13] and has been shown to be superior to other methods (such as support vector machine (SVM) [12]. Three layers feed-forward architecture was used for training ANN models with a back-propagation training algorithm. The input layer had 38 neurons (for each feature), whereas the second layer (hidden layer) had 5 neurons (details described below), and the third layer (output layer) consisted of only one output neuron to indicate the predictor output class label C_i ('−1' or '+1') for any given feature vector f_i. Both hidden and output layers used hyperbolic tangent sigmoid for transfer function. Training and validation of the models was performed with the Neural Network Toolbox available in Matlab R2013b (The Mathworks Inc., Natick, MA, USA).

Classifier performance was evaluated in terms of $F1$ score which is defined as

$$F1 = \frac{2 * Precision * Recall}{Precision + Recall} , \tag{4}$$

$$Precision = \frac{TP}{TP + FP} , \text{ and} \tag{5}$$

$$Recall = \frac{TP}{TP + FN} , \tag{6}$$

where TP is the number of true positives, FP is the number of false positives, and FN is the number of false negatives. For the leave-one-out cross-validation approach, data from one participant was used for validation (testing), whereas data from the remaining participants was used for training. This process was repeated for each participant.

An iterative approach was used to choose the number of neurons in the hidden layer. For this purpose, 30 visits were randomly selected (one visit from each participant) instead of the whole dataset to avoid over fitting. Number of neurons in the hidden layer was varied from 1 to 15 and for each neuron setting; 30-fold cross-validation was performed to compute the corresponding $F1$ score. For neural networks, the initial weights and biases are randomly assigned, which resulted in slightly different solution each time. To achieve generalizable results, the cross-validation procedure was repeated 10 times. An average of 10 iterations was computed to obtain a final $F1$-score for each fold. The final $F1$-score for each hidden neuron setting was computed by taking the average of all 30 visits. From the results, it was observed that the computed $F1$-score increases up to 5 neurons in the hidden layer, and the $F1$-score changes are small after adding more than 5 neurons. Therefore, 5 neurons were used for training final classification models.

ANN models classified epochs as "food intake" and "no-intake". A chew counting algorithm was used for epochs classified as food intake to estimate per-epoch chew counts ($CNT(i)$). For the i-th epoch, the chewing rate ($CR(i)$) was computed using Equation (1). For an epoch-based approach, cumulative estimated chew counts and average chewing rate for the nth visit were computed as $E_{CNT}(n) = \sum_{i=1}^{K} CNT(i)$ and $E_{CR}(n) = \frac{1}{K} \sum_{i=1}^{K} CR(i)$, respectively, where K represents the total number of epochs in a visit. Errors (both signed and unsigned) for chew counts were computed using Equations (2) and (3). For chew counting algorithms (in both semi- and fully automatic approaches), a 95% confidence interval was used to find the interval for the mean to check if the true mean error represents underestimation or over-estimation.

To compare the total number of chews estimated by the semi-automatic and fully automatic approaches with the manually annotated chew counts, a one-way analysis of variance (ANOVA) was used. The null hypothesis in this case was that the means of chew counts estimated (for all visits) by all approaches (manually annotated, semi-automatic and fully-automatic approaches) are the same, whereas the alternate hypothesis suggested that the means were different. An ANOVA was also performed for comparing the performance of the proposed method for different meal types such as breakfast, lunch, and dinner.

3. Results

The collected dataset consisted of a total of 5467 chewing sequences marked by human raters with a total of 62,001 chews (average chews per meal: 660 ± 267 chews). The average chewing rate for all meals from human annotation was 1.53 ± 0.22 chews per second. Table 1 shows meal parameters such as duration, number of bites, chews, swallows, and mass ingested grouped by type of meal.

For the semi-automatic approach, the algorithm estimated total chew count of ($\sum_{n=1}^{M} E_{CNT}(n)$) 58,666 (average chews per meal: 624 ± 278) with an average chewing rate of 1.44 ± 0.24 chews per second. The chew counting algorithm was able to achieve a mean absolute error (Equation (3)) of 10.4% ± 7.0% for the total number of chews compared to human annotated number of chews. The average signed error was (Equation (2)) 5.7% ± 11.2%. The 95% confidence interval (CI) for the mean for the signed error was CI (3.4%, 8.0%).

In the fully automatic approach, trained ANN models were able to detect food intake with an average $F1$ score of 91.0% ± 7.0% with the average precision and recall of 91.8% ± 9.0% and 91.3% ± 8.8% respectively, using leave-one-out cross-validation. Further application of the chew counting algorithm resulted in 59,862 total chews, 636 ± 294 average chews per meal, and an average chewing rate of 1.76 ± 0.31 chews per second. The mean absolute error was 15.0% ± 11.0%. In this case, the average signed error was 4.2% ± 18.2%. The 95% confidence interval (CI) for the mean for the signed error was CI (0.05%, 8.10%). Figure 5 shows the distribution of both mean absolute errors for the semi- and fully automatic approaches. Figure 6 shows the distribution of chew counts per meal for human annotated chews, the estimate chew counts from the chew counting algorithm for semi- and fully automatic approaches.

Table 2 shows the results of the ANOVA for comparing the mean chew counts among manually annotated, semi-automatic and fully-automatic approaches. Results of the statistical analysis showed no significant differences between the mean chew counts among different methods (*p*-value (0.68) > 0.05). Table 3 shows average errors in the chew count estimation of both approaches for breakfast, lunch, and dinner meals. Results of ANOVA show that there were no significant differences among different meal types (the semi-automatic approach *p*-value: 0.87 > 0.05; the fully-automatic approach *p*-value: 0.28 > 0.05).

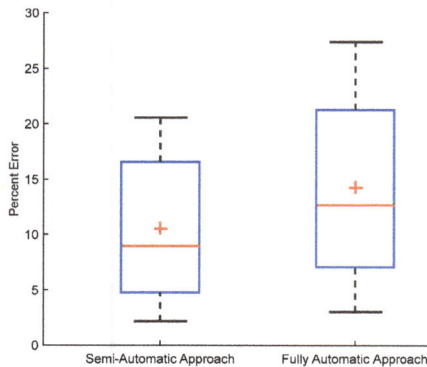

Figure 5. Distribution of mean absolute error of the chew counting algorithm for both semi- and automatic approaches.

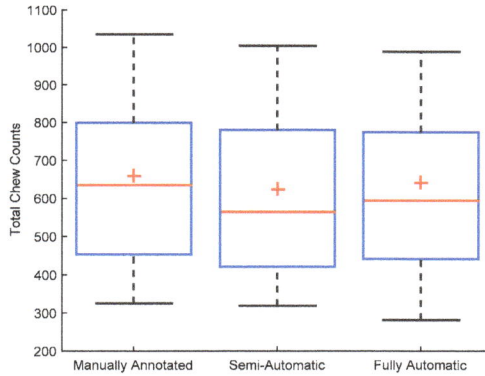

Figure 6. Box plots for total number per meal by human annotation; algorithm estimation with manually annotated, semi-automatic approach and fully automatic approaches.

Table 1. Details about the duration, number of bites, chews, swallow, and mass in grams for different meal types.

	Breakfast					Lunch				
	Duration s	Mass g	Bites #	Chews #	Swallows #	Duration s	Mass g	Bites #	Chews #	Swallows #
Mean	815.27	618.83	43.73	508.23	70.50	1011.06	675.50	41.33	748.31	78.53
STD	349.75	191.54	10.83	184.61	27.39	321.11	160.06	11.52	253.76	22.88
Min	406	299	28	204	40	485	381	23	267	44
Max	1964	914	69	1036	154	1839	1188	65	1352	131
Total	24,458	18,565	1312	15,247	2115	36,398	24,318	1488	26,939	2827
	Dinner					All Meals (Total)				
	Duration s	Mass g	Bites #	Chews #	Swallows #	Duration s	Mass g	Bites #	Chews #	Swallows #
Mean	1281.89	839.79	52.14	707.68	107.96	1029.24	706.35	45.32	659.59	84.73
STD	645.78	281.86	19.02	294.28	39.53	481.81	228.79	14.58	266.72	33.58
Min	460	198	25	181	55	406	198	23	181	40
Max	3052	1373	109	1441	209	3052	1373	109	1441	209
Total	35,893	23,514	1460	19,815	3023	96,749	66,397	4260	62,001	7965

Table 2. Results of one-way analysis of variance (ANOVA) for comparison between different chew counting approaches (manually annotated, semi-automatic and fully automatic).

Source of Variation	Sum of Squares	Degrees of Freedom	Mean Square	F	p-Value	F-Crit
Between Groups	60,737.45	2	30,368.73	0.39	0.68	3.03
Within Groups	21,857,591.03	279	78,342.62			
Total	21,918,328.49	281				

Table 3. Mean absolute errors for different meals for both semi- and fully automatic approaches.

Meal Type	Semi-Automatic	Fully Automatic
Breakfast	10.90% ± 7.59%	17.58% ± 12.95%
Lunch	10.30% ± 7.03%	12.65% ± 10.74%
Dinner	9.99% ± 6.63%	15.09% ± 9.19%
Overall	10.40% ± 7.03%	15.01% ± 11.06%

4. Discussion

This work presented a method for automatic recognition and quantification of chewing from piezoelectric sensor signals in terms of chewing counts and chewing rates. Results of the chew counting algorithm for both semi- and fully automated approaches suggest that the method proposed can provide objective and accurate estimation of chewing behavior. The method was tested on a comparatively large population and with a wide variety of food.

In the semi-automatic approach, the algorithm was used to estimate chew counts and chewing rates in manually annotated chewing segments and was able to achieve a mean absolute error of (Equation (3)) 10.4% ± 7.0%. For this approach, the mean signed error (Equation (2)) in comparison with human chew counts was 5.7% ± 11.2%. A 95% CI computed for signed error (3.43%, 8.02%) did not include zero with both limits being positive; therefore, the results show a trend of underestimation of the chew counts. A possible reason for this trend is the variability of food properties requiring different strength of chewing and variability in individual chewing patterns.

The fully automatic approach presents a more realistic use of the proposed method for automatic detection and quantification of chewing behavior. In the proposed method, ANN models for food intake detection preceded the use of the chew counting algorithm for epochs classified as food intake. This two-stage process resulted in a higher mean absolute error for chew counting (15.0% ± 11.0%) compared to the semi-automatic approach. The increase in error was expected as the error from both the classification stage and the chew counting stage accumulate. Other methods for chew count estimation from sensors have reported similar errors, e.g., in [9], the authors proposed the use of an accelerometer placed on the temporalis muscle and achieved a mean absolute error of 13.80 ± 8.29 for chew counts from four participants eating small pieces of watermelon. An acoustic-based chewing event counting approach using in-ear microphones presented in [8] reported an overall recall of 80% with precision at 60%–70% for three different food items. However, the approach presented in this work was tested on a larger population and with a wider variety of foods compared with the other reported results.

For cumulative chew counts, results of the one-way ANOVA (Table 2) show that the differences between the mean chew count for each experiment/visit were not statistically significant, i.e., there was not enough evidence to reject the null hypothesis (all means are equal) at the given *p*-value of 0.68. This shows that the proposed algorithm can provide chew counts that are close to the reference chew counts obtained by human raters from video observation. In this work, human annotated chew counts were used as the gold standard for algorithm development. Video-based annotation has been used by a number of researchers as the gold standard for algorithm development in food intake monitoring research [8,9,15]. The *p*-value also indicates that the trend of underestimation of the chew counts by the algorithm is not statistically significant and can be ignored based on the strong evidence (given by *p*-value). Table 3 shows that the mean absolute error was also independent of the meal type, i.e., breakfast, lunch, or dinner. This shows that the proposed method can be used for the estimation of chew counts for foods traditionally consumed for breakfast, lunch, or dinner without any sacrifice of the performance.

An average chewing rate for manually annotated chewing sequences was 1.53 chews per second, which is in the range of chewing frequencies reported in the literature (0.94–2.5 chews per second) [28]. The chewing rates estimated by the chew counting algorithm were also in the range of previously reported values. The fully automatic chew counting approach resulted in the highest chewing rate of 1.76 with the standard deviation of 0.31 chews per second. A potential reason is that the threshold for each epoch is a function of the signal amplitude in that particular epoch. Amplitude variations in epochs misclassified as food intake are smaller compared with the actual chewing; therefore, those particular epochs can result in higher chew counts. One possible improvement is to calculate global thresholds for each participant to avoid higher chew counts in epochs incorrectly classified as food intake. This will require the participant-dependent calibration of threshold computation.

In both semi- and fully automatic approaches, errors were computed for cumulative chew counts for each experiment/visit rather than the chews per known chewing segment. In the epoch-based approach, one chewing sequence may be divided into multiple epochs, and there is a possibility that a part of chewing sequence may be labeled as non-chewing epochs and vice versa (because of the boundary conditions mentioned above); therefore, a direct comparison between these two approaches was not possible. To tackle this issue, the comparison was performed for the total number of chews for each visit (meal).

One of the strengths of this study was the testing of the sensor, the signal processing, and the pattern recognition algorithms in the multi-meal experiments with a relatively larger population, where the daily eating behavior of people was replicated. Meals were kept as natural as possible. Participants were allowed to talk during the meals, and there were no restrictions on the body movements while sitting, such as head movements, coughing, etc. This enabled the inclusion of naturally occurring activities that are performed by individuals who are sitting, and it adds to the robustness of the classifications algorithm. Participants had the choice of selecting different meal times, i.e., breakfast, lunch, or dinner, and the choice of food items available at the cafeteria to make two different meals. While the food selection for a particular participant was somewhat limited, the study in general included a wide variety of food items with varying physical properties that may affect chewing. While not directly tested in this study, it is expected that the foods with different physical properties will have a different chew estimation error, which may average out across multiple food types over an extended period of time. This hypothesis will be tested in future studies.

All presented methods were developed as participant-independent group models that do not need individual calibration. Since each participant had different meal content and chewing patterns, the use of participant-dependent models (both for chew counting as well as for training classification models) may result in better accuracy and lower error, but will require calibration to each participant.

Overall, the presented method was able to detect and characterize the chewing behavior of a group of participants consisting of a wide range of adiposity, consuming a variety of food, and having a wide range of physical properties. This system needs to be further explored and tested in free living conditions. Sensor-derived chew counts can be used for creating models to estimate mass per bites [8] and energy intake in a meal [22]. One limitation of the fully automatic approach presented here is the use of a fixed size epoch, which contributes towards the increase in error. Rather than relying on the epoch-based approach, there is a need to develop algorithms that can first separate potential chewing sequences from non-chewing sequences (without using fixed window segments) and then use classification models to identifying them as food-intake or no-intake sequences.

Another limitation is that the sensor system used in this study relied on an adhesive sensor and wired connections to the data acquisition system. Since the reported study, the jaw sensor has been implemented as a Bluetooth-enabled wireless device that has been tested in free living conditions [13]. The attachment of the sensor to the skin using a medical adhesive has been shown to be simple and robust, similar to widely used adhesive bandages. In the free living study [13], the sensor did not experience any attachment issues and continuously remained on the body for approximately 24 h. Although not tested in this study, a previous study [13] demonstrated the robustness of the classifier to differentiate between chewing and other activities such as uncontrolled motion artifacts, e.g., walking, head motion, and other unscripted activities of daily living. The user burden imposed by the sensor was evaluated in [29]. Results of the study suggested that participants experienced a high level of comfort while wearing the adhesive sensor, and the presence of the sensor did not affect their eating. For long-term use (weeks and months), the sensor system may need further modifications to increase comfort and user compliance.

5. Conclusions

This work presented a method for the automatic detection and characterization of chewing behavior in terms of chew counts and chewing rates. A histogram-based peak detection algorithm was used to count chews in semi- and fully automatic approaches. For the semi-automatic approach, the method was able to achieve a mean absolute error of 10.4% ± 7.0%. In the fully automatic approach, sensor signals were first divided into 5-s epochs that were classified as chewing or non-chewing by an ANN. In the fully automatic approach, a classification accuracy of 91.0% and a mean absolute error of 15.0% ± 11.0% were achieved. These results suggest that the proposed method can be used to objectively characterize chewing behavior.

Acknowledgments: This work was partially supported by Nation Institute of Diabetes and Digestive and Kidney Diseases (grants number: R21DK085462 and R01DK100796). The authors have no potential conflicts of interest.

Author Contributions: Muhammad Farooq and Edward Sazonov conceived and designed the study; performed the experiments; analyzed the data; reviewed and edited the manuscript. All authors read and approved the manuscript.

Conflicts of Interest: The authors declare no conflict of interest.

Abbreviations

The following abbreviations are used in this manuscript:

ANN	artificial neural network
N	experiment/visit number
K	manually annotated chewing sequence
i	epoch number
CNT	chew counts
CR	chewing rate
$A_{CNT}(n)$	total annotated chew counts for n^{th} experiment/visit
$E_{CNT}(n)$	total estimated chew counts for n^{th} experiment/visit
|Error|	mean absolute error (unsigned)
Error	mean error (signed)

References

1. WHO. Obesity and Overweight. Available online: http://www.who.int/mediacentre/factsheets/fs311/en/ (accessed on 21 April 2011).
2. Fairburn, C.G.; Harrison, P.J. Eating disorders. *Lancet* **2003**, *361*, 407–416. [CrossRef]
3. Day, N.; McKeown, N.; Wong, M.; Welch, A.; Bingham, S. Epidemiological assessment of diet: A comparison of a 7-day diary with a food frequency questionnaire using urinary markers of nitrogen, potassium and sodium. *Int. J. Epidemiol.* **2001**, *30*, 309–317. [CrossRef] [PubMed]
4. Muhlheim, L.S.; Allison, D.B.; Heshka, S.; Heymsfield, S.B. Do unsuccessful dieters intentionally underreport food intake? *Int. J. Eat. Disord.* **1998**, *24*, 259–266. [CrossRef]
5. Thompson, F.E.; Subar, A.F. Chapter 1—Dietary Assessment Methodology. In *Nutrition in the Prevention and Treatment of Disease*, 3rd ed.; Coulston, A.M., Boushey, C.J., Ferruzzi, M., Eds.; Academic Press: Cambridge, MA, USA, 2013; pp. 5–46.
6. Champagne, C.M.; Bray, G.A.; Kurtz, A.A.; Monteiro, J.B.R.; Tucker, E.; Volaufova, J.; Delany, J.P. Energy Intake and Energy Expenditure: A Controlled Study Comparing Dietitians and Non-dietitians. *J. Am. Diet. Assoc.* **2002**, *102*, 1428–1432. [CrossRef]
7. Päßler, S.; Wolff, M.; Fischer, W.-J. Food intake monitoring: An acoustical approach to automated food intake activity detection and classification of consumed food. *Physiol. Meas.* **2012**, *33*, 1073–1093. [CrossRef] [PubMed]
8. Amft, O.; Kusserow, M.; Troster, G. Bite weight prediction from acoustic recognition of chewing. *IEEE Trans. Biomed. Eng.* **2009**, *56*, 1663–1672. [CrossRef] [PubMed]

9.   Wang, S.; Zhou, G.; Hu, L.; Chen, Z.; Chen, Y. CARE: Chewing Activity Recognition Using Noninvasive Single Axis Accelerometer. In Adjunct Proceedings of the 2015 ACM International Joint Conference on Pervasive and Ubiquitous Computing and the 2015 ACM International Symposium on Wearable Computers, New York, NY, USA, 7–11 September 2015; pp. 109–112.

10. Cadavid, S.; Abdel-Mottaleb, M.; Helal, A. Exploiting visual quasi-periodicity for real-time chewing event detection using active appearance models and support vector machines. *Pers. Ubiquitous Comput.* **2011**, *16*, 729–739. [CrossRef]

11. Kalantarian, H.; Alshurafa, N.; Le, T.; Sarrafzadeh, M. Monitoring eating habits using a piezoelectric sensor-based necklace. *Comput. Biol. Med.* **2015**, *58*, 46–55. [CrossRef] [PubMed]

12. Farooq, M.; Fontana, J.M.; Boateng, A.; McCrory, M.A.; Sazonov, E. A Comparative Study of Food Intake Detection Using Artificial Neural Network and Support Vector Machine. In Proceedings of the 12th International Conference on Machine Learning and Applications (ICMLA'13), Miami, FL, USA, 3–7 December 2013; pp. 153–157.

13. Fontana, J.M.; Farooq, M.; Sazonov, E. Automatic Ingestion Monitor: A Novel Wearable Device for Monitoring of Ingestive Behavior. *IEEE Trans. Biomed. Eng.* **2014**, *61*, 1772–1779. [CrossRef] [PubMed]

14. Fontana, J.M.; Farooq, M.; Sazonov, E. Estimation of feature importance for food intake detection based on Random Forests classification. In Proceedings of the 2013 35th Annual International Conference of the IEEE Engineering in Medicine and Biology Society (EMBC), Osaka, Japan, 3–7 July 2013; pp. 6756–6759.

15. Bedri, A.; Verlekar, A.; Thomaz, E.; Avva, V.; Starner, T. Detecting Mastication: A Wearable Approach. In Proceedings of the 2015 ACM on International Conference on Multimodal Interaction, New York, NY, USA, 9–13 November 2015; pp. 247–250.

16. Sazonov, E.; Fontana, J.M. A Sensor System for Automatic Detection of Food Intake through Non-Invasive Monitoring of Chewing. *IEEE Sens. J.* **2012**, *12*, 1340–1348. [CrossRef] [PubMed]

17. Spiegel, T.A. Rate of intake, bites, and chews-the interpretation of lean-obese differences. *Neurosci. Biobehav. Rev.* **2000**, *24*, 229–237. [CrossRef]

18. Li, J.; Zhang, N.; Hu, L.; Li, Z.; Li, R.; Li, C.; Wang, S. Improvement in chewing activity reduces energy intake in one meal and modulates plasma gut hormone concentrations in obese and lean young Chinese men. *Am. J. Clin. Nutr.* **2011**, *94*, 709–716. [CrossRef] [PubMed]

19. Zhu, Y.; Hollis, J.H. Increasing the number of chews before swallowing reduces meal size in normal-weight, overweight, and obese adults. *J. Acad. Nutr. Diet.* **2014**, *114*, 926–931. [CrossRef] [PubMed]

20. Nicklas, T.A.; Baranowski, T.; Cullen, K.W.; Berenson, G. Eating Patterns, Dietary Quality and Obesity. *J. Am. Coll. Nutr.* **2001**, *20*, 599–608. [CrossRef] [PubMed]

21. Lepley, C.; Throckmorton, G.; Parker, S.; Buschang, P.H. Masticatory performance and chewing cycle kinematics-are they related? *Angle Orthod.* **2010**, *80*, 295–301. [CrossRef] [PubMed]

22. Fontana, J.M.; Higgins, J.A.; Schuckers, S.C.; Bellisle, F.; Pan, Z.; Melanson, E.L.; Neuman, M.R.; Sazonov, E. Energy intake estimation from counts of chews and swallows. *Appetite* **2015**, *85*, 14–21. [CrossRef] [PubMed]

23. Farooq, M.; Chandler-Laney, P.C.; Hernandez-Reif, M.; Sazonov, E. Monitoring of infant feeding behavior using a jaw motion sensor. *J. Healthc. Eng.* **2015**, *6*, 23–40. [CrossRef] [PubMed]

24. Farooq, M.; Chandler-Laney, P.; Hernandez-Reif, M.; Sazonov, E. A Wireless Sensor System for Quantification of Infant Feeding Behavior. In Proceedings of the Conference on Wireless Health, New York, NY, USA, 2–4 June 2015.

25. Farooq, M.; Sazonov, E. Comparative testing of piezoelectric and printed strain sensors in characterization of chewing. In Proceedings of the 2015 37th Annual International Conference of the IEEE Engineering in Medicine and Biology Society (EMBC), Milan, Italy, 25–29 August 2015; pp. 7538–7541.

26. Fontana, J.M.; Lopez-Meyer, P.; Sazonov, E.S. Design of a instrumentation module for monitoring ingestive behavior in laboratory studies. In Proceedings of the 2011 Annual International Conference of the IEEE Engineering in Medicine and Biology Society, EMBC, Boston, MA, USA, 30 August–3 September 2011; pp. 1884–1887.

27. Sazonov, E.; Schuckers, S.; Lopez-Meyer, P.; Makeyev, O.; Sazonova, N.; Melanson, E.L.; Neuman, M. Non-invasive monitoring of chewing and swallowing for objective quantification of ingestive behavior. *Physiol. Meas.* **2008**, *29*, 525–541. [CrossRef] [PubMed]

28. Po, J.M.C.; Kieser, J.A.; Gallo, L.M.; Tésenyi, A.J.; Herbison, P.; Farella, M. Time-frequency analysis of chewing activity in the natural environment. *J. Dent. Res.* **2011**, *90*, 1206–1210. [CrossRef] [PubMed]
29. Fontana, J.M.; Sazonov, E.S. Evaluation of Chewing and Swallowing Sensors for Monitoring Ingestive Behavior. *Sens. Lett.* **2013**, *11*, 560–565. [CrossRef] [PubMed]

electronics

MDPI

Article

Robust and Accurate Algorithm for Wearable Stereoscopic Augmented Reality with Three Indistinguishable Markers

Fabrizio Cutolo [1,*], Cinzia Freschi [1], Stefano Mascioli [1], Paolo D. Parchi [1,2], Mauro Ferrari [1,3] and Vincenzo Ferrari [1,3,4]

[1] EndoCAS Center, Department of Translational Research and New Technologies in Medicine and Surgery, University of Pisa, Pisa 56124, Italy; cinzia.freschi@endocas.unipi.it (C.F.); stemsc19@gmail.com (S.M.); paolo.parchi@unipi.it (P.D.P.); mauro.ferrari@med.unipi.it (M.F.); vincenzo.ferrari@unipi.it (V.F.)
[2] 1st Orthopedic Division, University of Pisa, Pisa 56125, Italy
[3] Vascular Surgery Unit, Azienda Ospedaliero Universitaria Pisana, Pisa 56126, Italy
[4] Information Engineering Department, University of Pisa, Pisa 56122, Italy
* Correspondence: fabrizio.cutolo@endocas.unipi.it; Tel.: +39-050-995-689; Fax: +39-050-992-773

Academic Editors: Enzo Pasquale Scilingo and Gaetano Valenza
Received: 18 May 2016; Accepted: 9 September 2016; Published: 19 September 2016

Abstract: In the context of surgical navigation systems based on augmented reality (AR), the key challenge is to ensure the highest degree of realism in merging computer-generated elements with live views of the surgical scene. This paper presents an algorithm suited for wearable stereoscopic augmented reality video see-through systems for use in a clinical scenario. A video-based tracking solution is proposed that relies on stereo localization of three monochromatic markers rigidly constrained to the scene. A PnP-based optimization step is introduced to refine separately the pose of the two cameras. Video-based tracking methods using monochromatic markers are robust to non-controllable and/or inconsistent lighting conditions. The two-stage camera pose estimation algorithm provides sub-pixel registration accuracy. From a technological and an ergonomic standpoint, the proposed approach represents an effective solution to the implementation of wearable AR-based surgical navigation systems wherever rigid anatomies are involved.

Keywords: augmented reality; wearable displays; image-guided surgery; machine vision; camera calibration

1. Introduction

Augmented reality (AR) [1] is a ground-breaking technology in machine vision and computer graphics and may open the way for significant technological developments in the context of image-guided surgery (IGS). In AR-based applications, the key challenge is to ensure the highest degree of realism in merging computer-generated elements with live views of the surgical scene.

AR in IGS allows merging of real views of the patient with computer-generated elements generally consisting of patient-specific three-dimensional (3D) models of anatomy extracted from medical datasets (Figure 1). In this way, AR establishes a functional and ergonomic integration between surgical navigation and virtual planning by providing physicians with a virtual navigation aid contextually blended within the real surgical scenario [2].

In recent years, there has been a growing research interest in AR in medicine, which has driven a remarkable increase in the number of published papers. A PubMed search was performed of publications with the terms "augmented reality" OR "mixed reality" in the title or abstract. The first publication dated back to 1995 [3]. After 13 years, on 31 December 2008, the number of publications reached 255. During the last seven years, between 1 January 2009 and 30 April 2016, 647 papers were published, 168 of them in the past year. Nonetheless, only a few of the reported publications dealt with clinical validation of the technology described, and even fewer addressed its in vivo assessment. This is mostly due to the technological barriers encountered in the attempt to integrate similar AR systems into the surgical workflow.

Based on these considerations, the present work is aimed at developing strategies that could facilitate the profitable introduction of wearable AR systems to clinical practice.

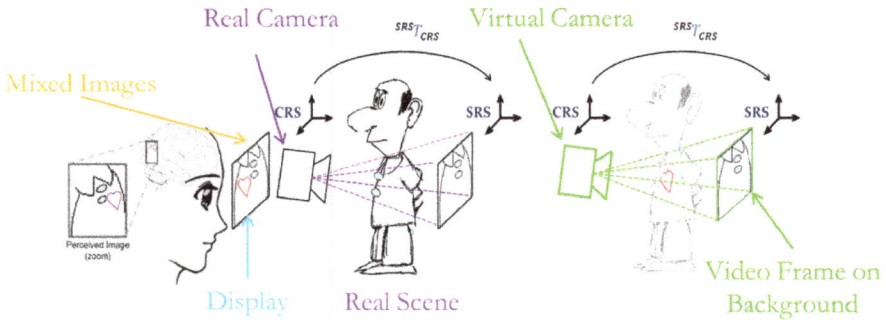

Figure 1. Augmented Reality video see-through paradigm: the 2D virtual image (heart) is mixed into an image frame of the real world grabbed by the external camera.

In the realm of AR-based IGS systems, various display technologies have been proposed. In light of avoiding abrupt changes to the surgical setup and workflow, historically the first AR-based surgical navigation systems were implemented on the basis of commonly used devices [4] such as surgical microscopes [5,6]. In laparoscopy, and generally in endoscopic surgery, the part of the environment where the surgeon's attention is focused during the surgical task (DVV's *Perception Location* [7]) is a stand-up monitor. Indeed, in such procedures, the surgeon operates watching endoscopic video images reproduced on the spatial display unit [8,9]. Therefore, the virtual information is usually merged with real-time video frames grabbed by the endoscope and presented on a stand-up monitor [10–12].

Alternative and promising approaches based on integral imaging (II) technology have been proposed [13,14]. II displays use a set of 2D elemental images from different perspectives to generate a full-parallax 3D visualization. Therefore, with II-based displays, a proper 3D overlay between virtual content and a real scene can be obtained. Certain embodiments of this technology have been specifically designed and tested for maxillofacial surgery and neurosurgery [15–19]. The II paradigm can provide the user with an egocentric viewpoint and a full-parallax augmented view in a limited viewing zone (imposed by the II display). However, wearable embodiments of II technology still require further development of both hardware and software aspects [20].

In general, the quality of an augmented reality (AR) experience, particularly in IGS systems, depends on how well the virtual content is blended with the surgical scene spatially, photometrically, and temporally [21]. In this regard, wearable AR systems offer the most ergonomic solution in those medical tasks that are manually performed under the surgeon's direct vision (open surgery, introduction of biopsy needle, palpation, etc.) because they minimize the extra mental effort required to switch focus between the real surgical task and the augmented view presented on the external display. Wearable AR systems based on head-mounted displays (HMDs) intrinsically provide the user with an egocentric viewpoint and do not limit freedom of movement around the patient [22–24]. Standard HMDs provide both binocular parallax and motion parallax and smoothly augment the user's perception of the surgical scene throughout the specific surgical procedure. At present, they are less obtrusive in the operating room (OR) than II systems. In HMDs, the see-through capability is provided through either a video or an optical see-through paradigm.

Typically, in optical see-through HMD systems, the user's direct view is augmented by the projection of virtual information either on semi-transparent displays placed in front of the eyes or directly onto the retina [25]. Accurate alignment between the direct view of the real scene and the virtual information is provided by real-time tracking of the visor and user-specific calibration that accounts for the change in relative position and orientation (pose) between display and eyes each time the user wears or moves the HMD [26,27]. Display-eye calibration is necessary to model intrinsically and extrinsically the virtual view frustum to the user's real one [28].

The video see-through solution is instead based on external cameras rigidly fixed in front of the HMD. In these systems, although the field of view is limited by the size of the camera optics and displays, a user-specific calibration routine is not necessary. Furthermore, in video see-through systems, the real scene and the virtual information can be synchronized, whereas in optical see-through devices, there is an intrinsic lag between immediate perception of the real scene and inclusion of the virtual elements. Therefore, at the current technological level, the use of video see-through systems is immediate, at least for those IGS applications that can tolerate slight delays between capture of the real scene by the cameras and its final presentation in augmented form.

Accurate alignment between the real scene and the virtual content is provided by tracking the HMD in relation to the real world (represented by matrix $^{SRS}T_{CRS}$ in Figure 1), which is usually performed by means of an external tracker [29].

In a previous work, we presented an early system based on a commercially available HMD equipped with two external cameras aligned to the user's eyes [23]. The see-through ability was created by combining 3D computer-generated models obtained by processing radiological images (e.g., CT or MRI) [30] with live views of the real patient. The distinctive feature of that AR system was that the pair of external cameras served both to capture the real scene and to perform stereo tracking.

As the authors, we share the conviction that the absence of an external tracker is a key element in enabling smooth and profitable integration of AR systems into the surgical workflow. Surgical navigation systems based on external infrared trackers have the major drawback of introducing unwanted line-of-sight constraints into the OR and of adding error-prone technical complexity to the surgical procedure [29]. Other tracking modalities are based on more complex surface-based tracking algorithms [12,31]. As an alternative to optical tracking, electromagnetic tracking systems are particularly suited for tracking hidden structures [32], but their accuracy and reliability are severely affected by the presence of ferromagnetic and/or conductive materials [33].

Standard video-based tracking methods featuring the use of large template-based markers provide highly accurate results in non-stereoscopic systems. Nonetheless, they are not suited for use in a surgical setting because they limit the surgeon's line of sight given their planar structure and they may occlude the visibility of the operating field.

In that early system, and as previously done in [10,34], real-time registration of the virtual content to the real images was achieved by localizing chromatically distinguishable spherical markers. The video marker-based registration method registers the virtual 3D space to the camera coordinate system (CRS) through real-time determination of the camera pose in the radiological coordinate system (SRS).

Small spherical markers do not seriously affect the line of sight and can be conveniently placed on the patient's skin with minimal logistic impact on the surgical workflow. With the objective of increasing system usability, the minimum set of markers (i.e., three) that could ensure a finite number of solutions to the camera pose estimation problem was chosen. The chromatic differences among the three markers and the stereo-camera setup enabled solution of the stereo correspondence problem and real-time computation of camera pose without the ambiguity of the general perspective-3-point (P3P) problem [35]. In practice, thanks to stereo tracking, the camera pose estimation problem can be reduced to determining the standard closed-form least-squares solution of the absolute orientation problem (AOP) given a set of three correspondences in the two 3D coordinate systems (CRS and SRS) [36]. The coordinates of the three markers in the CRS were recovered by applying stereo localization routines to the pairs of conjugate projections of the marker centroids taken from the image planes of the two cameras. Image coordinates of the marker centroids were determined by performing a feature extraction task using color segmentation and circular shape recognition. Hence, in the early system, robust feature extraction was crucial to providing accurate geometric registration.

Unfortunately, the shortcomings of the earlier approach were twofold: the non-fully controllable and/or inconsistent lighting conditions in the OR, and the intrinsic difficulty of robustly classifying three different colors using a standard thresholding technique. These shortcomings cannot be neglected if the system is to be integrated into the surgical workflow. Adoption of stringent thresholding criteria in the segmentation step may in fact result in inconsistent target identification because the connected regions tend to be poorly segmented. On the contrary, large thresholds may generate badly segmented regions or yield incorrect markers labelling.

In the present work, we shall present a tracking-by-detection solution that uses monochromatic markers and new marker labeling strategies to increase the robustness of the video-based tracking method under non-controllable lighting conditions.

In addition, the proposed solution overcomes another limitation of the earlier algorithm. As mentioned above, the 3D position of the markers in the CRS is estimated through stereoscopic triangulation routines applied to pairs of images acquired by the two external cameras. Nevertheless, the anthropomorphic geometry of the stereo setup can ensure adequate marker localization accuracy only at close distances. This localization error is inherent to the stereoscopic geometry and depends on the accuracy of the disparity estimate in the proposed feature extraction procedure and on the calibration errors in estimating the intrinsic and extrinsic camera parameters [37]. In Section 2.2.2, an example of such inaccuracy due to the anthropomorphic geometry of the stereo setup is reported. To cope with this limitation in this work we added a $P n P$-based optimization step, which refines the pose of both cameras separately and yields sub-pixel registration accuracy in the image plane.

Another interesting landmark-based mono-camera tracking solution has been proposed by Schneider et al. [38]. Their approach, based on an efficient and innovative 2D/3D point pattern matching algorithm, was specifically designed for computationally low-power devices and was proven to yield good results in terms of image registration accuracy and computational performance. Compared to that solution, our method needs fewer reference landmarks (i.e., three), whereas their single-view approach for estimating the camera pose cannot work if fewer than six landmarks can be seen. Use of a minimum set of three fiducial markers is in fact intended to limit the logistic payload for setup, and this aspect is key for facilitating the smooth integration of the system into the surgical workflow. The proposed solution tackles the ambiguity of the P3P problem through the stereoscopic settings of the video see-through system.

To the best of the authors' knowledge, no previous work in AR has addressed the image-to-patient registration problem and has achieved sub-pixel registration accuracy through a video marker-based method that uses only three chromatically indistinguishable markers.

2. Materials and Methods

This section is organized as follows. Section 2.1 provides a detailed description of the hardware and of the software libraries used to implement the proposed stereoscopic AR mechanism. Section 2.2 describes the new methods used to solve marker labeling and to obtain a first estimate of the camera pose in relation to the SRS. The same subsection also describes the optimization method that solves the perspective-3-point (P3P) problem and yields sub-pixel registration accuracy in the image plane. Finally, Section 2.3 explains the methodology used to evaluate registration accuracy.

2.1. System Overview

The aim of this work is to present a robust and accurate video-based tracking method suited for use in a clinical scenario. The solution is based on tracking three indistinguishable markers. The algorithm was developed for a HMD AR system, but it could be applied to other stereoscopic devices like binocular endoscopes or binocular microscopes. Reference hardware has been chosen to achieve a low-cost system by assembling off-the-shelf components and manufacturing custom-made parts. The custom-made video see-through HMD was made from a Z800 3D visor (eMagin, Hopewell Junction, NY, USA) (Figure 2). The HMD is provided with dual OLED panels and features a diagonal field of view (FoV) of 40°.

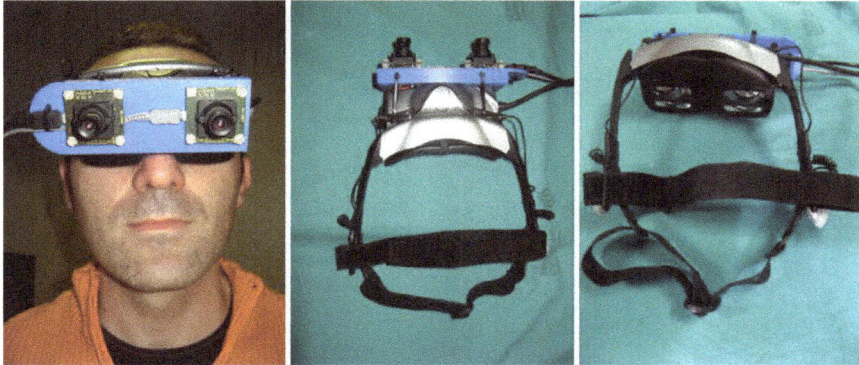

Figure 2. Video see-through head-mounted display (HMD) obtained by mounting two external cameras on top of a commercial 3D visor.

A plastic frame (ABS) was built through rapid prototyping to act as a support for the two external USB cameras equipped with 1/3″ image sensors UI-1646LE (IDS, Imaging Development Systems GmbH, Obersulm, Germany). By means of this support, the two cameras are mounted parallel to each other with an anthropometric interaxial distance ($\sim$7 cm) to provide a quasi-orthoscopic view of the augmented scene mediated by the visor. When the user looks at the real world while wearing the HMD, there are no appreciable differences between natural and visor-mediated views [39].

A toed-in camera configuration would be preferable for achieving better stereo overlap at close working distances, but if not coupled with simultaneous convergence of the optical display axes, this would go against the objective of this work: achievement of a quasi-orthostereoscopic AR HMD. As a matter of fact, another study by the authors has presented a different video see-through HMD that features the possibility of adjusting the degree of convergence of the stereo camera pair as a function of the working distance [40].

The Z800 HMD receives video frames from the computer via VGA cable and alternately transmits them to left and right internal monitors at 60 Hz in sync with the vsync signal. Therefore, the software, which renders and mixes the virtual model with the real frames, must set up and exchange left and the right views synchronously with the vsync signal as well. The proposed software application elaborates the grabbed video frames to perform real-time registration. Due to the computational complexity of the whole video see-through paradigm, a multithreaded application was implemented to distribute the operations among available processors to guarantee synchronization of the two views to be sent to the HMD. One thread sets up the AR views and ensures their synchronization, whereas the other performs video-based tracking.

A synthetic functional and logical description of the AR mechanism is as follows: real cameras grab video frames of the scene; video frames, after radial distortion compensation, are screened as backgrounds of the corresponding visor display; virtual anatomies, reconstructed offline from radiological images, are coherently merged to create the augmented scene. For coherent merging of real scenes and virtual content, the virtual content is observed by a couple of virtual viewpoints (virtual cameras) with projective parameters that mimic those of the real cameras and with poses that vary according to the real-time marker-based tracking method (Figure 3).

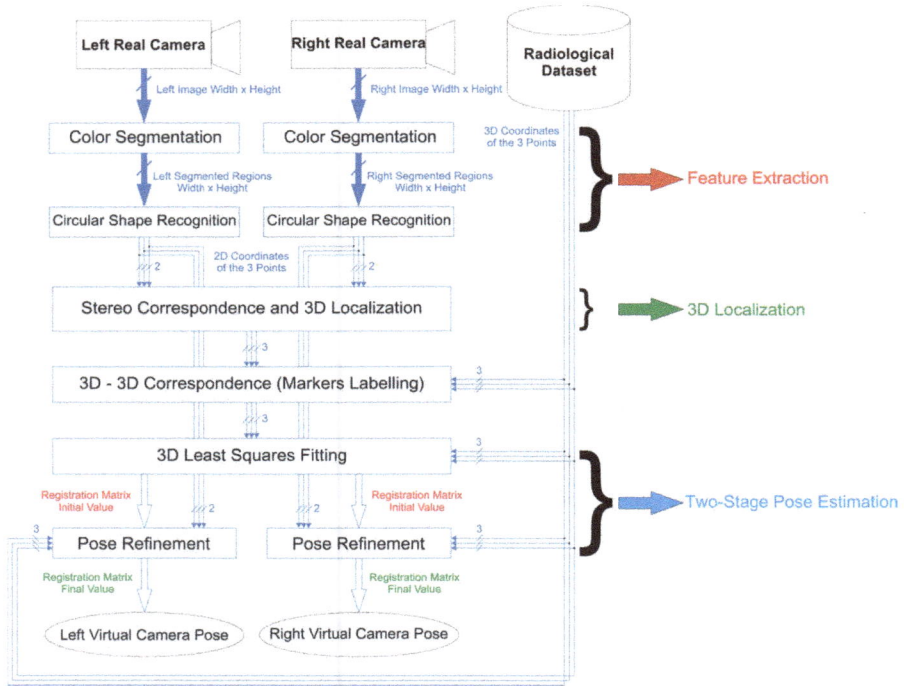

Figure 3. Localization and registration algorithm.

This AR mechanism was implemented in software libraries built in C++ on top of the multipurpose EndoCAS Navigator Platform modules [41]. Management of the virtual 3D scene was carried out through the OpenSG 1.8 open-source software framework (www.opensg.org). As for the machine vision routines needed to implement the video-based tracking method, the Halcon 7.1 library (MVTec Software Gmbh, Munich, Germany, 2008) was used. The whole application was implemented to be compatible with several 3D displays (working either with side-by-side or alternate frames) and with all cameras for which DirectShow drivers by Microsoft are available. The configurable software framework is described in more detail in [42].

In a video see-through system, to achieve an accurate and robust fusion between reality and virtuality, the virtual scene must be rendered so that the following three conditions are satisfied:

1. The virtual camera projection models $\approx$ to the real ones.
2. The relative pose between the two virtual cameras of the stereo setup $\approx$ to the real one.
3. The pose of the virtual anatomies/surgical tools $\approx$ to the real ones.

The first condition implies that the virtual camera viewing frustums are to be modeled on the real ones in terms of image size, focus length, and center of projection (intrinsic calibration). At the same time, the second condition implies that the relative pose between the two virtual cameras of the stereo setup must be set equal to the pose between the two real cameras (extrinsic calibration).

These two calibration routines can be performed offline by implementing Zhang's calibration routine [43] (in this research, Halcon libraries were used for this task). The nonlinear part of the internal camera model (due to lens radial distortion) was taken into account by compensating for the distortion over the grabbed images before rendering them onto the background of the left and right visor displays.

Finally, the pose of the virtual elements in the virtual scene must be set equal to the real pose between the real anatomies/tools and the physical camera. This latest condition was satisfied by using a video marker-based tracking method that will be described in the following subsections.

2.2. 3D Localization and Tracking Algorithm

The poses of the two cameras relative to the anatomy and vice versa are determined by tracking passive colored markers constrained to the surgical scene in defined positions. The proposed video-based tracking solution relies on stereo localization of three monochromatic markers and is robust to inconsistent lighting conditions. 3D coordinates of the markers in the left CRS are retrieved by applying stereo *3D Localization* routines on pairs of conjugate projections of the markers' centroids onto the image planes of the two cameras. Image coordinates of the marker centroids are determined by a feature extraction task performed using *Color Segmentation* and *Circular Shape Recognition*.

2.2.1. Feature Extraction, Stereo Correspondence, and Marker Labeling

As an overall concept, color segmentation based on thresholding must ensure a robust tradeoff between illumination invariance and absence of segmentation overlaps among differently colored regions. Adoption of stringent thresholding criteria may result in inconsistent target identification because the connected regions may be poorly segmented. On the contrary, large thresholds may generate badly segmented regions or yield incorrect marker labeling in the case of multicolored markers. This drawback is emphasized by the use of cheap and/or small cameras equipped with Bayer filter color sensors. Such sensors provide inferior color quality and lower signal-to-noise ratio than those based on three sensors and a trichroic beam splitter prism for each pixel (3-CCD sensing). Use of monochromatic markers makes it possible to achieve higher robustness in the *Feature Extraction* step and in the presence of non-controllable and inconsistent lighting conditions because incorrect labelling is intrinsically avoided.

To cope partially with the limitation of using visible light as an information source, *Color Segmentation* was performed in the HSV (hue, saturation, value) color space. HSV is a human-oriented representation of the distribution of the electromagnetic radiation energy spectrum [44]. HSV enables a sufficiently robust segmentation of objects that undergo non-uniform levels of illumination intensity, shadows, and shading [45,46]. The assumption is that light intensity primarily affects the value (V) channel, whereas the hue (H), and to a lesser extent the saturation (S) channels are less influenced by illumination changes [46]. The chromatic choice for the markers must lean towards highly saturated colors, as was done in [47]. In this way, segmentation based on thresholding becomes more selective: it can be performed with a high cutoff value in the S-channel.

After *Color Segmentation*, three broader connected regions with a circular shape factor >0.5 are identified on both images. Then, the centroids of the selected regions are determined. These image points correspond to the projections of the marker centroids on the image planes of the two cameras. Figure 4 shows the results of *Color Segmentation*. After *Circular Shape Recognition*, the 2D projections of the three marker centroids on the left and right images are known.

The *Stereo Correspondence* problem is solved with a method based on minimizing an energy term computed by applying standard projective rules to all possible permutations of matches between the feature-point triplets on the image pair. In more detail, knowing the internal parameters and the relative pose between the two cameras, it is possible to determine the 3D position of a point from its projections on the left and right cameras (stereo triangulation). The 3D position of the point in the CRS can be approximated as the middle of the shortest segment joining the two projection lines. The distance between the two projection lines (DPL) is correlated with the localization error and depends on working distance, inter-camera distance, calibration quality, and identification accuracy of the conjugate image points. By working with a set of indistinguishable markers, it is not possible to localize the markers in the CRS without ambiguity because the correspondence between projected points on the left and right cameras (known as conjugate points) is unknown. The algorithm calculates the position of the three marker centroids together with the associated DPL for each of the six possible permutations of possible conjugate point matches. Hence, the solution for the stereo correspondence problem is assumed to be the one that minimizes the sum of the three DPLs over the six permutations. Once the right correspondence has been determined, the positions of the three marker centroids in the CRS are given, and the *Stereo Correspondence* and *3D Localization* steps are complete.

Figure 5 shows the results of the *Stereo Correspondence* step on a pair of sample images. Note that after this step, the correspondence between each of the projected marker centroids on the two images is known, but the marker labels (i.e., the *3D-3D Correspondence*) remain unknown.

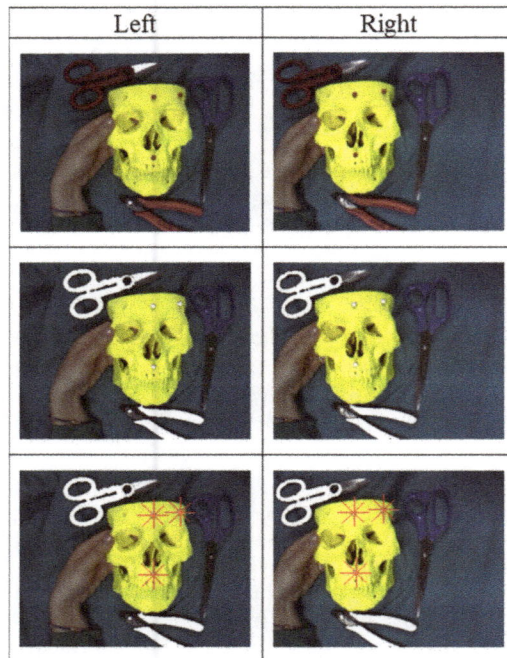

Figure 4. Results of the *Color Segmentation* and *Circular Shape Recognition* steps. In the first row, left and right camera native frames are shown. The second row shows the results of *Color Segmentation*, and the third row shows the results of *Circular Shape Recognition*.

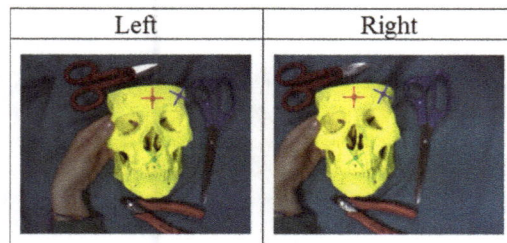

Figure 5. *Stereo Correspondence*: The correspondence between the three points on the left and right images is solved using multiple stereoscopic triangulation routines on the six possible permutations of the three points. In the images, the correct correspondences between the three points are shown.

Therefore, before solving the registration problem, the *3D-3D Correspondence* problem must be determined, which involves finding the proper set of corresponding points in the CRS and SRS. The *3D-3D Correspondence* between the two sets of 3D points is solved by a geometric procedure that takes account of the similarity of the triangles formed by such points. This approach requires that the distances between markers not be equal.

2.2.2. Two-Stage Pose Estimation

The rigid transformation between the two reference systems, namely the camera pose and the SRS, is encapsulated by matrix $\overline{R}|\overline{T}$. Pose estimation is performed using a two-stage method, with the first step being solving the AOP by standard *3D Least-Squares Fitting* of the two point sets through SVD [36]. Figure 6 shows the visual results of the first registration step between the two reference systems. As shown in the figure, due to stereo localization inaccuracies, the image registration resulting from the AOP solution may be inaccurate.

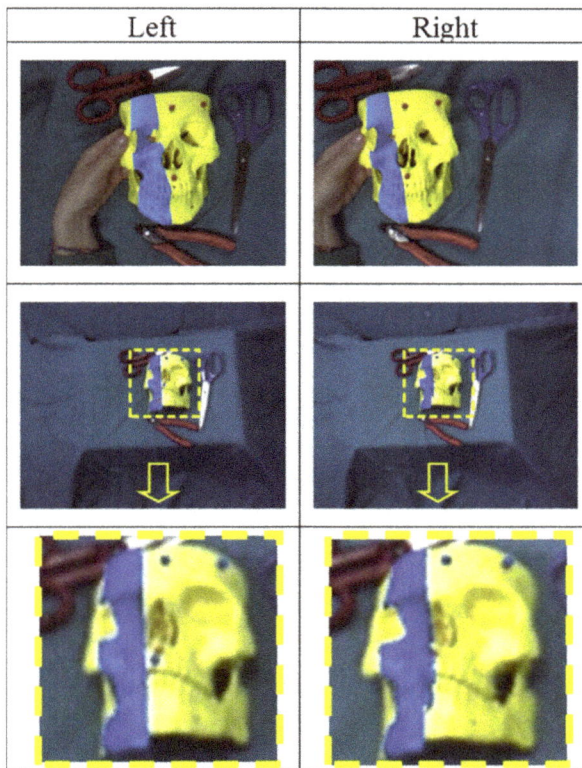

Figure 6. Geometric registration through SVD: Geometric registration solved by a *Least-Squares Fitting* method that provides a first rough alignment between the virtual information and the real scene. As shown in the first row, at close distances (about 40 cm), geometric registration is sufficiently accurate in the presence of calibrated cameras and with reliable disparity estimates. As shown in the second row, far from the scene (~100 cm), alignment accuracy rapidly degrades. The third row shows a zoomed detail of the second row.

Because of the geometry of the stereo setup, the limited focal length, and the degradation of the stereo camera calibration, adequate accuracy of *3D Localization* of the markers at greater distances cannot be ensured. The major error component in *3D Localization* is along the optical axis (z-axis) and increases with the square of the distance. The depth resolution is calculated as in [37,48]:

$$\Delta Z \cong \frac{z^2}{fb} \Delta d. \tag{1}$$

As an example, let us assume that we have determined a fixed and ideally error-free estimate of the focal length and the baseline (in the described system, $f \cong 4.8$ mm, $b \cong 70$ mm). Given a disparity accuracy of ± 1 pixel and a sensor diagonal of $1/3''$ ($\Delta d \cong 7.2$ μm), the depth resolution ΔZ is approximately ± 5 mm at a working distance of 50 cm. At 100 cm, this error increases to approximately ± 21 mm (see the last row of Figure 6). Therefore, because of data noise and the geometry of the stereo setup, the SVD solution of the AOP cannot yield a sufficiently accurate result in terms of geometric registration. On this basis, this paper proposes a methodology for refining the estimates of both camera poses to increase the accuracy of the video-based tracking technique.

The general problem of determining the pose of a calibrated camera with respect to a scene or object given its intrinsic parameters and a set of *n* world-to-image point correspondences was first formally introduced in the computer vision community by Fishler and Bolles in 1981 [49] using the term "Perspective-*n*-Point" problem (P*n*P). The P*n*P problem pertains to several areas of interest and is key to many fields like computer vision, robotics, and photogrammetry. In the transformation-based definition given in [49], the P*n*P problem aims to estimate the camera pose given a set of correspondences between *n* 3D points (known as "control points") and their 2D projections in the image plane [50]. If the number of corresponding points is <six, which is the most common and practical situation, the P*n*P problem generally does not guarantee the uniqueness of the solution. The P3P problem entails the smallest subset of control points that yields a finite number of solutions. In computer-vision applications, study of the multi-solution phenomenon for the closed-form methods has become very popular because of the "pivotal role played" by the P3P problem within the set of problems with a large number of uncertainties [51].

Regardless of the number of control points, the P*n*P problem can be faced with mainly two categories of methods: closed-form methods and iterative optimization methods [52]. Closed-form methods are usually faster, but often do not provide a unique solution and are usually less accurate and more susceptible to noise [35,53–57]. Iterative optimization approaches are based on minimizing a chosen cost function and, if a good initial guess of the solution is provided, determine the closest solution [58,59]. In our case, the initial guess is provided by the SVD solution of the AOP. Therefore, we added a P*n*P-based iterative optimization step (with *n* = 3) to the pose estimation routine. The optimization problem can be formalized as:

$$\overline{R}|\overline{T} = \arg\min \sum_{i=1}^{3} d(p_i, \hat{p}_i)^2 = \arg\min \sum_{i=1}^{3} \| p_i - \hat{p}_i(K, \hat{R}, \hat{T}, P_i) \|^2, \tag{2}$$

where the residual function $d(p_i, \hat{p}_i)$ represents the absolute distance, on the image plane, between the measured projections p_i after compensation of the radial distortion, and the calculated projections $\hat{p}_i$; $\hat{p}_i$ are computed by applying the transformation matrix $\hat{R}|\hat{T}$ and the projection matrix K to the control points P_i^{SRS}; and $\overline{R}|\overline{T}$ is the unknown transformation matrix with six dof (three rotational and three translational). Hence, knowing K, all the P_i^{SRS}, and p_i', $\overline{R}|\overline{T}$ can be calculated by minimizing the sum of the squared residuals. This optimization problem is solved using a library routine by Halcon. The iterative routine is applied to both left and right camera frames and provides more accurate image registration for the left and right views. Figure 7 shows the results of the *Pose Refinement* step applied over the images in the last row of Figure 6.

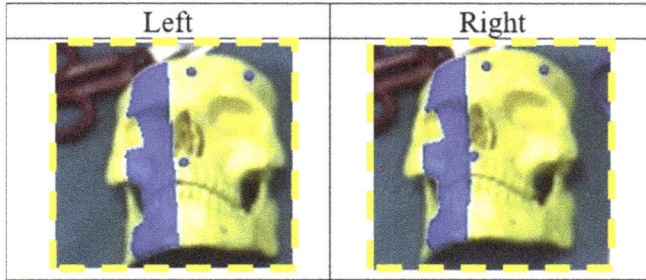

Figure 7. *Pose Refinement*: Virtual information was perfectly aligned also at greater distances (approximately 100 cm) after the *Pose Refinement* step. The frames shown on the figures constitute a refinement of the augmented frames in the last row of Figure 6.

2.3. Evaluation of Registration Accuracy

Two experiments were performed to assess registration accuracy. The first aimed to evaluate the 2D visualization alignment between virtual and real information, as done in [60]. The goal of the second experiment was to estimate the error committed by the user in a target-reaching task, and hence a testing strategy similar to that proposed in [17] was used. For each trial, the errors on both channels (right and left) before and after the *Pose Refinement* step were measured.

The experimental setup consisted of a plastic board with dimensions (160 × 100 mm) intended to reproduce the area of a typical surgical field of intervention; this panel, covered by a layer of white cardboard, included reference holes close to the vertices in known positions. For this specific test, three red plastic spheres with a diameter of 5.92 mm (measured by a digital caliper with a resolution of 0.01 mm) were used as markers. The colored markers were arranged on top of the reference holes on the panel. The 3D coordinates of the marker centroids in the board reference system (i.e., SRS) were known.

The first experiment calculated the 2D target visualization error (TVE2D) expressed in pixels. The TVE2D represents the mean offset between real objects and their virtual reproductions on the image plane.

To this end, ten validation points in the form of black crosses were printed over the white cardboard in known positions. To assess the accuracy of the AR registration, the HMD was placed at four different positions with different distances and orientations in relation to the SRS (distances ranging between 300 and 900 mm).

For each AR view, TVE2D was measured between the centroids of the black crosses (real objects) and of the red crosses (virtual objects), as shown in Figure 8.

The second experiment was aimed at empirically estimating the error committed by the user in the task of reaching a planned target point over a planar surface under AR guidance; this error was called the 3D target reaching error (TReachE3D). The user, under AR guidance, was asked to mark, using a thin pen over the white cardboard, the center of the virtual crosses showed on the displays. The test was repeated at four distances between 300 and 900 mm. After each test, the cardboard was scanned using a desktop scanner. Reached points were visually determined and expressed in the SRS (knowing the reference hole positions in the scanned image). Finally, the distances between reached and correct/planned points were computed.

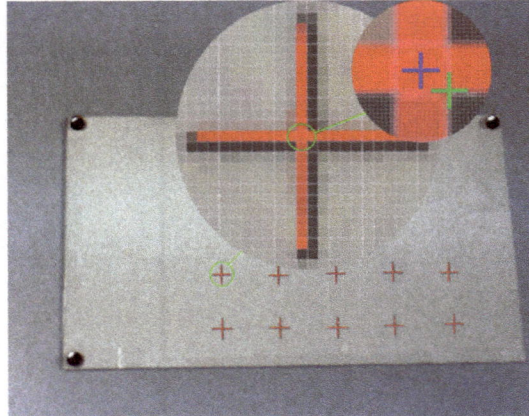

Figure 8. Evaluation of the TVE: image frame from the left display at a distance of 300 mm between the HMD and the validation board. The two circles show a zoomed detail of the frame with the centroids of the virtual (red) and real (black) cross highlighted respectively in blue and green. The black virtual spheres align exactly with the real red markers.

3. Results

Table 1 presents the results of the first validation experiment for the left and right cameras before and after the *Pose Refinement* step. Errors were measured on the image plane and are expressed in pixels.

Table 1. Mean and standard deviation of TVE2D over 10 validation points for both sides before and after *Pose Refinement*.

Camera	TVE2D Before Pose Refinement	TVE2D After Pose Refinement
Left Camera	1.72 pixel (±0.71)	0.86 pixel (±0.53)
Right Camera	1.48 pixel (±0.58)	0.88 pixel (±0.67)

Knowing the intrinsic parameters of the two cameras, it is also possible to estimate the visualization error in space in mm (TVE3D) at fixed distances [10,60]. TVE3D is calculated by inverting the projection equation:

$$\text{TVE3D} \cong \frac{\text{TVE2D}}{k} \frac{Z_C}{f} \tag{3}$$

where Z_C is the estimated working distance, f represents the focal length estimated in the calibration phase and corresponding to 4.8 mm, and k is the scaling factor of the image sensor (number of pixels per unit of length). In the present case, $1/k$ can be calculated from the image sensor specifications and corresponds approximately to 7.2 μm. The mean TVE3D at 700 mm for the left camera was 1.8 mm without *Pose Refinement* and was decreased to 0.9 mm by minimizing the reprojection error in the *Pose Refinement* step.

Figure 9 shows the results of the second experiment, which provided an estimate of spatial accuracy in 3D space. The *Pose Refinement* step increased the accuracy. Mean (μ) errors ± standard deviations (σ) at 700 mm for the left camera were 2.30 ± 0.91 and 1.00 ± 0.56 mm respectively before and after the *Pose Refinement* step. It is interesting to highlight the tendency of the error, without *Pose Refinement*, to drift upward approximately with the square of the distance between camera and validation board, following the same trend as the localization error. Finally, the computational payload for the entire algorithm was also evaluated. Depending on the scene and the environmental lighting, the average running time ranged between 10 and 15 ms. Running time was evaluated using a standard PC with a quad-core i7-3770@3.4GHz processor and 8GB RAM. The graphics card used was a GeForce GT 620 (NVIDIA, Santa Clara, CA, USA). In any case, the time required for the localization and registration thread was less than 33 ms (working at 60 Hz).

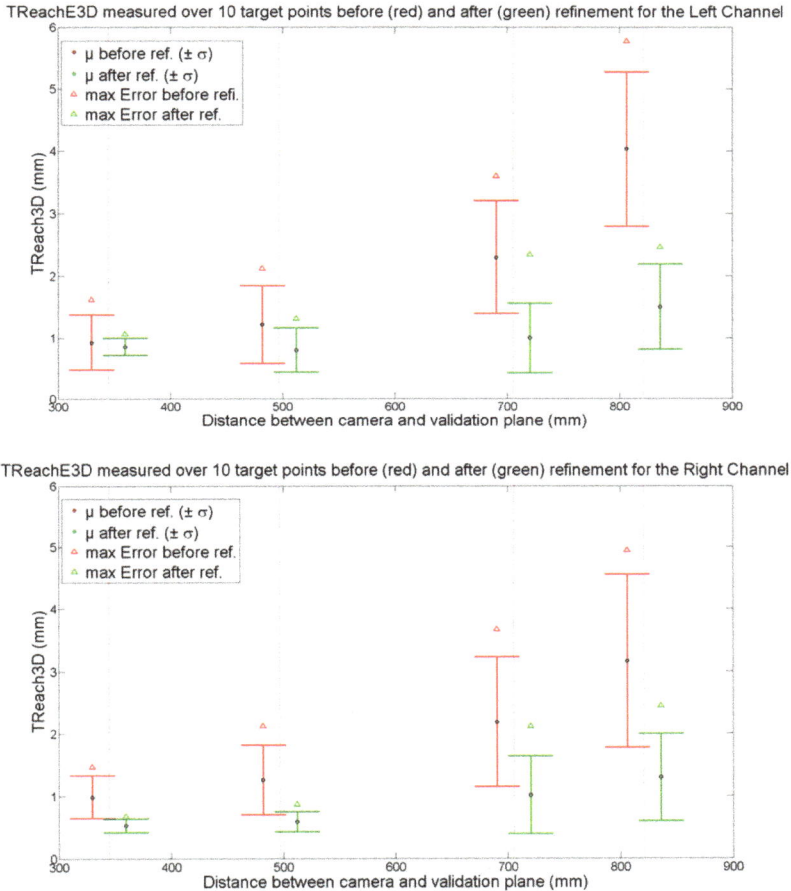

Figure 9. Target reaching error in space (TReach3D) made by the user in the task of reaching target crosses on a board positioned at different distances under AR guidance. Red and green values, referring respectively to TReach3D before and after the *Pose Refinement* step, are slightly shifted along the *x*-axis for readability reasons.

4. Discussion and Conclusions

This paper has introduced new video-based tracking methods suitable for stereoscopic AR video see-through systems. The proposed solution is aimed at providing accurate camera pose estimation and is based on tracking three indistinguishable markers. The algorithm was developed for a wearable AR system, but it might also be applied to other stereoscopic devices like binocular endoscopes or binocular microscopes.

The proposed algorithm avoids the need for an external tracker to detect the relative pose between the cameras and the real scene. Coherent superimposition of virtual information onto real images is achieved through a video marker-based tracking method.

Video-based tracking methods need a robust estimation of the physical camera projective model, i.e., intrinsic and extrinsic camera parameters. In the proposed AR application, the estimate of the intrinsic camera parameters is the result of a standard off-line calibration process, whereas the extrinsic camera parameters are determined online. Solutions for simultaneous on-line estimation of both intrinsic and extrinsic camera parameters have been proposed [61]. In any case, the limitation of using fixed intrinsic parameters does not generally affect the overall usability of wearable video see-through AR systems. Camera zooming, which implies a change in intrinsic camera parameters, would in fact cause an unnatural sensation to the user because of the resulting incoherence between changes in motion perceived by the user in the displayed images (due to changes in camera field of view) and actual head motions. The proposed solution takes into account considerations of system applicability in a clinical scenario.

Colored spheres were chosen as markers. The reason for this choice was that small spherical markers can be conveniently placed around the surgical area without compromising the surgeon's field of view. The use of a minimum set of three fiducial markers is also intended to limit the logistic payload for setup.

Use of monochromatic markers makes it possible to achieve high robustness in the feature extraction step and also in the presence of non-controllable and/or inconsistent lighting conditions. This choice has required marker labeling methods. The proposed algorithm solves both stereo and 3D–3D correspondence problems before registration. The stereo correspondence problem is solved by applying multiple stereoscopic triangulation routines on pairs of images simultaneously grabbed from the two cameras. 3D–3D correspondence is determined by a geometrical procedure.

Furthermore, the proposed algorithm provides sub-pixel registration accuracy between real and virtual scenes thanks to a PnP-based optimization step. The strategy for refining each camera pose does not need a perfectly calibrated stereoscopic system.

A key factor in performing highly accurate measurements with stereoscopic systems is to know with extreme confidence the relative pose between the two cameras. A relevant drawback of using wearable trackers is represented by the non-ideal stability in the constraints between the two stereo cameras, which may cause a potential change in their relative pose while the visor is being used [40]. Such systems need frequent calibration to cope with degradation of the stereo calibration over time.

Pose refinement provides sub-pixel video registration accuracy and can compensate for potential loss of accuracy in the estimate of the relative pose between the two cameras. The accuracy and robustness of the proposed wearable AR stereoscopic video see-through display pave the way for the introduction of such technology in clinical practice.

One way to translate this solution into clinical practice and to provide radiological images for patient registration is the following: virtual anatomies are reconstructed offline from radiological images [30]; the positions of the three fiducial points are identified on the 3D model of the anatomy (e.g., by applying radiopaque markers on the patient before acquiring a CT image, or by considering physical landmarks as references); before the surgical procedure, three monochromatic markers, whose centroids must be in the same position as the three fiducial points, are anchored onto the patient. This approach is well known and used in other IGS systems.

The proposed solution has already been used in a study in maxillofacial surgery that was published in 2014 [62]. The study was focused on in vitro validation of the proposed stereoscopic video see-through AR system as an aid for manual repositioning of facial bone fragments. The AR visualization modality used in the clinical study, which provides an ergonomic interaction paradigm within the augmented scene, draws its inspiration from and tries to mimic physically the paradigm on which the PnP problem is formulated. This task-oriented AR visualization modality has been more thoroughly described in a recently published manuscript [63].

More recently, the video-based tracking method has been positively validated in vitro to aid trocar insertion during a percutaneous procedure in orthopedic surgery [64].

Supplementary Materials: The following are available online at www.mdpi.com/2079-9292/5/3/59/s1. Video S1: Video_Demo_Maxillofacial_Application.

Acknowledgments: This work was funded by the Italian Ministry of Health grant, SThARS (Surgical training in identification and isolation of deformable tubular structures with hybrid Augmented Reality Simulation, 6/11 2014–5/11/2017), Grant "Ricerca finalizzata e Giovani Ricercatori 2011–2012", Young Researchers, Italian Ministry of Health.

Author Contributions: F.C. and V.F. designed and developed the wearable AR system and proposed the localizer-free video-based tracking mechanism. F.C., S.M., and C.F. implemented the AR software framework. M.F., P.D.P., and F.C. set the requirements for the new wearable solution for AR-based image-guided surgery. The main goal of this analysis was to insert the proposed system within the context of state-of-the-art solutions for AR-based image-guided surgery. F.C. and V.F. analyzed and discussed the results.

Conflicts of Interest: The authors declare no conflict of interest.

References

1. Milgram, P.; Kishino, F. A taxonomy of mixed reality visual-displays. *IEICE Trans. Inf. Syst.* **1994**, *E77-D (12)*, 1321–1329.

2. Kersten-Oertel, M.; Jannin, P.; Collins, D.L. The state of the art of visualization in mixed reality image guided surgery. *Comput. Med. Imaging Graph.* **2013**, *37*, 98–112. [CrossRef] [PubMed]

3. Cuchet, E.; Knoplioch, J.; Dormont, D.; Marsault, C. Registration in neurosurgery and neuroradiotherapy applications. *J. Image Guid. Surg.* **1995**, *1*, 198–207. [CrossRef]

4. Roberts, D.W.; Strohbehn, J.W.; Hatch, J.F.; Murray, W.; Kettenberger, H. A frameless stereotaxic integration of computerized tomographic imaging and the operating microscope. *J. Neurosurg.* **1986**, *65*, 545–549. [CrossRef] [PubMed]

5. Edwards, P.J.; Hawkes, D.J.; Hill, D.L.G.; Jewell, D.; Spink, R.; Strong, A.; Gleeson, M. Augmentation of reality using an operating microscope for otolaryngology and neurosurgical guidance. *J. Image Guid. Surg.* **1995**, *1*, 172–178. [CrossRef]

6. Edwards, P.J.; Hill, D.L.G.; Hawkes, D.J.; Spink, R.; Colchester, A.C.F.; Strong, A.; Gleeson, M. Neurosurgical guidance using the stereo microscope. In *Computer Vision, Virtual Reality Robotics in Medicine*; Ayache, N., Ed.; Springer: Berlin/Heidelberg, Germany, 1995; Volume 905, pp. 555–564.

7. Kersten-Oertel, M.; Jannin, P.; Collins, D.L. Dvv: Towards a taxonomy for mixed reality visualization in image guided surgery. *Med. Imaging Augment. Real.* **2010**, *6326*, 334–343.

8. Freysinger, W.; Gunkel, A.R.; Thumfart, W.F. Image-guided endoscopic ENT surgery. *Eur. Arch. Oto-Rhino-Laryngol.* **1997**, *254*, 343–346. [CrossRef]

9. Caversaccio, M.; Giraldez, J.G.; Thoranaghatte, R.; Zheng, G.; Eggli, P.; Nolte, L.P.; Ballester, M.A.G. Augmented reality endoscopic system (ARES): Preliminary results. *Rhinology* **2008**, *46*, 156–158. [PubMed]

10. Baumhauer, M.; Simpfendorfer, T.; Muller-Stich, B.P.; Teber, D.; Gutt, C.N.; Rassweiler, J.; Meinzer, H.P.; Wolf, I. Soft tissue navigation for laparoscopic partial nephrectomy. *Int. J. Comput. Assist. Radiol. Surg.* **2008**, *3*, 307–314. [CrossRef]

11. Ieiri, S.; Uemura, M.; Konishi, K.; Souzaki, R.; Nagao, Y.; Tsutsumi, N.; Akahoshi, T.; Ohuchida, K.; Ohdaira, T.; Tomikawa, M.; et al. Augmented reality navigation system for laparoscopic splenectomy in children based on preoperative ct image using optical tracking device. *Pediatr. Surg. Int.* **2012**, *28*, 341–346. [CrossRef] [PubMed]

12. Haouchine, N.; Dequidt, J.; Berger, M.O.; Cotin, S. Deformation-based augmented reality for hepatic surgery. *Stud. Health Technol. Inf.* **2013**, *184*, 182–188.

13. Liao, H.; Nakajima, S.; Iwahara, M.; Kobayashi, E.; Sakuma, I.; Yahagi, N.; Dohi, T. Intra-operative real-time 3-D information display system based on integral videography. In Proceedings of the 4th International Conference on Medical Image Computing and Computer-Assisted Intervention—Miccai 2001, Utrecht, The Netherlands, 14–17 October 2001; Niessen, W., Viergever, M., Eds.; Springer: Berlin/Heidelberg, Germany, 2001; Volume 2208, pp. 392–400.

14. Liao, H.; Hata, N.; Nakajima, S.; Iwahara, M.; Sakuma, I.; Dohi, T. Surgical navigation by autostereoscopic image overlay of integral videography. *IEEE Trans. Inf. Technol. Biomed.* **2004**, *8*, 114–121. [CrossRef] [PubMed]

15. Iseki, H.; Masutani, Y.; Iwahara, M.; Tanikawa, T.; Muragaki, Y.; Taira, T.; Dohi, T.; Takakura, K. Volumegraph (overlaid three-dimensional image-guided navigation). Clinical application of augmented reality in neurosurgery. *Stereotact. Funct. Neurosurg.* **1997**, *68*, 18–24. [CrossRef] [PubMed]

16. Narita, Y.; Tsukagoshi, S.; Suzuki, M.; Miyakita, Y.; Ohno, M.; Arita, H.; Saito, Y.; Kokojima, Y.; Watanabe, N.; Moriyama, N.; et al. Usefulness of a glass-free medical three-dimensional autostereoscopic display in neurosurgery. *Int. J. Comput. Assist. Radiol. Surg.* **2014**, *9*, 905–911. [CrossRef] [PubMed]

17. Liao, H.; Inomata, T.; Sakuma, I.; Dohi, T. Surgical navigation of integral videography image overlay for open MRI-guided glioma surgery. *Med. Imaging Augment. Real.* **2006**, *4091*, 187–194.

18. Liao, H.E.; Inomata, T.; Sakuma, I.; Dohi, T. 3-D augmented reality for mri-guided surgery using integral videography autostereoscopic image overlay. *IEEE Trans. Biomed. Eng.* **2010**, *57*, 1476–1486. [CrossRef] [PubMed]

19. Suenaga, H.; Tran, H.H.; Liao, H.G.; Masamune, K.; Dohi, T.; Hoshi, K.; Mori, Y.; Takato, T. Real-time in situ three-dimensional integral videography and surgical navigation using augmented reality: A pilot study. *Int. J. Oral Sci.* **2013**, *5*, 98–102. [CrossRef] [PubMed]

20. Ferrari, V.C.E. Wearable augmented reality light field optical see-through display to avoid user dependent calibrations: A feasibility study. In Proceedings of the IEEE Science and Information Conference, SAI 2016, London, UK, 13–15 July 2016; pp. 1211–1216.

21. Sielhorst, T.; Feuerstein, M.; Navab, N. Advanced medical displays: A literature review of augmented reality. *J. Disp. Technol.* **2008**, *4*, 451–467. [CrossRef]

22. Birkfellner, W.; Figl, M.; Huber, K.; Watzinger, F.; Wanschitz, F.; Hummel, J.; Hanel, R.; Greimel, W.; Homolka, P.; Ewers, R.; et al. A head-mounted operating binocular for augmented reality visualization in medicine-design and initial evaluation. *IEEE Trans. Med. Imaging* **2002**, *21*, 991–997. [CrossRef] [PubMed]

23. Ferrari, V.; Megali, G.; Troia, E.; Pietrabissa, A.; Mosca, F. A 3-D mixed-reality system for stereoscopic visualization of medical dataset. *IEEE Trans. Biomed. Eng.* **2009**, *56*, 2627–2633. [CrossRef] [PubMed]

24. Sielhorst, T.; Bichlmeier, C.; Heining, S.M.; Navab, N. Depth perception—A major issue in medical AR: Evaluation study by twenty surgeons. In Proceedings of the 9th International Conference on Medical Image Computing and Computer-Assisted Intervention—MICCAI 2006, Copenhagen, Denmark, 1–6 October 2006; Volume 4190, pp. 364–372.

25. Rolland, J.P.; Fuchs, H. Optical versus video see-through mead-mounted displays in medical visualization. *Presence Teleoperators Virtual Environ.* **2000**, *9*, 287–309. [CrossRef]

26. Kellner, F.; Bolte, B.; Bruder, G.; Rautenberg, U.; Steinicke, F.; Lappe, M.; Koch, R. Geometric calibration of head-mounted displays and its effects on distance estimation. *IEEE Trans. Vis. Comput. Graph.* **2012**, *18*, 589–596. [CrossRef] [PubMed]

27. Genc, Y.; Tuceryan, M.; Navab, N. Practical solutions for calibration of optical see-through devices. *Int. Symp. Mixed Augment. Real. Proc.* **2002**, 169–175.

28. Plopski, A.; Itoh, Y.; Nitschke, C.; Kiyokawa, K.; Klinker, G.; Takemura, H. Corneal-imaging calibration for optical see-through head-mounted displays. *IEEE Trans. Vis. Comput. Graph.* **2015**, *21*, 481–490. [CrossRef] [PubMed]

29. Navab, N.; Heining, S.M.; Traub, J. Camera augmented mobile C-arm (CAMC): Calibration, accuracy study, and clinical applications. *IEEE Trans. Med. Imaging* **2010**, *29*, 1412–1423. [CrossRef] [PubMed]

30. Ferrari, V.; Carbone, M.; Cappelli, C.; Boni, L.; Melfi, F.; Ferrari, M.; Mosca, F.; Pietrabissa, A. Value of multidetector computed tomography image segmentation for preoperative planning in general surgery. *Surg. Endosc.* **2012**, *26*, 616–626. [CrossRef] [PubMed]

31. Marmulla, R.; Hoppe, H.; Muhling, J.; Eggers, G. An augmented reality system for image-guided surgery. *Int. J. Oral Maxillofac. Surg.* **2005**, *34*, 594–596. [CrossRef] [PubMed]

32. Ferrari, V.; Viglialoro, R.M.; Nicoli, P.; Cutolo, F.; Condino, S.; Carbone, M.; Siesto, M.; Ferrari, M. Augmented reality visualization of deformable tubular structures for surgical simulation. *Int. J. Med. Robot.* **2015**, *12*, 231–240. [CrossRef] [PubMed]

33. Franz, A.M.; Haidegger, T.; Birkfellner, W.; Cleary, K.; Peters, T.M.; Maier-Hein, L. Electromagnetic tracking in medicine—A review of technology, validation, and applications. *IEEE Trans. Med. Imaging* **2014**, *33*, 1702–1725. [CrossRef] [PubMed]
34. Kanbara, M.; Okuma, T.; Takemura, H.; Yokoya, N. A stereoscopic video see-through augmented reality system based real-time vision-based registration. In Proceedings of the IEEE Virtual Reality Conference, New Brunswick, NJ, USA, 18–22 March 2000; pp. 255–262.
35. Haralick, R.M.; Lee, C.N.; Ottenberg, K.; Nolle, M. Review and analysis of solutions of the 3-point perspective pose estimation problem. *Int. J. Comput. Vis.* **1994**, *13*, 331–356. [CrossRef]
36. Arun, K.S.; Huang, T.S.; Blostein, S.D. Least-squares fitting of 2 3-D point sets. *IEEE Trans. Pattern Anal. Mach. Intell.* **1987**, *9*, 699–700. [CrossRef]
37. Kyto, M.; Nuutinen, M.; Oittinen, P. Method for measuring stereo camera depth accuracy based on stereoscopic vision. *Three-Dimens. Imaging Interact. Meas.* **2011**, *7864*. [CrossRef]
38. Schneider, A.; Baumberger, C.; Griessen, M.; Pezold, S.; Beinemann, J.; Jurgens, P.; Cattin, P.C. Landmark-based surgical navigation. *Clin. Image-Based Proc. Transl. Res. Med. Imaging* **2014**, *8361*, 57–64.
39. Cutolo, F.; Parchi, P.D.; Ferrari, V. Video see through ar head-mounted display for medical procedures. In Proceedings of the 17th IEEE International Symposium on Mixed and Augmented Reality (ISMAR), Munich, Germany, 10–12 September 2014; pp. 393–396. [CrossRef]
40. Ferrari, V.; Cutolo, F.; Calabro, E.M.; Ferrari, M. Hmd video see though AR with unfixed cameras vergence. In Proceedings of the 17th IEEE International Symposium on Mixed and Augmented Reality (ISMAR), Munich, Germany, 10–12 September 2014; pp. 265–266. [CrossRef]
41. Megali, G.; Ferrari, V.; Freschi, C.; Morabito, B.; Turini, G.; Troia, E.; Cappelli, C.; Pietrabissa, A.; Tonet, O.; Cuschieri, A.; et al. Endocas navigator platform: A common platform for computer and robotic assistance in minimally invasive surgery. *Int. J. Med. Robot. Comput. Assist. Surg.* **2008**, *4*, 242–251. [CrossRef] [PubMed]
42. Cutolo, F.; Siesto, M.; Mascioli, S.; Freschi, C.; Ferrari, M.; Ferrari, V. Configurable software framework for 2D/3D video see-through displays in medical applications. In *Augmented Reality, Virtual Reality, and Computer Graphics: Third International Conference, Avr 2016, Lecce, Italy, 15–18 June 2016. Proceedings, Part II*; De Paolis, T.L., Mongelli, A., Eds.; Springer International Publishing: Cham, Switzerland, 2016; pp. 30–42.
43. Zhang, Z.Y. A flexible new technique for camera calibration. *IEEE Trans. Pattern Anal. Mach. Intell.* **2000**, *22*, 1330–1334. [CrossRef]
44. Boker, S. *The Representation of Color Metrics and Mappings in Perceptual Color Space*; The University of Virginia, 1995. Available online: http://people.virginia.edu/~smb3u/ColorVision2/ColorVision2.html (accessed on 13 September 2016).
45. Kyriakoulis, N.; Gasteratos, A. Light-invariant 3D object's pose estimation using color distance transform. In Proceedings of the 2010 IEEE International Conference on Imaging Systems and Techniques (IST), Thessaloniki, Greece, 1–2 July 2010; pp. 105–110.
46. Loukas, C.; Lahanas, V.; Georgiou, E. An integrated approach to endoscopic instrument tracking for augmented reality applications in surgical simulation training. *Int. J. Med. Robot. Comput. Assist. Surg.* **2013**, *9*, E34–E51. [CrossRef] [PubMed]
47. Diotte, B.; Fallavollita, P.; Wang, L.J.; Weidert, S.; Thaller, P.H.; Euler, E.; Navab, N. Radiation-free drill guidance in interlocking of intramedullary nails. *Lect. Notes Comput. Sci.* **2012**, *7510*, 18–25.
48. Chang, C.C.; Chatterjee, S. Quantization-error analysis in stereo vision. In Proceedings of the Conference Record of The Twenty-Sixth Asilomar Conference on Signals, Systems and Computers, Pacific Grove, CA, USA, 26–28 October 1992; pp. 1037–1041.
49. Fischler, M.A.; Bolles, R.C. Random sample consensus—A paradigm for model-fitting with applications to image-analysis and automated cartography. *Commun. ACM* **1981**, *24*, 381–395. [CrossRef]
50. Wu, Y.H.; Hu, Z.Y. PnP problem revisited. *J. Math. Imaging Vis.* **2006**, *24*, 131–141. [CrossRef]
51. Zhang, C.X.; Hu, Z.Y. Why is the danger cylinder dangerous in the P3P problem. *Zidonghua Xuebao/Acta Autom. Sin.* **2006**, *32*, 504–511.
52. Garro, V.; Crosilla, F.; Fusiello, A. Solving the PnP problem with anisotropic orthogonal procrustes analysis. In Proceedings of the 2012 Second Joint 3DIM/3DPVT Conference: 3D Imaging, Modeling, Processing, Visualization & Transmission 2012, Zurich, Switzerland, 13–15 October 2012; pp. 262–269.
53. Quan, L.; Lan, Z.D. Linear N-point camera pose determination. *IEEE Trans. Pattern Anal. Mach. Intell.* **1999**, *21*, 774–780. [CrossRef]

54. Fiore, P.D. Efficient linear solution of exterior orientation. *IEEE Trans. Pattern Anal. Mach. Intell.* **2001**, *23*, 140–148. [CrossRef]

55. Gao, X.S.; Hou, X.R.; Tang, J.L.; Cheng, H.F. Complete solution classification for the perspective-three-point problem. *IEEE Trans. Pattern Anal. Mach. Intell.* **2003**, *25*, 930–943.

56. Ansar, A.; Daniilidis, K. Linear pose estimation from points or lines. *IEEE T Pattern Anal.* **2003**, *25*, 578–589. [CrossRef]

57. Lepetit, V.; Moreno-Noguer, F.; Fua, P. Epnp: An accurate o(N) solution to the pnp problem. *Int. J. Comput. Vis.* **2009**, *81*, 155–166. [CrossRef]

58. Haralick, R.M.; Joo, H.; Lee, C.N.; Zhuang, X.H.; Vaidya, V.G.; Kim, M.B. Pose estimation from corresponding point data. *IEEE Trans. Syst. Man Cyber.* **1989**, *19*, 1426–1446. [CrossRef]

59. Lu, C.P.; Hager, G.D.; Mjolsness, E. Fast and globally convergent pose estimation from video images. *IEEE Trans. Pattern Anal. Mach. Intell.* **2000**, *22*, 610–622. [CrossRef]

60. Muller, M.; Rassweiler, M.C.; Klein, J.; Seitel, A.; Gondan, M.; Baumhauer, M.; Teber, D.; Rassweiler, J.J.; Meinzer, H.P.; Maier-Hein, L. Mobile augmented reality for computer-assisted percutaneous nephrolithotomy. *Int. J. Comput. Assist. Radiol. Surg.* **2013**, *8*, 663–675. [CrossRef] [PubMed]

61. Taketomi, T.; Okada, K.; Yamamoto, G.; Miyazaki, J.; Kato, H. Camera pose estimation under dynamic intrinsic parameter change for augmented reality. *Comput. Graph.-UK* **2014**, *44*, 11–19. [CrossRef]

62. Badiali, G.; Ferrari, V.; Cutolo, F.; Freschi, C.; Caramella, D.; Bianchi, A.; Marchetti, C. Augmented reality as an aid in maxillofacial surgery: Validation of a wearable system allowing maxillary repositioning. *J. Cranio-Maxillofacial Surg.* **2014**, *42*, 1970–1976. [CrossRef] [PubMed]

63. Cutolo, F.; Badiali, G.; Ferrari, V. Human-PnP: Ergonomic ar interaction paradigm for manual placement of rigid bodies. In *Augmented Environments for Computer-Assisted Interventions*; Linte, C., Yaniv, Z., Fallavollita, P., Eds.; Springer International Publishing: Cham, Switzerland, 2015; Volume 9365, pp. 50–60.

64. Cutolo, F.; Carbone, M.; Parchi, P.D.; Ferrari, V.; Lisanti, M.; Ferrari, M. Application of a new wearable augmented reality video see-through display to aid percutaneous procedures in spine surgery. In *Augmented Reality, Virtual Reality, and Computer Graphics: Third International Conference, Avr 2016, Lecce, Italy, 15–18 June 2016. Proceedings, Part II*; De Paolis, T.L., Mongelli, A., Eds.; Springer International Publishing: Cham, Switzerland, 2016; pp. 43–54.

electronics

MDPI

Article

Miniaturized Blood Pressure Telemetry System with RFID Interface

Michele Caldara [1], Benedetta Nodari [1,*], Valerio Re [1] and Barbara Bonandrini [2]

[1] Department of Engineering, University of Bergamo, Viale Marconi 5, Dalmine (BG) 24044, Italy; michele.caldara@unibg.it (M.C.); valerio.re@unibg.it (V.R.)

[2] Istituto di Ricerche Farmacologiche "Mario Negri", Via Stezzano 87, Bergamo 24126, Italy; barbara.bonandrini@marionegri.it

* Correspondence: benedetta.nodari@unibg.it; Tel.: +39-035-205-2159

Academic Editors: Enzo Pasquale Scilingo and Gaetano Valenza
Received: 23 May 2016; Accepted: 22 August 2016; Published: 30 August 2016

Abstract: This work deals with the development and characterization of a potentially implantable blood pressure telemetry system, based on an active Radio-Frequency IDentification (RFID) tag, International Organization for Standardization (ISO) 15693 compliant. This approach aims to continuously measure the average, systolic and diastolic blood pressure of the small/medium animals. The measured pressure wave undergoes embedded processing and results are stored onboard in a non-volatile memory, providing the data under interrogation by an external RFID reader. In order to extend battery lifetime, RFID energy harvesting has been investigated. The paper presents the experimental characterization in a laboratory and preliminary in-vivo tests. The device is a prototype mainly intended, in a future engineered version, for monitoring freely moving test animals for pharmaceutical research and drug safety assessment purposes, but it could have multiple uses in environmental and industrial applications.

Keywords: blood; embedded; energy harvesting; implantable; in-vivo; ISO 15693; low-power; pressure; RFID

1. Introduction

In recent years, implantable smart sensors and Wireless Sensor Network (WSN) technologies have been considered key research areas for both computer science and electronics. Reliable physiological parameters monitoring with miniaturized smart sensor nodes can enhance healthcare applications and, at the same time, to improve the patient quality of life. Implantable telemetry systems take advantage of continuous electronic components miniaturization, progresses in sensor capability, diffusion of wireless data transfer technologies and power consumption reduction. The synergy between Micro Electro-Mechanical Systems (MEMS), a microcontroller and RFID technologies allows to extend the sensor capabilities by adding embedded computational power and wireless interface with the lowest possible supply consumption, enabling in this way the achievement of accurate and low-cost wireless systems. Telemetry is a well-established method of monitoring physiological functions in awake and freely moving laboratory animals, while minimizing stress artifacts. Currently, such systems are employed in pharmacological research to measure physiological signals such as blood pressure, heart rate, blood flow, electrocardiogram, respiratory rate, sympathetic nerve activity, body temperature etc., in a wide range of animal species: small animals such as rats, mice, gerbils and hamsters; and medium animals, such as dogs, rabbits, monkeys, guinea pigs, and pigs, etc. [1]. Nowadays, the small size of implantable monitoring devices and their extended battery lifetime permit in-vivo tests for several days, with no necessity of manipulation, leading to an improvement of experiments capabilities [2]. Most of animal implantable telemetry systems are composed by a small biocompatible enclosure containing the sensor, the electronic device and the battery plugged to a catheter filled with physiological solution. The catheter end is inserted into an artery and it is fixed with biocompatible glue. The commercial systems are usually designed to accommodate intraperitoneal placement. Typically, the animal cage is placed near the system receiver for allowing data transmission. The main wireless technologies for transmitting data are definitely Bluetooth, Wi-Fi, General Packet Radio Service, Enhanced Data rates for Gsm Evolution, ZigBee and Near Field Communication (NFC). NFC protocols are developed on the radiofrequency identification RFID standards, in particular exploiting the industrial, scientific and medical (ISM) radio band at 13.56 MHz ± 7 kHz, which is worldwide available. Nowadays, the use of the RFID in telemetry and telemedicine is increasing considerably in the biomedical and health domains. This is due to the fact that the solutions for human and animal implants are advantageous in terms of exposure to Electro Magnetic fields, since in the range 1–20 MHz human tissues do not significantly attenuate the electromagnetic waves. Moreover, the external reader provides the power needed to establish the communication [3]. The system disclosed in this paper is an active RFID sensing tag, compliant with the ISO 15,693 protocol, capable to continuously measure and process the arterial blood pressure data, providing results to an external reader. The electronics are intended to be placed in a subcutaneous region of small/medium animals, with the goal of achieving a minimal invasive implant and of reducing the tag-reader distance. The paper describes the system in terms of hardware and firmware, the laboratory characterization and preliminary in-vivo test. Long-term animal studies will be performed with the next version of the system, after further miniaturization and proper encapsulation.

2. RFID Pressure Sensing System

2.1. Requirements

The aim of such a work is to develop a prototype capable to continuously monitoring blood pressure, with the requirement to be comparable to the state-of-the-art in terms of dimensions and functionalities. For this reason, the device specifications have been dictated by state-of-the-art implantable systems, whose main features are summarized in Table 1. The system, based on Near Field Communication (NFC) technology and embedded processing, has been conceived in order to be potentially implantable in test animals (rats), used in pharmaceutical research field and human [2]. It requires only a revision of the electronics toward further miniaturization.

Table 1. State-of-the-art of telemetry system for small animals and human.

Sensor	Use	Sensor Dimensions	Range	Resolution & Accuracy
DataSCI (HD-S21)	Blood pressure (small animals)	5.9 cc; 8 gr;	−20–300 mmHg	±3 mmHg; −0.25 mmHg/month;
TSE (Stellar)	Blood pressure (small animals)	6 cc; 11 gr; 16×30 mm^2;	0–300 mmHg	-
Millar (TRM54P)	Blood pressure (small animals)	12 gr;	−20–300 mmHg	±2 mmHg; <4 mmHg/month
Mitter (G2 HR)	Heart rate (small animals)	11 gr; 15.5×6.5 mm^2;	120–780 BPM	1.5%
Developed system	Blood pressure (small animals)	8 gr; 5.6cc; $30 \times 17.5 \times 11$ mm^3	0–300 mmHg	±3 mmHg;
ENDOCOM	Blood pressure (Human)	15×18.5 mm^2	-	-
Cardio MEMS	Blood pressure (Human)	$2 \times 3.4 \times 15$ mm^3	-	-

The device main requirements concern the small volume (in the range 4.4–6 cm^3), the weight (in the range 7.6–12 gr), the pressure accuracy (±3 mmHg) and the capability of continuously measuring blood pressure in the range 0–300 mmHg. For an accurate determination of the maximum and the minimum pressure values, considering that the maximum heart rate of rats can be up to 400 bpm (6.67 Hz), the pressure signal should be sampled at least a factor ten of the maximum rat's heart rate (i.e., 66.67 Hz). This permits a good reconstruction of the pressure wave after acquisition and to better identify the minimum and maximum values. Moreover, the typical experiment duration in pharmacological research is about 10 days, implying ultra-low power electronics, which permit the use of a small and lightweight battery. In order to have an efficient and optimal system, the blood pressure wave is measured and processed on-board, thus only the maximum and minimum values are stored on a local memory of the device, leaving the freedom of a subsequent data upload via RFID, without the risk of losing acquired data. Despite this, if required, the device is able to transmit the full waveform data. The embedded processing can open new approaches in conducting experiments, minimizing the data transfer time and better exploiting the local memory usage.

2.2. Sensor Platform Description

The architecture of the system is depicted in Figure 1a. The pressure sensor is coupled to a catheter, filled with biocompatible saline solution, with the purpose to sense the blood pressure signal once it is inserted into an artery. In order to prevent the clots inside the catheter, the saline solution could be mixed with anticoagulant. The pressure is thus transferred to the opposite end of the catheter where the pressure sensor membrane is placed. An ultra-low-power microcontroller processes instantaneous pressure data and it stores the results on a non-volatile memory. The board bill of material, minimized to reduce area occupancy, includes the pressure sensor, the low-power microcontroller (STM32L162RD, STMicroelectronics, Geneva, CH, Switzerland), a non-volatile memory (M24LR64E-R EEPROM 64 kbit, STMicroelectronics) with a double interface (I2C and RFID), and an energy harvesting pin. The latter being capable of providing microcontroller analysis results via RFID. The system includes also an optimized power management chain and a Li-poly rechargeable battery to power the device (PWB1389). Thanks to the electronic device's low power-consumption, the battery capacity is only 90 mAh, with volume of only $25 \times 11 \times 3.5$ mm^3 and an approximate weight of 2 gr. While the battery voltage decreases from 4.1 V to 3 V during the procedure, the buck boost regulator is able to maintain a constant supply value of 3.4 V. Cascaded to the buck boost, a dedicated Low DropOut regulator (LDO) is used to supply the sensor and the Electrically Erasable Programmable Read-Only Memory (EEPROM) at 3.3 V only when a measurement or memory communication is needed, additionally providing a noiseless supply for the sensors.

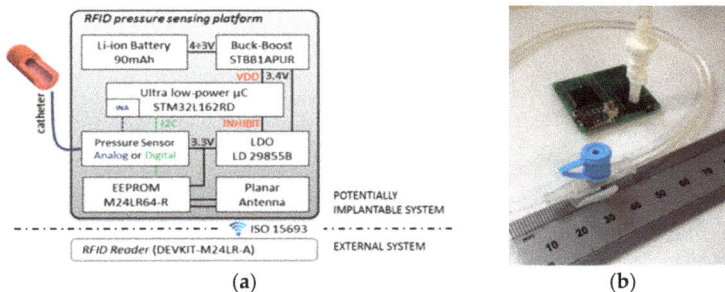

Figure 1. (**a**) Sensor platform block diagram. The potentially implantable system is above the dotted line; (**b**) Laboratory setup for characterization; vertically stacked mechanical arrangement suitable for implant.

The developed device, depicted in Figure 1b, is assembled on a two layers 34×30 mm^2 FR4 (Flame Retardant 4) printed circuit board, designed in order to be divided in two parts, which can be stacked vertically with the battery. Such a configuration creates a "multilayer" system with total dimension of $30 \times 17.5 \times 11$ mm^3, a volume of 5.6 cm^3 and a weight of 8 gr. Such compactness is comparable with some of the most diffused telemetering systems cited in Table 1 and it makes the system potentially implantable in the abdominal cavity of the animal, once coated with a biocompatible material. By replacing an actual components case with smaller packages it can dramatically reduce the present device volume, with a minimal impact on the cost. The choice of the actual components case has been dictated by the need of test and debugging. In order to further reduce the device volume and to improve robustness, the interconnection between the catheter and the sensor should also be improved. The Printed Circuit Board (PCB) has been designed with the capability of alternatively mounting two different temperature-compensated pressure sensors. The first one is analog (MPX2300DT1, Freescale, Austin, TX, USA) and the second one is digital (NPA-700M-005D, General Electric, Fairfield, CT, USA), both selected for their pressure accuracy of ± 3 mmHg in the range 0–260 mmHg. Table 2 summarizes the main features of both sensors; they have been characterized experimentally with the purpose of determining the optimal solution in terms of power consumption and resolution.

Table 2. Digital and analog pressure sensor main features.

Features	Digital Sensor	Analog Sensor
Pressure range	0–260 mmHg	0–300 mmHg
Dimension	9×11 mm^2	9×6 mm^2
Sensitivity	50 counts/mmHg	5 µV/V/mmHg
Accuracy	$\pm 1.5\%$	$\pm 1.5\%$
Interface	I2C	Analog differential
Max ODR *	833 Hz	ADC ** sample rate
Supply current (at 10 V)	1.5 mA	1 mA

* Output Data Rate; ** Analog to Digital Conversion.

The RFID reader (DEVKIT-M24LR-A, STMicroelectronics) is placed under the cage of the animal and it provides the features summarized in Table 3.

Table 3. RFID reader main features.

Features	DEVKIT-M24LR-A
Antenna dimensions	337×237 mm^2
Operating frequency	13.56 MHz
Max transmitting power	1 W
Interface	I2C and RF

2.3. Firmware

Typically the drug assessment experiments need to monitor the arterial blood pressure every minute for several days, so it is not needed to monitor each cardiac cycle. For this reason, the firmware has been implemented in such a way that the microcontroller acquires pressure sensor data for 5 s. Then it applies an algorithm for the maximum and minimum detection and computes the average systolic and diastolic pressure values. Finally, the microcontroller stores the results on the non-volatile memory. The LDO which provides a supply for the sensor and the EEPROM is disabled until the next acquisition occurs. After this phase, in order to maximize the device operation time, the microcontroller enters into a stand-by mode for 55 s and successively it wakes-up thanks to the internal real-time clock (RTC). Since the EEPROM collects all the data permanently, it is not necessary to have an active communication link between the sensor and the RFID reader for all the experiment duration. When the external reader identifies the tag memory, all the data history is downloaded and erased for the next acquisitions.

The system embedded firmware includes the algorithm for the identification of the blood pressure peaks, essential for analyzing the effects of pharmacological treatments on test animals. Peaks and troughs, corresponding to systolic and diastolic pressure respectively, characterize the blood pressure signal. Systolic phase occurs during ventricles contraction, while diastolic phase take places at the beginning of the cardiac cycle, when the ventricles are filled. The identification of peaks is a common problem in the analysis of physiological signals; often it is necessary to detect peaks in real time, but the task is frequently complicated by baseline wander and other type of interference [4]. The search for systolic and diastolic pressure of every cardiac cycle has been implemented with the Todd-Andrews algorithm. The choice of such an algorithm is due to the benefits in terms of local peaks, baseline wandering and dynamic threshold immunity. It mainly consists of three steps: First, it looks for the maximum on the rising edge of the signal by comparing each sample in the array with the previous one. Then it calculates the difference between every sample of the falling edge and the identified maximum. If the result of the subtraction is greater than a fixed threshold, the global maximum has been found. The threshold can be computed dynamically on the amplitude of the previous signal period. The same process is applied on the falling edge to find the minimum value. Figure 2 shows the algorithm validation that has been done manually varying the input signal pressure with a sphygmomanometer.

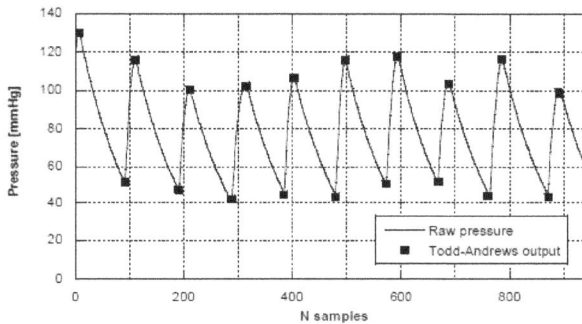

Figure 2. Algorithm validation, performed varying pressure and representing raw data, obtained from the digital pressure sensor and algorithm max/min determined values.

2.4. Power Consumption

The device's power consumption has been characterized on each component at different operating modalities by replacing the battery with a power supply (Agilent E363A, Santa Clara, CA, USA), remotely controlled in order to track the sourced current. The battery lifetime is approximately 67 h for the system using digital sensor and about 63 h with the analog sensor. The difference between them is given by the power consumption of the microcontroller ADC needed only for analog sensor signal acquisition (see Table 4). In order to satisfy the specification of a 10 day lifetime, it turns out to power the device with a Li-poly battery with an increased capacity (300 mAh could be suitable) or to take advantage of the EEPROM energy harvesting function, that it has been characterized in a following paragraph. The current consumption of 0.7 mA during system stand-by mode is only due to the voltage regulator quiescent current and RTC clock activated.

Table 4. Device supply currents.

Pressure Sensor Mounted	Average Current Provided by the Battery (V = 3.7 V)			
	Irun (ΔT = 5 s)		Isleep (ΔT = 55 s)	Iaverage (ΔT = 60 s)
	Isensor + EEPROM	Itotal		
Analog	3.41 mA	9.49 mA	0.7 mA	1.43 mA
Digital	3.53 mA	8.51 mA	0.7 mA	1.35 mA

2.5. The RFID Interface

RFID technology for human and animal implants is generally based on passive or active devices and makes it possible to achieve read ranges in the order of 10 s of centimeters, a very short range once compared to the other wireless technologies, but suitable to the majority of the applications. Basically, an RFID interface uses communication via electromagnetic waves to exchange data between an interrogator (also known as the reader) and an object (transponder or tag). The communication must respect given standards, as the protocol ISO 15693, in which the 13.56 MHz carrier electromagnetic wave is ASK (Amplitude Shift Keying) modulated for data transmission [5,6]. ASK wave is 10% or 100% modulated, obtaining a data rate of 1.6 kbps using the 1/256 pulse coding mode or a maximum data rate of 26 kbps using the 1/4 pulse coding mode. An integrated circuit for storing and processing data and an antenna for receiving and transmitting them, generally compose the transponder. The electromagnetic field generated by the interrogator provides the power to the transponder for the data communication, achieving a wireless link with no power consumption on the transponder side. Unlike other wireless technologies, the obtainable data rate is quite limited, suggesting a one-shot measurement operation or an on-board data processing in order to optimize the transmission time. From the review of state-of-the-art RFID devices, it can be observed that inductive coupling operating in HF (13.56 MHz and below) frequency range is presently the best method to wirelessly send power and data from off-body interrogator to RFID device implanted inside body. Since in the range 1–20 MHz the EM waves are not significantly attenuated by human tissues [7,8].

The width of the PCB has been constrained by the size of the antenna, designed with a square shape as a two layers planar coil on the PCB. In fact, to allow the implantable system to interact with the external reader through RFID communication, the antenna should have an outer diameter of 15.7 mm and an internal one of 5.1 mm. Moreover, considering the spacing and width of the convolutions of 10 mils, the antenna should have 12 turns [9]. The shape and the dimensions used for the antenna allow also to obtain a value of a measured self-inductance equal to 4.6 μH (4.7 μH theoretical), in order to have a resonant circuit at 13.56 MHz. The RFID antenna self-inductance has been measured with an Impedance Analyzer (Agilent 4395A, Santa Clara, CA, USA) on a frequency span up to 500 MHz (see Figure 3). A self-resonance at 36 MHz, due to the antenna parasitic capacitance (4 pF), is clearly visible in Figure 3. The operating point at 13.56 MHz is also emphasized.

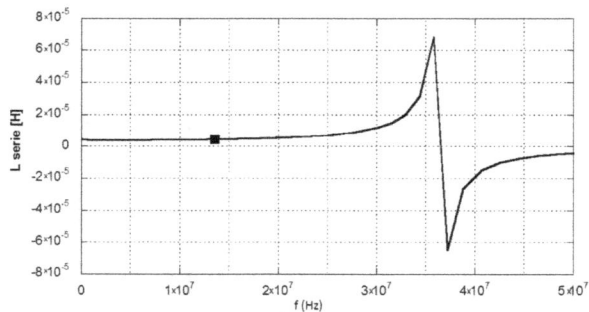

Figure 3. Radio Frequency Identification antenna measured self-inductance. The marker indicates the RFID operating frequency at 13.56 MHz.

3. System Characterization and Experimental Results

The device has been initially characterized in the laboratory and then evaluated on test rats. The system functionality tests were carried out connecting a catheter to the device pressure sensor and imposing air pressure changes with the sphygmomanometer. The pressure has been monitored with both the RFID sensor tag and a reference manometer (SICK PBS), which it has an analog

4–20 mA interface (see Figure 4). For comparison purposes the firmware was modified in order to acquire the pressure signal for 20 s and store raw data on the RFID tag memory. A software application was implemented in order to view on a PC the signal pressure, uploaded using the RFID reader. The communication distance between the interrogator and the transponder has been successfully tested up to 11 cm, which does not change interposing a hand or other material (i.e., wood, plastic, etc.) between the reader and the device [10]. In fact, thanks to RFID advantages, the tissues interposed between do not significantly attenuate the electromagnetic waves during the tag-reader communication.

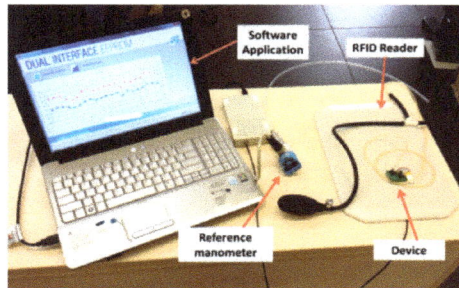

Figure 4. System setup functional sensor.

3.1. System with Digital Pressure Sensor

Figure 5a depicts the simultaneous acquisition of the pressure wave from the reference pressure sensor and the proposed device with digital pressure sensor mounted. The reference sensor resolution is 3 mmHg$_{rms}$, whereas the device measured resolution is 2.5 mmHg$_{rms}$ for the digital version. The differences between the two traces are comprised between ± 5 mmHg. During the system characterization it has also been tested the maximum output data rate (833 Hz) of the I2C sensor interface, which provides a direct connection to the microcontroller.

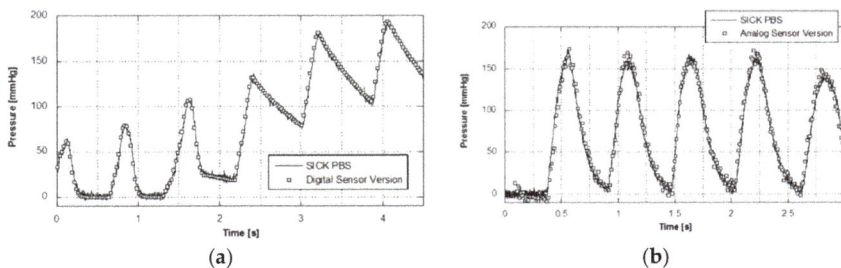

Figure 5. (a) Comparison between reference pressure sensor and the RFID tag with digital pressure sensor; (b) Comparison between reference pressure sensor and the RFID tag with analog pressure sensor.

3.2. System with Analog Pressure Sensor

The analog pressure sensor, based on a bridge configuration, is connected to an instrumentation amplifier (INA) obtained by using three operational amplifiers, which are embedded in the microcontroller. This solution is optimal in terms of space occupancy, since only passive elements are needed to acheive full front-end amplification. The same tests done with the digital pressure sensor have been performed even with the system with the analog sensor configuration. The measured resolution of the analog pressure sensor is 5 mmHg$_{rms}$ and the difference between the two acquisition traces is less than ±10 mmHg (see Figure 5b).

Figure 6 shows the measured transfer function of the instrumentation amplifier, featuring a gain of 38.5 dB and a bandwidth of 10 kHz. The INA output is acquired by the microcontroller ADC running at 1 kS/s.

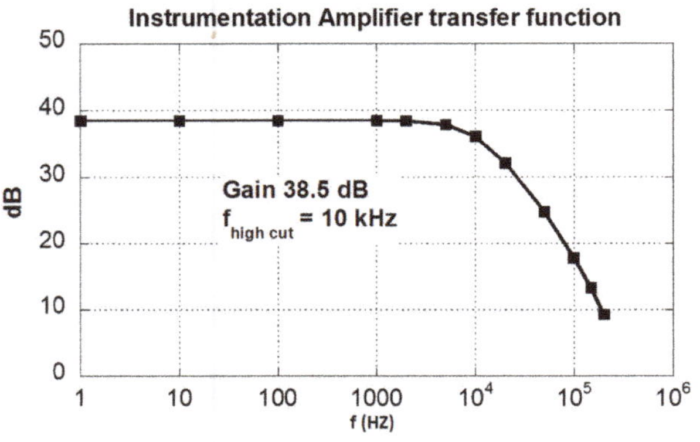

Figure 6. Instrumentation amplifier transfer function.

3.3. In-Vivo Tests

An adult male Sprague-Dawley rat (Charles River Laboratories International Inc., Wilmington, MA, USA) weighing 400–500 gr was used for the in-vivo tests of the proposed system, after the laboratory characterization. After anesthesia with isoflurane, the left femoral artery was isolated and cannulated with a PE50 catheter. Animal care and treatment were conducted in conformity with the institutional guidelines, in compliance with national (DL n. 116/1992, Circ. 8/1994) and international (EEC Dir. 86/609, OJL 358, Dec 1987; NIH Guide for the Care and Use of Laboratory Animals, US NRC, 1996) laws and policies. The catheter end was split and connected both to the device with digital pressure sensor mounted, placed aside the animal and to a reference pressure sensor (Deltran II, Utah Medical Products, Midvale, UT, USA), as depicted in Figure 7. The synchronous acquisitions lasting 20 s, depicted in Figure 8, are in accordance within 1 mmHg. The Fast Fourier transform of the time-base blood pressure signal is pointed out in Figure 9, showing that the blood pressure wave, having an average frequency of 5.76 Hz (346 bpm), is modulated at 0.83 Hz by the animal breath.

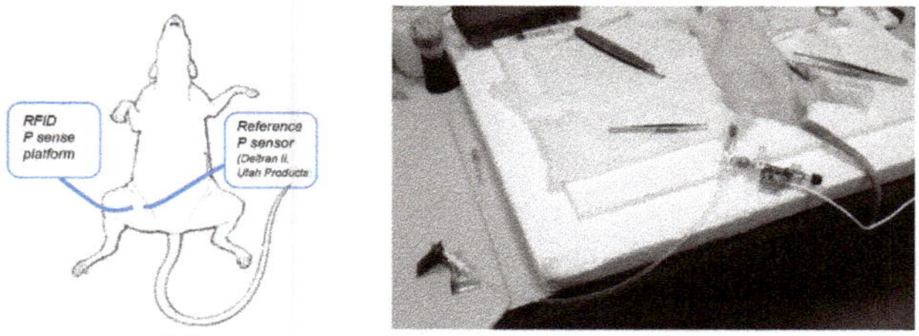

Figure 7. (a) Representation of in-vivo test set up; (b) In-vivo experiment.

(a)

(b)

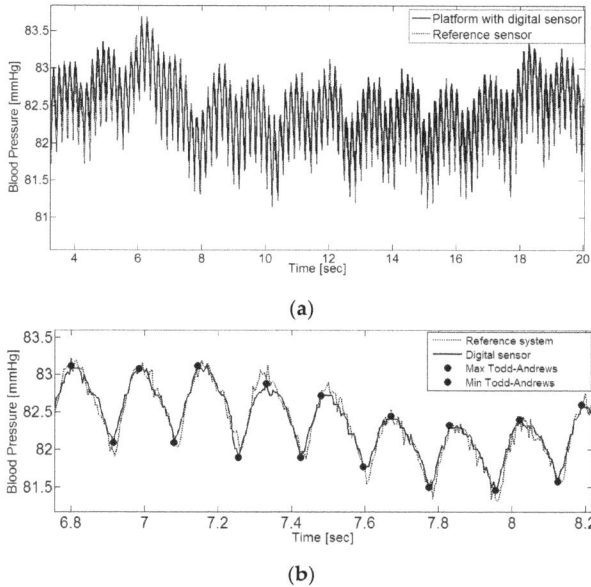

Figure 8. (**a**) Synchronous recording (~20 s) of the femoral artery blood pressure with the device mounting the digital sensor and a reference measurement system; (**b**) zoom of the traces showing systolic and diastolic peak detection, independent of baseline wandering.

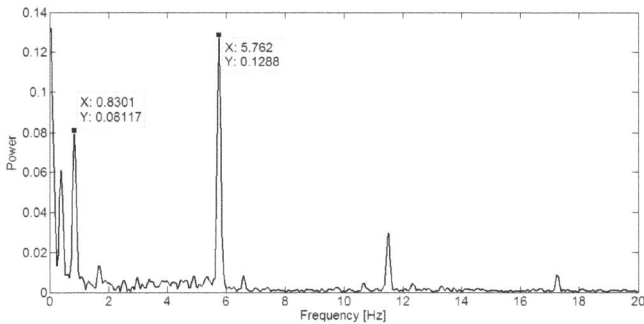

Figure 9. Fast Fourier transform spectrum of the signal depicted in Figure 8.

Table 5 reports the blood pressure maximum, minimum and mean values of blood pressure acquired by both the prototype and the reference sensor, during 20 s acquisition. The results obtained with the presented device are in accordance with the reference instrument within 0.12%.

Table 5. Maximum, minimum and mean values of blood pressure acquired by digital and reference sensor during 20 s acquisition.

Value	Digital Sensor	Reference Sensor	Error
Mean	82.57 mmHg	82.47 mmHg	0.12%
Max	83.89 mmHg	82.69 mmHg	1.43%
Min	81.14 mmHg	81.16 mmHg	0.02%

The same in-vivo test has been performed on the device mounting the analog pressure sensor, in order to compare the performances. The system was able to follow the average blood pressure but not to detect the peaks, due to its lower resolution with respect to the digital sensor version, as shown in Figure 10. The dominant frequencies in this case are not detectable as in the results obtained with the digital pressure sensor.

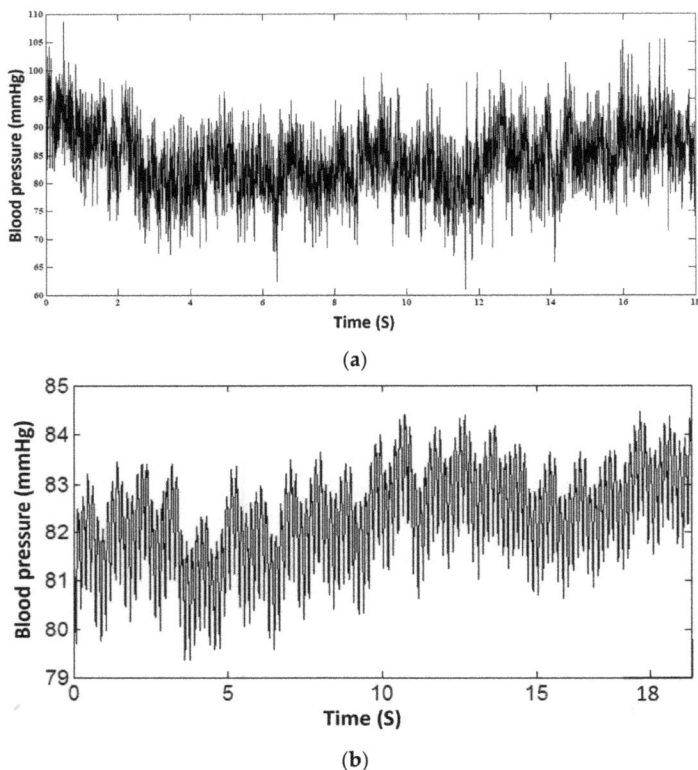

(a)

(b)

Figure 10. (**a**) Synchronous recording (~20 s) of the femoral artery blood pressure with the device mounting the analog sensor; (**b**) pressure wave acquired by the reference sensor.

3.4. Energy Harvesting Investigation

In order to maximize the device operation time keeping the 90 mAh small-sized battery, the EEPROM Energy Harvesting (EH) functionality has been investigated. The general purpose of the EH mode is to deliver a part of the unnecessary RF power received by the EEPROM on the RF input pin, in order to recharge the battery. When the external reader enables the EH mode and the RF field strength is sufficient, an unregulated DC voltage is provided on a pin of the EEPROM. Figure 11 shows the characterization results in terms of available power and supplied current by the EEPROM with different loads, in function of the reader-tag distance. The EH output current could be used to recharge the battery during the system sleep time, by adding a low-power step up regulator. Up to 5 mm distance between the reader and the RFID tag, the output power (about 9.5 mW) is maximized with a 1 kΩ load and it is independent on the distance itself. It has been estimated that the supplied current by the EEPROM Energy Harvesting function during the 55 s sleep time could double the battery lifetime (130 h).

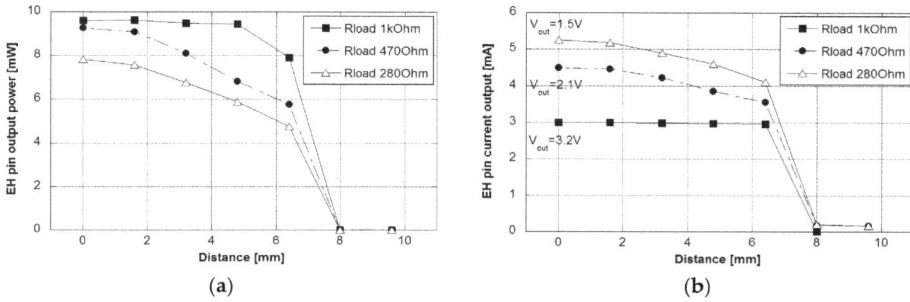

Figure 11. Electrically Erasable Programmable Read-Only Memory energy harvesting characterization as a function of the distance from the reader; in terms of power (**a**) and current (**b**); power can be extracted up to 6.5 mm distance.

4. Conclusions

This paper described the development, the laboratory characterization, and preliminary in-vivo results of a potentially implantable blood pressure-sensing platform with an RFID interface. It shows a meaningful use of RFID interface to measure the vital signals. The electronic system exhibits performances and functionalities that make it suitable for implantable blood pressure telemetry. Two different pressure sensors were characterized, the first providing a digital output and the second an analog one. The system configuration with the digital pressure sensor produced the best performances in terms of power consumption (1.35 mA average current, 3.7 V supply) and resolution (0.4 mmHg$_{rms}$). The in-vivo measurement, conducted implanting the catheter and not the electronics, provides results in accordance within 1 mmHg with those obtained with a reference sensor. Finally, a characterization of the Energy Harvesting function provided by the EEPROM with RFID interface has demonstrated the possibility to double the battery lifetime. The system exhibits volume, dimensions, and weight comparable with small-animals state-of-the-art telemetry systems, introducing some new features as the embedded processing, low cost PCB RFID antenna and energy harvesting functionality. In particular, the system is able to measure the pressure wave and, after embedded processing, to store the calculated blood pressure values onboard. Such data are available for subsequent interrogation by an external RFID reader. Nevertheless, the connection between the pressure sensor and the catheter is still to be optimized. As a consequence, the device is presently suitable for medium-size animals; the work is presently focusing on further tag miniaturization, achievable by choosing smaller packages for the electronics components, and on the biocompatible device encapsulation. Moreover, further long-term animal studies are planned to be performed in order to fully validate the system. It is worth emphasizing that the developed pressure sensing system is fitting for many applications, such as environmental or structural monitoring and human telemetry.

Author Contributions: V.R. and M.C. conceived and designed the experiments; B.N. performed the experiments; B.N. and M.C. analyzed the data; B.B. contributed to in-vivo tests; B.N. and M.C. wrote the paper.

Conflicts of Interest: The authors declare no conflict of interest.

References

1. Braga, V.A.; Burmeister, M. Applications of Telemetry in Small Laboratory Animals for Studying Cardiovascular Diseases. In *Modern Telemetry*; InTech: Rijeka, Croatia, 2011.
2. Rey, M.; Weber, E.W.; Hess, P.D. Simultaneous pulmonary and systemic blood pressure and ECG interval measurement in conscious, freely moving rats. *J. Am. Assoc. Lab. Anim. Sci.* **2012**, *51*, 231–238. [PubMed]

3. Romain, O.; Mazeyrat, J.; Garda, P.; Talleb, H.; Lautru, D.; Wong, M.-F.; Wiart, J.; Hanna, V.F.; Lagrée, P.-Y.; Bonneau, M.; et al. ENDOCOM: Implantable wireless pressure sensor for the follow-up of abdominal aortic aneurysm stented. *IRBM* **2011**, *32*, 163–168.

4. Todd, B.S.; Andrews, D.C. The identification of peaks in physiological signal. *Comput. Biomed. Res.* **1998**, *32*, 322–335. [CrossRef] [PubMed]

5. *Identification Cards—Contactless Integrated Circuit Cards—Vicinity Cards—Part 2: Air Interface and Initialization*; ISO/IEC 15693-2; International Organization for Standardization: Geneva, Switzerland.

6. *Identification Cards—Contactless Integrated Circuit(s) Cards—Vicinity Cards—Part 3: Anti-Collision and Transmission Protocol*; ISO/IEC 15693-3; International Organization for Standardization: Geneva, Switzerland.

7. Valdastri, P.; Menciassi, A.; Dario, P. Transmission power requirements for novel Zigbee implants in the gastrointestinal tract. *IEEE Trans. Biomed. Eng.* **2008**, *55*, 1705–1710. [CrossRef] [PubMed]

8. Aubert, H. RFID technology for human implant devices. *C. R. Phys.* **2011**, *12*, 675–683. [CrossRef]

9. Zhao, J. A new calculation for designing multilayer planar spiral inductors. *EDN (Electr. Des. News)* **2010**, *55*, 37–40.

10. Caldara, M.; Nodari, B.; Re, V. Development of a potentially implantable pressure sensing platform with RFID interface. In Proceedings of the 2013 IEEE Sensors, Baltimore, MD, USA, 3–6 November 2013.

electronics

MDPI

Article

Skin Admittance Measurement for Emotion Recognition: A Study over Frequency Sweep

Alberto Greco [1,*], Antonio Lanata [1], Luca Citi [2], Nicola Vanello [1], Gaetano Valenza [1] and Enzo Pasquale Scilingo [1]

[1] Department of Information Engineering and Research Center "E. Piaggio", School of Engineering, University of Pisa, 56122 Pisa, Italy; a.lanata@centropiaggio.unipi.it (A.L.); nicola.vanello@unipi.it (N.V.); g.valenza@iet.unipi.it (G.V.); e.scilingo@centropiaggio.unipi.it (E.P.S.)

[2] School of Computer Science and Electronic Engineering, University of Essex, Colchester CO4 3SQ, UK; lciti@essex.ac.uk

* Correspondence: alberto.greco@centropiaggio.unipi.it; Tel.: +39-050-221-7462

Academic Editor: Mostafa Bassiouni
Received: 26 June 2016; Accepted: 1 August 2016; Published: 4 August 2016

Abstract: The electrodermal activity (EDA) is a reliable physiological signal for monitoring the sympathetic nervous system. Several studies have demonstrated that EDA can be a source of effective markers for the assessment of emotional states in humans. There are two main methods for measuring EDA: endosomatic (internal electrical source) and exosomatic (external electrical source). Even though the exosomatic approach is the most widely used, differences between alternating current (AC) and direct current (DC) methods and their implication in the emotional assessment field have not yet been deeply investigated. This paper aims at investigating how the admittance contribution of EDA, studied at different frequency sources, affects the EDA statistical power in inferring on the subject's arousing level (neutral or aroused). To this extent, 40 healthy subjects underwent visual affective elicitations, including neutral and arousing levels, while EDA was gathered through DC and AC sources from 0 to 1 kHz. Results concern the accuracy of an automatic, EDA feature-based arousal recognition system for each frequency source. We show how the frequency of the external electrical source affects the accuracy of arousal recognition. This suggests a role of skin susceptance in the study of affective stimuli through electrodermal response.

Keywords: electrodermal activity; wearable system; pattern recognittion; skin conductance; skin admittance

1. Introduction

Emotions play a fundamental role in the daily life and a possible continuous monitoring of them could be very beneficial in understanding and managing personal well-being promoting mental healthy state [1,2]. In the last decades, several studies have been proposed to increase knowledge on emotion recognition, and, consequently, develop an automatic emotional detection systems. Even though a scientific definition of emotion is still controversial, many models of emotions have been developed [3–7]. Among them, the circumplex model of affect [3] describes emotions as a combination of an arousal level (i.e., the intensity of the emotion perception) and a valence level (i.e., the degree of pleasantness). According to these emotional models, many human–machine systems have been developed in order to automatically recognize humans' affective and mood states [8–19] by interpreting physiological changes as a response to an external triggering event. To this extent, several psychophysiological features extracted from different peripheral biosignals have been widely used in the literature [20–23]. Some of the most commonly used physiological signs in affective recognition are: electrocardiogram and related heart rate variability series, electrodermal activity (EDA), respiration, muscle activity, peripheral temperature, eye gaze, as well as brain activity [24–29].

Among these, EDA has been widely used to assess the arousal level in humans because of its ability to quantify changes in the sympathetic nervous system (SNS) [30,31]. EDA signal is comprised of a low-frequency tonic component, and a higher frequency phasic component [30,32]. Anatomically, EDA changes are due to the sudomotor nerve activity (SMNA), which is part of the SNS, that directly control the eccrine sweat glands [30]. Therefore, EDA can be easily monitored through voltage/current measures between two fingers, where there is a higher concentration of the eccrine sweat glands with respect to other body sites [30].

In this context and aiming at continuously monitoring the EDA signal in a ecological scenario, wearable monitoring systems are the most interesting and advantageous devices. In fact, wearable sensors are greatly valued due to their comfort, portability, non-invasiveness, and their wireless communication capabilities with either a computer, a mobile embedded system or other wearable sensors [33–37].

Referring to EDA measure, three standard methodologies are usually employed. The first one is called endosomatic measurement. This is rarely used and consists in measuring, directly on the skin, the potential difference between two skin sites, in a passive way. It does not need special amplifiers and coupling electrical circuits. Although it is a quite unknown bioprocess, it is accepted that changes in skin potential during sympathetic activity may be provoked by the sodium reabsorption across the duct walls and the consequential change of the ionic potential in the sweat ducts [30].

The other two methods are based on an exosomatic approach, i.e., a small external current is directly injected into the skin. Consequently, two different methods are used: the direct current method (DC) and the alternating current method (AC) at different frequency levels. Generally, they perform a measure of resistance (impedance) or conductance (admittance: "conductance + j susceptance"), where j is the imaginary unit. When DC source is used, resistance and conductance are inversely related. Instead, when AC source is used, the inverse paradigm between resistance and conductance is not valid, but can be applied to the complex impedance and complex admittance. Moreover, in the AC regime, the impedance is represented by a circuit comprised of a resistor-capacitor in series, instead the admittance is represented by a parallel circuit (of a resistor and a capacitor). The Parallel equivalent should be preferred since ionic conduction and polarization are in parallel in biological tissues. More in detail, the skin can be considered as a dielectric whose permittivity and conductivity are complex due to free charges and the AC losses of bound charges [38].

Although AC methods allow measuring capacity changes in the electrodermal responses, DC procedures are the most implemented [39]. Within DC methodology, even if much effort has been spent to standardize, an agreement concerning the use of either constant-voltage or constant-current scheme has not yet been achieved. Therefore, different designs can be found. In constant-voltage sources, the conductance of the skin can be directly measured as an output of the circuit without the need of any further transformation. However, constant-current sources provide more stability and exhibit less tolerance, but, in this method, much attention must be payed to possible damage to the sweat ducts due to the injected current through a small area of the skin [34].

Unfortunately, a very low number of studies have been published on the difference among AC and DC stimulation in EDA measurement. One of the most interesting works [39] showed that, when using an AC source at 88 Hz, the major contribution to EDA is given by the conductance term and not by the susceptance. More specifically, the authors found that, at the instant of the conductance response, there were no susceptance responses, and this could indicate the absence of a significant capacitance in the sweat ducts. Moreover, to the best of our knowledge, in the current literature, no specific wearable systems, which are able to perform both DC and AC electrical stimulations, have already been reported. As a matter of fact, in a previous study, we have already proposed a fabric-based sensorized glove [40]. However, this system was also able to continuously record the EDA using only a DC source.

Starting from the above considerations and from the recommendations provided in [41–43], we propose a novel textile wearable system, which is able to perform an exosomatic EDA measurement using both AC and DC methods. Using this novel device, our study aims at investigating whether the admittance contribution at different frequency sources (in the range from DC to 1 kHz of AC) could affect the ability of EDA of inferring the central state during emotional stimulation. To this end, we designed an experimental paradigm using visual affective stimuli, and we developed an automatic arousal recognition system, in order to test potential differences in inferring the arousal state of the two approaches.

2. Materials and Methods

Exosomatic EDA measures changes in the electrical conductance (DC method) or admittance (AC method) due to both the psycho-physiological state of a person and the response to an external event. The EDA signal has a frequency bandwidth of 0–2 Hz [44] and, for a correct analysis, should be decomposed in two sub-components containing different and complementary information: a low-frequency tonic component, which reflects the subject's general psycho-physiological state and its autonomic regulation [30], and a phasic component, which is the superposed higher-frequency change directly related to an external stimulus [45]. A frequent issue in the decomposition process consists in the overlapping of consecutive phasic responses, which occur in the case of inter-stimulus intervals less than around 10–20 s [46,47].

Measurement of EDA changes provides evidence of eccrine sweat gland activity. The eccrine sweat glands are innervated by sympathetic fibers, and, in normal ambient temperatures, palmar, finger (or plantar) glands reflect responses to psychological rather than thermoregulatory stimuli [48]. Therefore, EDA is considered as an ideal way to monitor the autonomic nervous system and, more specifically, its sympathetic branch [30].

In this section, we first report on the EDA acquisition system prototype and then briefly describe details on the cvxEDA models, which is presented in [49]. Note that this method is able to discern overlapping consecutive electrodermal responses (EDRs), likely to be present in the case of an inter-stimulus interval shorter than the EDR recovery time.

2.1. Multi-Frequency Sensorized Glove

EDA is acquired by a glove where integrated textile electrodes were placed at the distal phalanges of the index and middle fingers (Figure 1). Textile electrodes, provided by Smartex s.r.l. [40] (Pisa, Italy), are made up of 80% polyester yarn knitted with 20% steel wire, with a dimension of 1×2 cm. In one of our previous studies [40], we performed a comparison of textile sensors with Ag/AgCl electrodes demonstrating comparable performance. More specifically, reported results on electrode characterization, performed by means of the voltage-current characteristics, and its electric impedance showed that textile electrode achieves a good electrical and thermal coupling with biological site. Moreover, the use of a wearable textile system exhibits several advantages in terms of portability and usability for long-term monitoring, and gives minimal constraints. This latter characteristic is very significant when the system is used in an ecological environment.

The analog front-end of the designed electronics, which is responsible for measuring the DC and AC exosomatic EDA, is based on a variable-gain current-to-voltage operational amplifier. The electric current injected into the skin is variable and programmable (from 0 to 1 kHz), and for this purpose we used the AD9833 provided by Analog Device [50] (Norwood, MA, USA). This chip is a low power, programmable waveform generator, which is used to switch up the frequency of the skin electrical stimulation among 0 (i.e., DC), 10, 100 and 1000 Hz [39]. Moreover, a low-pass filter (cutoff frequency of 3 Hz) and a further amplification stage were applied to the raw EDA data before the successive digitalization step.

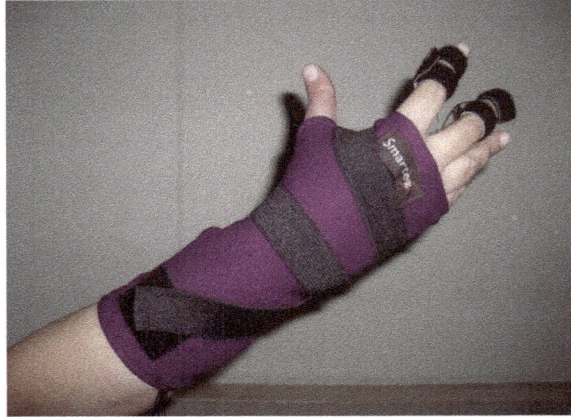

Figure 1. Sensorized glove for the acquisition of the Electrodermal Activity (EDA).

The preprocessed EDA signal was digitally converted with a sampling frequency of 15 kHz, thanks to the 12-bit analog-to-digital converter built in the Texas Instrumen (Dallas, TX, USA) MSP430 microcontroller (Figure 2). The MSP430fxx family of microcontrollers are designed to be low cost and, specifically, low power consumption embedded applications. It is a very popular choice especially in wireless networking systems and it is built around a 16-bit RISC (Reduced instruction set computing) CPU. In our prototype, we used the MSP430x6xx Series, which are able to run up to 25 MHz, have up to 512 KB flash memory and up to 66 KB RAM. Moreover, this series includes an innovative power management module for optimal power consumption [51].

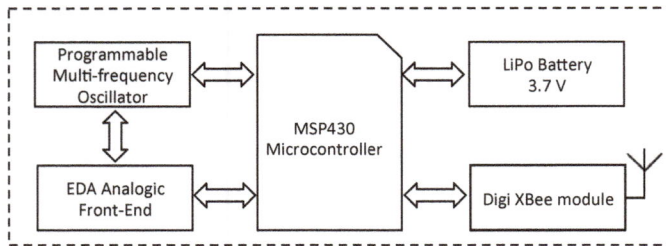

Figure 2. Block scheme of the electronic circuit.

Moreover, wireless communication was implemented by an Xbee module (Minnetonka, MN, USA) connected to the USART (Universal Synchronous Receiver-Transmitter) of the MSP430. Specifically, it was used to exchange data between the transceiver and a dedicated multi-platform software application. Finally, a lithium-polymer battery with a voltage of 3.7 V and a capacity of around 750 mAh was chosen as power supply [34]. The large capacitance of the battery allows a long-term-continuous monitoring, and it is an essential characteristic to stream the data wirelessly. An external circuit was developed to support the rechargeable battery through a USB port. Finally, a voltage regulator is responsible for supplying 3.0 volts from the battery to all the components of the device.

2.2. EDA Processing Using cvxEDA Algorithm

CvxEDA proposed a representation of the phasic responses as the output of a linear time-invariant system to a sparse non-negative driver signal. The model assumes that the observed EDA (y) is the sum of the phasic activity (r), a slow tonic component (t), and an additive independent and identically distributed zero-average Gaussian noise term ϵ:

$$y = r + t + \epsilon. \tag{1}$$

Physiologically-plausible characteristics (temporal scale and smoothness) of the tonic input signal can be achieved by means of a cubic spline with equally-spaced knots every 10 s, an offset and a linear trend term:

$$t = B\ell + Cd, \tag{2}$$

where B is a tall matrix whose columns are cubic B-spline basis functions, ℓ is the vector of spline coefficients, C is an $N \times 2$ matrix (where N is the length of the EDA time series) with $C_{i,1} = 1$ and $C_{i,2} = i/N$, d is a 2×1 vector with the offset and slope coefficients for the linear trend.

The phasic component is the result of a convolution between the SMNA, p, and an impulse response $h(t)$ shaped like a biexponential Bateman function [52–54]:

$$h(t) = (e^{-\frac{t}{\tau_1}} - e^{-\frac{t}{\tau_2}})\, u(t) \tag{3}$$

where τ_1 and τ_2 are, respectively, the slow and the fast time constants of the phasic curve shape, and $u(t)$ is the unitary step function.

Taking the Laplace transform of Equation (3) and then its discrete-time approximation with sampling time δ (using a bilinear transformation), we obtain an autoregressive moving average (ARMA) model (see details in [49]) that can be represented in matrix form as

$$H = M^{-1}A, \tag{4}$$

where M and A are tridiagonal matrices with the MA and AR coefficients along the diagonals. Using an auxiliary variable q such that

$$q = A^{-1}p, \quad r = Mq, \tag{5}$$

we write the final observation model as

$$y = Mq + B\ell + Cd + \epsilon. \tag{6}$$

Given the EDA model Equation (6), the goal is to identify the maximum a posteriori (MAP) neural driver SMNA (p) and tonic component (t) parametrized by $[q, \ell, d]$, for the measured EDA signal (y). CvxEDA rewrites the MAP problem as a constrained minimization Quadratic Programming (QP) convex problem (see details in [49,55]):

$$\begin{aligned}
\text{minimize} \quad & \frac{1}{2}\|Mq + B\ell + Cd - y\|_2^2 + \alpha\|Aq\|_1 + \frac{\gamma}{2}\|\ell\|_2^2 \\
\text{subj. to} \quad & Aq \geq 0.
\end{aligned} \tag{7}$$

This optimization problem can be re-written in the standard QP form and solved efficiently using one of the many sparse-QP solvers available. After finding the optimal $[q, \ell, d]$, the tonic component t can be derived from Equation (2) while the sudomotor nerve activity driving the phasic component can be easily found as $p = Aq$.

The objective function Equation (7) to be minimized is a quadratic measure of misfit or prediction error between the observed data and the values predicted by the model. Moreover, the prior knowledge about the spiking sparse nature and nonnegativity of the SMNA (p) and the smoothness of the tonic component are accounted for by the regularizing terms and the constraint.

The strength of the penalty is regulated by α and γ terms. A sparser estimate is yielded by large values of α. Concerning γ, higher values mean a stronger penalization of ℓ, i.e., a smoother tonic curve. Of note, fixed values of $\tau_1 = 0.7s$, $\tau_2 = 3.0s$, $\alpha = 0.008$ and $\gamma = 0.01$, which were chosen during previous exploratory tests on separate data, were employed throughout this analysis.

CvxEDA algorithm is implemented in Matlab language and the software is available online [56].

2.3. Experimental Protocol

Forty healthy subjects were enrolled in the experiment, aged 26 ± 4 (18 females). All subjects gave written informed consent before taking part in the study, which was approved by the local Ethics Committee. The experiment was designed as following:

- initial resting phase of 1 min;
- maximal expiration task phase of about 1 min;
- affective visual stimulation phase of 2 min;
- final resting phase of 1 min;

(The two elicitation phases will be described in detail in the next sub-sections.) Subjects were comfortably seated in an acoustically insulated room in front of a computer screen while their EDA was recorded using the presented acquisition system.

Textile electrodes were placed for 10 min before the acquisition for achieving a stable skin/dry-electrode coupling and limiting the temporal and thermic effect [57].

Note that the group of 40 healthy subjects was split into four subgroups, each of which comprised of 10 subjects. Each subgroup were acquired with a different exosomathic method such as DC (group 1), AC with a frequency of 10 Hz (group 2), AC with a frequency of 100 Hz (group 3) and AC with a frequency of 1 kHz (group 4).

2.3.1. Maximal Expiration Task

In this session of the experiment, all of the 40 subjects performed a forced maximal expiration task [58], in which they were asked to breathe out with the maximum possible intensity in order to trigger the SNS-mediated expiration reflex.

After the initial resting state session, the subjects breathe normally and rest in front of the computer monitor for about 20 s. Then, they had to perform a deep expiration twice with an inter-stimulus interval of about 20 s, after a neutral visual input on the screen.

The use of the forced expiration task is justified by the need of having a stimulus whose EDA response was as reliable and objective as possible. In fact, previous studies have demonstrated that this stimulation is a reliable way to evoke phasic responses unaffected by emotional change with better reproducibility, less habituation, and more stable waveform patterns than other experimental paradigms (including electrical) [58]. In this way, the presence of at least one phasic response after each stimulus was ascertained. Therefore, we could investigate whether the cvxEDA algorithm was able to identify each phasic response for each acquisition method.

2.3.2. Affective Visual Stimulation

In the second elicitation session, each group of 10 participants was stimulated by projecting on a screen images selected from the official IAPS (International Affective Picture System) database [59]. The IAPS dataset is a collection of images ranked in terms of arousal (i.e., intensity of perception) and valence (pleasantness of perception). This protocol session is designed to assess the pattern recognition system ability on each data group (i.e., of each method) to correctly classify stimulations with different arousal content and provide meaningful information about SNS activation.

The slideshow timeline consists of three neutral images, six aroused images and three other neutral images. Each image was shown for 10 s.

2.4. EDA Analysis and Classification Procedure

For each dataset, the convex-optimization-based EDA model (cvxEDA) described in Section 2.2 was applied to each time series after a Z-score normalization (this is not a mandatory step before applying the cvxEDA algorithm, but an increase in the speed of the QP-solver). Concerning the Impulse Response Function (IRF) parameters considered for this study, values of $\tau_2 = 0.7\,$s, $\tau_1 = 0.7\,$s, $\alpha = 0.4$ and $\gamma = 0.01$ were employed throughout this analysis, according to previous exploratory tests on separate data.

In the respiratory stimulation dataset, the presence of an estimated burst of SMNA activity was verified in each 5 s time window following a stimulus onset, in order to prove the model's ability to correctly detect partially overlapped phasic responses.

As summarized in Table 1, we segmented each signal in correspondence to each IAPS image time window, and we extracted several features from both the tonic and phasic component.

Table 1. List of features extracted from Electrodermal Activity (EDA) phasic and tonic components.

Feature	Description
Npeak	number of significant SMNA peaks wrw
AUC	Area under curve of reconstructed phasic signal wrw (μSs)
peak	maximum amplitude of significant peaks of SMNA signal wrw [1] (μS)
MeanTonic	Mean value of the tonic component within each image time window (μS)

wrw= within response window (i.e., 5 s after stimulus onset).

Classification Procedure

The feature set, extracted from each single IAPS image, was used as the input of a pattern recognition algorithm in order to classify the two arousal levels, according to the IAPS rates. The supervised classification of the feature set was implemented following a Leave-One-Subject-Out procedure (LOSO) applied to a K-nearest neighbors (K-NN)-based classifier. For each of the N iterations (where N is the total number of participants), the whole dataset was split into a training set including $(N-1)$ subjects and a test set including the cvxEDA feature values of the the remaining subject N_{th}. Moreover, for each iteration of the LOSO scheme, a feature selection procedure was performed in order to identify the combination of parameters that resulted in the highest recognition accuracy within the training set examples. Each selected feature constituted a single dimension of the feature space. The LOSO pattern recognition procedure is illustrated in Figure 3.

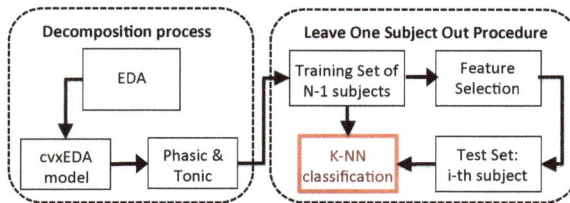

Figure 3. Overall block scheme of the proposed valence recognition system. The EDA is processed in order to extract the phasic and tonic components using the cvxEDA algorithm. According to the protocol timeline, several features are extracted. The K-nearest neighbors (K-NN) classifier is engaged to perform the pattern recognition by adopting a leave-one-subject-out procedure.

3. Results

According to the cvxEDA model, all EDA data (Figure 4a) were decomposed into two signals, a sparse component p and a smooth component t. Of note, we interpret p as the activity of the sudomotor nerve (Figure 4b), and t as the tonic level (Figure 4c)).

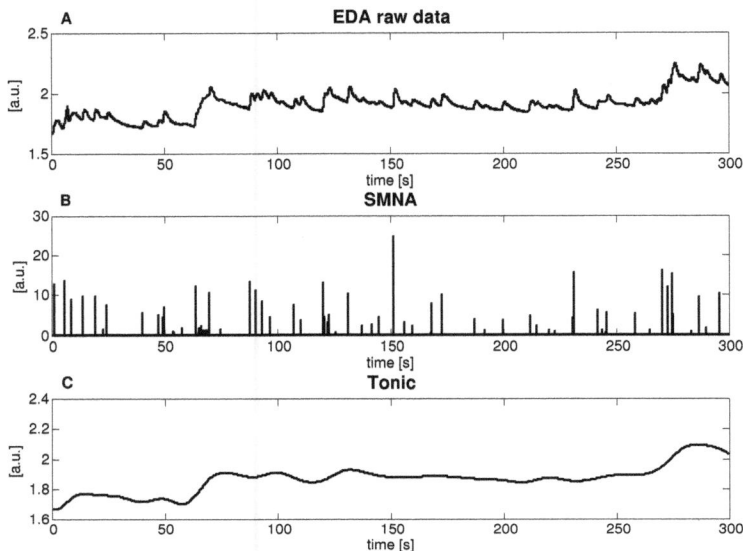

Figure 4. Application of the cvxEDA decomposition procedure to the EDA signal recorded (i.e., admittance module) for a representative subject. (**A**) raw EDA signal, Z-score normalized; (**B**) estimated sparse phasic driver component p; (**C**) estimated slow tonic component t.

3.1. Maximal Expiration Task Results

We performed both a visual and a statistical inspection of time series to verify whether the effectiveness of the protocol in eliciting phasic responses was confirmed for all different kinds of acquisition method (DC and AC).

After the application of the cvxEDA model, we considered a time windows of 5 s after the onset of each expiration task, and we looked for peaks of the SMNA signal (in fact, the phasic response is defined as the part of the signal arises within a predefined response window of 1–5 s [30,45]). Of note, due to the stimulus intervals of about 20 s, no overlap between consecutive responses occurred.

Results of an intersubject analysis showed that cvxEDA was able to correctly detect the corresponding phasic peak response over 97.5% of the respiratory stimuli. Moreover, a visual inspection of the small percentage of cases that were not correctly identified revealed a very low signal-to-noise ratio of the signal.

3.2. Automatic Arousal Recognition Results

Results of the arousal-level-classification-procedure on the four datasets, namely, DC, AC 10 Hz, AC 100 Hz, AC 1 kHz, are shown in Tables 2–5. The recognition accuracy is reported in the form of a confusion matrix. An element r_{ij} of the confusion matrix indicates a percentage of mismatches, i.e., how many times a pattern belonging to class i was erroneously classified as belonging to class j. Terms r_{ij} on the main diagonal of the confusion matrix correspond to correct classifications.

Both DC and AC measures did not show very high average recognition accuracy. However, it is worth noting that, using 100 Hz of the frequency current source, we obtain an average accuracy significantly higher than in the other cases. More specifically, using DC, 10 Hz and 1 kHz, the average accuracy was in the range of 62.5% to 63.34%, whereas at 100 Hz, the pattern recognition system showed an accuracy of 71.67%.

Table 2. Confusion matrix of Neutral vs. Arousal images using an cvxEDA feature set extracted with a Direct Current (DC) source.

K-NN (DC)	Neutral	Arousal
Neutral	63.33%	35.00%
Arousal	36.67%	65.00%

K-NN means K-nearest neighbors.

Table 3. Confusion matrix of Neutral vs. Arousal images using cvxEDA feature set extracted with Alternating Current (AC) source at 10 Hz.

K-NN (AC 10 Hz)	Neutral	Arousal
Neutral	65.00%	38.33%
Arousal	35.00%	61.67%

Table 4. Confusion matrix of Neutral vs. Arousal images using cvxEDA feature set extracted with AC source at 100 Hz.

K-NN (AC 100 Hz)	Neutral	Arousal
Neutral	68.33%	25.00%
Arousal	31.67%	75.00%

Table 5. Confusion matrix of Neutral vs. Arousal images using cvxEDA feature set extracted with AC source at 1 kHz.

K-NN (AC 1 kz)	Neutral	Arousal
Neutral	63.33%	38.33%
Arousal	36.67%	61.67%

4. Discussions and Conclusions

In this study, we proposed a novel wearable EDA acquisition system prototype. It consisted of a sensorized glove provided with textile electrodes at the fingertips able to acquire the exosomatic EDA using both DC and AC methods. In order to test the usability of the novel sensorized glove and to investigate about possible differences between DC and AC stimulation (i.e., 10 Hz, 100 Hz, 1 kHz), we designed an experimental paradigm where 40 healthy subjects were stimulated by means of a mechanical expiration task and visual affective stimuli selected from the IAPS database. From this collection of pictures ranked in terms of arousal and valence level, two groups of images were selected: a group of neutral images and a group of negative aroused images.

The EDA signals were analyzed by means of the cvxEDA model [49]. The cvxEDA algorithm is based on the three concepts of sparsity, Bayesian statistics and convex optimization. It provides a decomposition of the EDA in its two components, i.e., phasic and tonic, and estimates the sudomotor nerve activity that control the eccrine sweat process, giving a window into the sympathetic nerve activity.

Results from the application of the cvxEDA algorithm showed no differences in the identification of the phasic peak response after the deep respiration stimulus among the DC and three AC methods. In fact, over 97% of the peaks were identified in the SMNA signal in the time response window of 5 s after the stimulus onset (i.e., directly evoked by the stimulus [30,45]). We could conclude that all of the investigated methods for the exosomatic measurement of the EDA reliably measure the phasic responses to eliciting stimuli.

Considering the four groups of data separately (i.e., DC, AC 10 Hz, AC 100 Hz, AC 1 kHz), in the second part of the experiment, we investigated possible differences in inferring the arousal state. Specifically, we performed a classification procedure of the arousal levels in the four groups of signals. Results showed that an alternating current method at 100 Hz could improve the arousal recognition accuracy up to 71% (while other acquisition modalities did not overcome an average accuracy of 63.5%). These results suggested that not only the skin conductance plays an important role in the electrodermal affective response, but also the susceptance (i.e., imaginary part of the skin impedance) may contain relevant information about the SNS. Moreover, this relationship between AC frequency and recognition accuracy is strongly nonlinear due to the nonlinear relationship between skin impedance, and amplitude and frequency of the external electrical source [60]. Specifically, it is well-known that the current density under a surface plate electrode could be non-uniform, and electrode surfaces present fractal properties creating local areas of different current densities. The onset of non-linearity may therefore be gradual, and start very early at very limited areas on the electrode surface (e.g., it has been shown that very weak non-linearity is measurable at voltages than 100 mV). Hence, it may be difficult to differentiate between the non-linearity of the electrode processes and the tissue processes [61].

We are aware that works stated that the role of the susceptance is less important with respect to the conductance at low frequency [39], but our results seem to indicate that a significant difference in EDA results are frequency dependent even more when they are not mechanical but emotionally evoked.

In other words, we assume that it could be feasible that emotional stimuli may involve a capacitive component in the medium under investigation that has a bigger contribution at 100 Hz.

Moreover, we should take into account that Ohm's law, given by $J = \sigma E$, in such a medium could be not valid, and it may be useful to treat σ as a complex quantity in order to incorporate dielectric losses and frequency dependence, therefore defining σ as: $\sigma = \sigma' + j\sigma''$.

Future works will investigate the real and imaginary components of the admittance in the analysis of the EDA dynamics by involving time varying methods that could highlight the nonlinear nature of the electrodermal response.

Acknowledgments: This work was partially supported by the European Commission–Horizon 2020 Program under Grant 689691 "NEVERMIND" (NEurobehavioural predictiVE and peRsonalised Modelling of depressIve symptoms duriNg primary somatic Diseases with ICT-enabled self-management procedures).

Author Contributions: A.G., A.L., and E.P.S. conceived and designed the experiments; A.G. performed the experiments and analyzed the data; A.G. and L.C. developed the analysis tools; G.V., N.V. and E.P.S. contributed to the result interpretation; A.G., A.L., G.V. and E.P.S. wrote the paper.

Conflicts of Interest: The authors declare no conflict of interest.

References

1. Gross, J.J.; Muñoz, R.F. Emotion regulation and mental health. *Clin. Psychol. Sci. Pract.* **1995**, *2*, 151–164.
2. Berking, M.; Wupperman, P. Emotion regulation and mental health: recent findings, current challenges, and future directions. *Curr. Opin. Psychiatry* **2012**, *25*, 128–134.
3. Posner, J.; Russell, J.A.; Peterson, B.S. The circumplex model of affect: An integrative approach to affective neuroscience, cognitive development, and psychopathology. *Dev. Psychopathol.* **2005**, *17*, 715–734.
4. Schlosberg, H. Three dimensions of emotion. *Psychol. Rev.* **1954**, *61*, 81.
5. Ortony, A.; Clore, G.L.; Collins, A. *The Cognitive Structure of Emotions*; Cambridge University Press: Cambridge, MA, USA, 1990.

6. Lisetti, C.L.; Gmytrasiewicz, P. Can a rational agent afford to be affectless? A formal approach. *Appl. Artif. Intell.* **2002**, *16*, 577–609.

7. Reisenzein, R.; Hudlicka, E.; Dastani, M.; Gratch, J.; Hindriks, K.; Lorini, E.; Meyer, J.J.C. Computational modeling of emotion: Toward improving the inter-and intradisciplinary exchange. *IEEE Trans. Affect. Comput.* **2013**, *4*, 246–266.

8. Greco, A.; Valenza, G.; Nardelli, M.; Bianchi, M.; Citi, L.; Scilingo, E.P. Force–Velocity Assessment of Caress-Like Stimuli Through the Electrodermal Activity Processing: Advantages of a Convex Optimization Approach. *IEEE Trans. Hum. Mach. Syst.* **2016**, 1–10.

9. Calvo, R.A.; D'Mello, S. Affect detection: An interdisciplinary review of models, methods, and their applications. *IEEE Trans. Affect. Comput.* **2010**, *1*, 18–37.

10. Valenza, G.; Greco, A.; Citi, L.; Bianchi, M.; Barbieri, R.; Scilingo, E. Inhomogeneous Point-Processes to Instantaneously Assess Affective Haptic Perception through Heartbeat Dynamics Information. *Sci. Rep.* **2016**, *6*, 1–14.

11. Valenza, G.; Nardelli, M.; Gentili, C.; Bertschy, G.; Kosel, M.; Scilingo, E.P. Predicting Mood Changes in Bipolar Disorder through Heartbeat Nonlinear Dynamics. *Biomed. Health Inform.* **2016**, *20*, 1034–1043.

12. Rukavina, S.; Gruss, S.; Hoffmann, H.; Tan, J.W.; Walter, S.; Traue, H.C. Affective Computing and the Impact of Gender and Age. *PLoS ONE* **2016**, *11*, e0150584.

13. Valenza, G.; Citi, L.; Gentili, C.; Lanata, A.; Scilingo, E.P.; Barbieri, R. Point-process nonlinear autonomic assessment of depressive states in bipolar patients. *Methods Inf. Med.* **2014**, *53*, 296–302.

14. Lanata, A.; Greco, A.; Valenza, G.; Scilingo, E.P. A pattern recognition approach based on electrodermal response for pathological mood identification in bipolar disorders. In Proceedings of the 2014 IEEE International Conference on Acoustics, Speech and Signal Processing (ICASSP), Florence, Italy, 4–9 May 2014; pp. 3601–3605.

15. Valenza, G.; Lanatà, A.; Scilingo, E.P.; De Rossi, D. Towards a smart glove: Arousal recognition based on textile electrodermal response. In Proceedings of the 2010 Annual International Conference of the IEEE Engineering in Medicine and Biology, Buenos Aires, Argentina, 31 August–4 September 2010; pp. 3598–3601.

16. Cowie, R.; Douglas-Cowie, E.; Tsapatsoulis, N.; Votsis, G.; Kollias, S.; Fellenz, W.; Taylor, J.G. Emotion recognition in human-computer interaction. *IEEE Signal Process. Mag.* **2001**, *18*, 32–80.

17. Mazzei, D.; Greco, A.; Lazzeri, N.; Zaraki, A.; Lanata, A.; Igliozzi, R.; Mancini, A.; Stoppa, F.; Scilingo, E.P.; Muratori, F. Robotic social therapy on children with autism: preliminary evaluation through multi-parametric analysis. In Proceedings of the 2012 International Conference on Privacy, Security, Risk and Trust (PASSAT), and 2012 International Confernece on Social Computing (SocialCom), Amsterdam, The Netherlands, 3–5 September 2012; pp. 955–960.

18. Betella, A.; Zucca, R.; Cetnarski, R.; Greco, A.; Lanatà, A.; Mazzei, D.; Tognetti, A.; Arsiwalla, X.D.; Omedas, P.; De, R.D. Inference of human affective states from psychophysiological measurements extracted under ecologically valid conditions. *Using Neurophysiol. Signals Reflect Cognit. Affect. State* **2015**, *66*, doi:10.3389/fnins.2014.00286.

19. Anagnostopoulos, C.N.; Iliou, T.; Giannoukos, I. Features and classifiers for emotion recognition from speech: A survey from 2000 to 2011. *Artif. Intell. Rev.* **2015**, *43*, 155–177.

20. Lanatà, A.; Valenza, G.; Greco, A.; Gentili, C.; Bartolozzi, R.; Bucchi, F.; Frendo, F.; Scilingo, E.P. How the Autonomic Nervous System and Driving Style Change With Incremental Stressing Conditions During Simulated Driving. *IEEE Trans. Intell. Transp. Syst.* **2015**, *16*, 1505–1517.

21. Valenza, G.; Greco, A.; Gentili, C.; Lanata, A.; Sebastiani, L.; Menicucci, D.; Gemignani, A.; Scilingo, E. Combining electroencephalographic activity and instantaneous heart rate for assessing brain-heart dynamics during visual emotional elicitation in healthy subjects. *Philos. Trans. R. Soc. A* **2016**, *374*, 20150176.

22. Kim, J.; André, E. Emotion recognition based on physiological changes in music listening. *IEEE Trans. Pattern Anal. Mach. Intell.* **2008**, *30*, 2067–2083.

23. Koelstra, S.; Muhl, C.; Soleymani, M.; Lee, J.S.; Yazdani, A.; Ebrahimi, T.; Pun, T.; Nijholt, A.; Patras, I. Deap: A database for emotion analysis; using physiological signals. *IEEE Trans. Affect. Comput.* **2012**, *3*, 18–31.

24. Jang, E.; Rak, B.; Kim, S.; Sohn, J. Emotion classification by machine learning algorithm using physiological signals. *Proc. Comput. Sci. Inform. Technol. Singap.* **2012**, *25*, 1–5.

25. Healey, J.A.; Picard, R.W. Detecting stress during real-world driving tasks using physiological sensors. *IEEE Trans. Intell. Transp. Syst.* **2005**, *6*, 156–166.

26. Koji, N.; Nozawa, A.; Ide, H. Evaluation of emotions by nasal skin temperature on auditory stimulus and olfactory stimulus. *IEEJ Trans. Electron. Inform. Syst.* **2004**, *124*, 1914–1915.

27. Lanata, A.; Valenza, G.; Scilingo, E.P. Eye gaze patterns in emotional pictures. *J. Ambient Intell. Hum. Comput.* **2013**, *4*, 705–715.

28. Lanata, A.; Scilingo, E.P.; De Rossi, D. A multimodal transducer for cardiopulmonary activity monitoring in emergency. *IEEE Trans. Inform. Technol. Biomed.* **2010**, *14*, 817–825.

29. Krupa, N.; Anantharam, K.; Sanker, M.; Datta, S.; Sagar, J.V. Recognition of emotions in autistic children using physiological signals. *Health Technol.* **2016**, doi:10.1007/s12553-016-0129-3.

30. Boucsein, W. *Electrodermal Activity*; Springer Science & Business Media: Berlin/Heidelberg, Germany, 2012.

31. Greco, A.; Valenza, G.; Lanata, A.; Rota, G.; Scilingo, E. Electrodermal Activity in Bipolar Patients during Affective Elicitation. *IEEE J. Biomed. Health Inform.* **2014**, *18*, 1865–1873.

32. Olausson, H.; Cole, J.; Rylander, K.; McGlone, F.; Lamarre, Y.; Wallin, B.G.; Krämer, H.; Wessberg, J.; Elam, M.; Bushnell, M.C.; et al. Functional role of unmyelinated tactile afferents in human hairy skin: Sympathetic response and perceptual localization. *Exp. Brain Res.* **2008**, *184*, 135–140.

33. Hanson, M.A.; Powell, H.C., Jr.; Barth, A.T.; Ringgenberg, K.; Calhoun, B.H.; Aylor, J.H.; Lach, J. Body area sensor networks: Challenges and opportunities. *Computer* **2009**, *42*, 58.

34. Martínez-Rodrigo, A.; Zangróniz, R.; Pastor, J.M.; Fernández-Caballero, A. Arousal level classification in the ageing adult by measuring electrodermal skin conductivity. In *Ambient Intelligence for Health*; Springer: New York, NY, USA, 2015; pp. 213–223.

35. Lee, Y.; Lee, B.; Lee, M. Wearable sensor glove based on conducting fabric using electrodermal activity and pulse-wave sensors for e-health application. *Telemed. E-Health* **2010**, *16*, 209–217.

36. Patel, S.; Park, H.; Bonato, P.; Chan, L.; Rodgers, M. A review of wearable sensors and systems with application in rehabilitation. *J. Neuroeng. Rehabilit.* **2012**, *9*, 1.

37. Garbarino, M.; Lai, M.; Bender, D.; Picard, R.W.; Tognetti, S. Empatica E3—A wearable wireless multi-sensor device for real-time computerized biofeedback and data acquisition. In Proceedings of the 2014 EAI 4th International Conference on Wireless Mobile Communication and Healthcare (Mobihealth), Athens, Greece, 3–5 November 2014; pp. 39–42.

38. Martinsen, O.G.; Grimnes, S. *Bioimpedance and Bioelectricity Basics*; Academic press: Cambridge, MA, USA, 2011.

39. Martinsen, Ø.; Grimnes, S.; Sveen, O. Dielectric properties of some keratinised tissues. Part 1: Stratum corneum and nailin situ. *Med. Biol. Eng. Comput.* **1997**, *35*, 172–176.

40. Lanatà, A.; Valenza, G.; Scilingo, E.P. A novel EDA glove based on textile-integrated electrodes for affective computing. *Med. Biol. Eng. Comput.* **2012**, *50*, 1163–1172.

41. Martinsen, O.; Grimnes, S. On using single frequency electrical measurements for skin hydration assessment. *Innov. Technol. Biol. Méd.* **1998**, *19*, 395–400.

42. Martinsen, Ø.G.; Grimnes, S. Facts and myths about electrical measurement of stratum corneum hydration state. *Dermatology* **2001**, *202*, 87–89.

43. Martinsen, Ø.G.; Grimnes, S.; Nilsen, J.K.; Tronstad, C.; Jang, W.; Kim, H.; Shin, K.; Naderi, M.; Thielmann, F. Gravimetric method for in vitro calibration of skin hydration measurements. *IEEE Trans. Biomed. Eng.* **2008**, *55*, 728–732.

44. Ishchenko, A.; Shev'ev, P. Automated complex for multiparameter analysis of the galvanic skin response signal. *Biomed. Eng.* **1989**, *23*, 113–117.

45. Benedek, M.; Kaernbach, C. Decomposition of skin conductance data by means of nonnegative deconvolution. *Psychophysiology* **2010**, *47*, 647–658.

46. Breska, A.; Maoz, K.; Ben-Shakhar, G. Interstimulus intervals for skin conductance response measurement. *Psychophysiology* **2011**, *48*, 437–440.

47. Dawson, M.E.; Schell, A.M.; Filion, D.L. 7 the Electrodermal System. *Handb. Psychophysiol.* **2007**, *159*, 200–223.

48. Christie, M.J. Electrodermal activity in the 1980s: A review. *J. R. Soc. Med.* **1981**, *74*, 616.

49. Greco, A.; Valenza, G.; Lanata, A.; Scilingo, E.P.; Citi, L. cvxEDA: A Convex Optimization Approach to Electrodermal Activity Processing. *IEEE Trans. Biomed. Eng.* **2016**, *63*, 797–804.

50. Analog Device: AD9833 Low Power, Programmable Waveform Generator. Available online: http://www.analog.com/en/products/rf-microwave/direct-digital-synthesis-modulators/ad9833.html (accessed on 25 July 2016).

51. Texas Instrument: MSP430 Ultra-Low-Power Microcontrollers. Available online: http://www.ti.com/lsds/ti/microcontrollers_16-bit_32-bit/msp/overview.page (accessed on 25 July 2016).

52. Garrett, E.R. The Bateman function revisited: A critical reevaluation of the quantitative expressions to characterize concentrations in the one compartment body model as a function of time with first-order invasion and first-order elimination. *J. Pharmacokinet. Biopharm.* **1994**, *22*, 103–128.

53. Alexander, D.; Trengove, C.; Johnston, P.; Cooper, T.; August, J.; Gordon, E. Separating individual skin conductance responses in a short interstimulus-interval paradigm. *J. Neurosci. Methods* **2005**, *146*, 116–123.

54. Benedek, M.; Kaernbach, C. A continuous measure of phasic electrodermal activity. *J. Neurosci. Methods* **2010**, *190*, 80–91.

55. Greco, A.; Lanata, A.; Valenza, G.; Scilingo, E.P.; Citi, L. Electrodermal activity processing: A convex optimization approach. In Proceedings of the 2014 36th IEEE Annual International Conference of the Engineering in Medicine and Biology Society (EMBC), Chicago, IL, USA, 26–30 August 2014; pp. 2290–2293.

56. cvxEDA. Algorithm for the Analysis of Electrodermal Activity (EDA) Using Convex Optimization. Available online: https://www.mathworks.com/matlabcentral/fileexchange/53326-cvxeda (accessed on 25 July 2016).

57. Searle, A.; Kirkup, L. A direct comparison of wet, dry and insulating bioelectric recording electrodes. *Physiol. Meas.* **2000**, *21*, 271.

58. Kira, Y.; Ogura, T.; Aramaki, S.; Kubo, T.; Hayasida, T.; Hirasawa, Y. Sympathetic skin response evoked by respiratory stimulation as a measure of sympathetic function. *Clin. Neurophysiol.* **2001**, *112*, 861–865.

59. Lang, P.; Bradley, M.; Cuthbert, B. International affective picture system (IAPS): Digitized photographs, instruction manual and affective ratings. In *Technical Report A-6*; University of Florida: Gainesville, FL, USA, 2005.

60. Mørkrid, L.; Qiao, Z.G. Continuous estimation of parameters in skin electrical admittance from simultaneous measurements at two different frequencies. *Med. Biol. Eng. Comput.* **1988**, *26*, 633–640.

61. Sawan, M.; Laaziri, Y.; Mounaim, F.; Elzayat, E.; Corcos, J.; Elhilali, M. Electrode-Tissues interface: Modeling and experimental validation. *Biomed. Mater.* **2007**, *2*, S7.

electronics

MDPI

Article

An Embedded Sensing and Communication Platform, and a Healthcare Model for Remote Monitoring of Chronic Diseases

Sergio Saponara [1,2,*], Massimiliano Donati [2], Luca Fanucci [1,2] and Alessio Celli [2]

[1] Dipartimento Ingegneria della Informazione-Università di Pisa; via G. Caruso 16, 56122 Pisa, Italy; luca.fanucci@unipi.it
[2] IngeniArs srl, via Ponte a Piglieri 8, 56121 Pisa, Italy; massimiliano.donati@ingeniars.com (M.D.); alessio.celli@ingeniars.com (A.C.)
* Correspondence: sergio.saponara@iet.unipi.it; Tel.: +39-050-2217602

Academic Editors: Enzo Pasquale Scilingo and Gaetano Valenza
Received: 18 June 2016; Accepted: 25 July 2016; Published: 4 August 2016

Abstract: This paper presents a new remote healthcare model, which, exploiting wireless biomedical sensors, an embedded local unit (gateway) for sensor data acquisition-processing-communication, and a remote e-Health service center, can be scaled in different telemedicine scenarios. The aim is avoiding hospitalization cost and long waiting lists for patients affected by chronic illness who need continuous and long-term monitoring of some vital parameters. In the "1:1" scenario, the patient has a set of biomedical sensors and a gateway to exchange data and healthcare protocols with the remote service center. In the "1:N" scenario the use of gateway and sensors is managed by a professional caregiver, e.g., assigned by the Public Health System to a number N of different patients. In the "point of care" scenario the patient, instead of being hospitalized, can take the needed measurements at a specific health corner, which is then connected to the remote e-Health center. A mix of commercially available sensors and new custom-designed ones is presented. The new custom-designed sensors range from a single-lead electrocardiograph for easy measurements taken by the patients at their home, to a multi-channel biomedical integrated circuit for acquisition of multi-channel bio signals, to a new motion sensor for patient posture estimation and fall detection. Experimental trials in real-world telemedicine applications assess the proposed system in terms of easy usability from patients, specialist and family doctors, and caregivers, in terms of scalability in different scenarios, and in terms of suitability for implementation of needed care plans.

Keywords: wireless biomedical sensors; healthcare embedded platform; chronic health patient monitoring; biomedical data gateway; e-Health service center

1. Introduction

One of the main trends in biomedical applications is developing telemedicine systems for the remote monitoring of people affected by chronic diseases [1–16]. Particularly in developed countries such as the United States, Canada, Europe, Japan, South Korea, and Australia, the increasing percentage of elderly people and the need for public health systems (PHS) to cut the budget for hospitalization are fostering the rise of a new healthcare paradigm: hospitalization should be reserved only for patients with acute syndromes that can be solved in a short period. The healthcare model for patients affected by chronic illness, and needing continuous and long-term monitoring of some vital parameters should be based on telemedicine. According to a medical protocol established by a doctor, some biomedical parameters of the patient are periodically measured at home or in a point of care (e.g., a pharmacy) by the patients themselves, their relatives, or a professional caregiver (e.g., a nurse paid by the PHS or by medical insurance). The biomedical signals to be measured depend on the specific illness and may include measurements of ECG (ElectroCardioGraphy), blood pressure, body temperature and weight, oxygen saturation level in the blood (SpO$_2$), chest impedance, hearth rate and breath rate, and glycemia. These are the main parameters relevant for the three main chronic illnesses in western countries: Chronic Heart Failure (CHF), Chronic Obstructive Pulmonary Disease (COPD), and diabetes. These types of chronic illness affect approximately 15 million people in Europe, with an incidence of 3.6 million new cases every year; and the trend is the same in the United States [4]. Moreover, in this work we add proper motion sensors to measure the posture of the patient, which influences the measure of some biomedical parameters. Indeed, false alarms can be generated if the vital signs are acquired in a non-correct position of the patient. As an additional service, motion sensors are also useful for fall detection and consequent alarm generation. Fall detection is one of the main causes of home accidents for elderly people. Telemedicine is also a key technology for overcoming the problem of remote regions with low population density, where hospitals can be far from the town where people live (e.g., internal and/or mountain zones of Europe or the United States).

It is worth noting that the technologies discussed in this paper can enable telemedicine to reduce the hospitalization of patients, but cannot decide which patients will be hospitalized and when. The decision about how many patients (and which ones) are acute and should be hospitalized, and how many patients (and which ones), although affected by a chronic illness, are non-acute and should be monitored remotely will depend on the medical protocol defined by specialist doctors, and on the budget constraints of the PHS. If during the remote monitoring the patient's biomedical parameters get worse, according to the medical protocol established by the doctors, the patient can be re-hospitalized.

In the rest of the paper, Section 2 presents the state of the art and highlights the main contributions of this work. Section 3 presents the remote healthcare model and the embedded processing/communication platform. Section 4 is focused on the communication between the gateway at the client side and the e-Health service center at the server side. Section 5 deals with biomedical sensors using COTS (Commercial off the Shelf) components and three new custom-designed biomedical devices. The new custom-designed sensors include a single-lead ECG for easy CHF measurements taken by the patients at their home; a multi-channel biomedical ASIC (Application Specific Integrated Circuit) for acquisition of multi-channel ECG, EEG (ElectroEncephaloGraphy) or EMG (ElectroMioGraphy), blood pressure, and body temperature; and a new motion sensor for patient posture estimation and fall detection. Experimental trials are addressed in Section 6. Conclusions are drawn in Section 7.

2. Review of the State of the Art and Main Contributions of the Work

Several state-of-the-art wearable sensors and telemedicine platforms [3,4,17–39] have been proposed in the literature, but a successful and universal healthcare model is still missing. The main reason is that most works are only focused on a specific sub-part of the system, or on a specific type of disease. For example, [17,19–23,27–29] are focused on integrated smart sensors. Many studies (e.g., [18,25,30,31]) are focused only on the acquisition and communication gateway or on the remote server connected to the hospital information system (HIS). Moreover, [3,4,17,25,33–39] and [26,27] deal only with the monitoring of patients affected by heart disease and diabetes, respectively. Studies [19–23] are focused only on posture estimation and fall detection in patients. In [28–30] only contactless detection measurement of breath rate and/or heart rate is presented. Furthermore, most of these works come from academia and fail to address the qualification and certification issues of real-world biomedical applications. Most of the above works present just a new sensor, without any integration of real-world telemedicine scenarios that are characterized by multiple actors: patients and their relatives, professional caregivers (family doctor, nurse, specialist doctor), call centers, which are operating at home, or the hospital or points of care such as a pharmacy or a residence for elderly people.

Most current biomedical monitoring platforms try to exploit the computational and communication capabilities of smartphones with touchscreen user interface, or even with smartwatches [21,24,26,30–32]. This way, state-of-the-art works are missing one of the key features of a telemedicine service: easy usability of the interface for users, who are mainly elderly people. Smartphones and smartwatches are more suited for wellness applications targeting younger people.

To address these issues, this paper presents a new remote healthcare model, exploiting wireless biomedical sensors that can be scaled to different telemedicine scenarios:

- The "1:1" scenario, where each of the patient has a set of biomedical sensors and an embedded acquisition, processing, and communication platform (hereafter called a gateway) to exchange data and healthcare protocols with a remote service center and/or HIS for telemedicine, where a doctor is connected.
- The "1:N" scenario where the embedded acquisition and communication telemedicine platform is manged by a nurse, e.g., assigned by the PHS to a number N of different patients. The nurse is visiting and taking care of data acquisition from a set of N patients. The relevant biomedical data are then transmitted to the remote service center and/or HIS. In the "1:N" scenario it is the nurse that is moving and visiting patients at their homes.
- The "point of care" scenario where a local building, e.g., a pharmacy, or a point of care in a school or a residence for elderly people, hosts the embedded acquisition and communication telemedicine platform and the set of sensors. The patients, instead of being hospitalized, with increased cost and waiting lists, can take the needed measurements at a specific point of care, which is then connected to the remote service center and/or HIS. In this scenario the patients are moving toward the point of care where a nurse supervises the biomedical measurement acquisitions to be collected and transmitted.

In this paper, different from the state of the art, the whole value chain is implemented from the health care model at the top, down to the technical implementations of sensors, data acquisition and communication platform, and integration with the service center and HIS. The work is the result of the collaboration between academia (the University of Pisa) and industry (IngeniArs S.r.l.), the latter being responsible for integration in different real-world telemedicine scenarios, taking care of all actors involved, including certification and qualification issues.

The communication between the home gateway and the HIS is based on approved medical protocols such as HL7 CDA (Clinical Document Architecture). The communication is physically running on wireless technologies available everywhere, like 3G or 4G cellular network, or satellite connections, wired technologies like ADSL/VDSL (Asymmetric or Very-high speed Digital Subscriber Line), or Fiber to the Cabinet/Home (FTTC/FTTH).

The platform can be used to implement a predefined measuring protocol, i.e., a care plan assigned remotely by the family or specialist doctor. Extra-protocol measurements can be taken by the patient or the caregiver in case of necessity or can be requested remotely by the doctor. A mixture of COTS sensors and custom ones specifically designed by the research group are presented and used in this work. With respect to previous publications of the authors in [4,17,25,28], this work is extended in terms of:

- Support of any chronic illness and not only CHF as in [4,17,25].
- Analysis of the system-level telemedicine model with differentiation of the biomedical monitoring kit according to the above scenarios: "1:1", "1:N," and "point of care," which is missing in [4,17,25,28].
- Development of new custom sensors, particularly the single-lead ECG one for easy self-measurements at home, whereas [4,25,28] were mainly based on commercial sensors.
- Integration of motion sensors for fall detection and patient's posture analysis, missing in [4,17,25,28].

This work also presents real-world results from experimental trials carried out in the field of several European and regional research and health projects such as Health@Home (EU Ambient Assisted Living program), RIS and RACE (EU-Tuscany Region FESR program), and Domino (Tuscany region PHS project).

3. Remote Healthcare Model and Embedded Sensing/Communication Platform

According to Figure 1, the proposed telemedicine model includes several elements to cover the different sub-systems belonging to a distributed health care system. The key blocks in Figure 1 are:

- A monitoring kit (embedded sensing and communication platform, oval A with green borders in Figure 1, whose description is detailed in Sections 3.1–3.3) which, depending if the "1:1" or "1:N" scenario is implemented, can be used by patients for self-measurement or by caregivers, e.g., nurses, during planned home visits.
- A totem for the monitoring of the biomedical parameters to be installed at point of cares (e.g., pharmacies, residences for elderly people, or other healthcare points); this is the embedded sensing and communication platform indicated with oval B with green borders in Figure 1, whose description is detailed in Sections 3.1 and 3.4).
- Management platform of the electronic health record and of the home-care plan (e-Health center detailed in Section 4), integrated with the HIS, and available to specialist doctors directly or through operators of a service center. Optionally, through the service center the data of the electronic health record can also be made available to the family doctor. Alarms can be automatically generated by the embedded sensing/communication units (thanks to local signal processing capability) at home or at the point of care, or by a caregiver analyzing the data. Automatically generated alarms should be validated by a caregiver. An alarm, generated or validated by a caregiver, is communicated to emergency units for a fast re-hospitalization of the patient and, optionally, to a pre-selected list of relatives/friends.

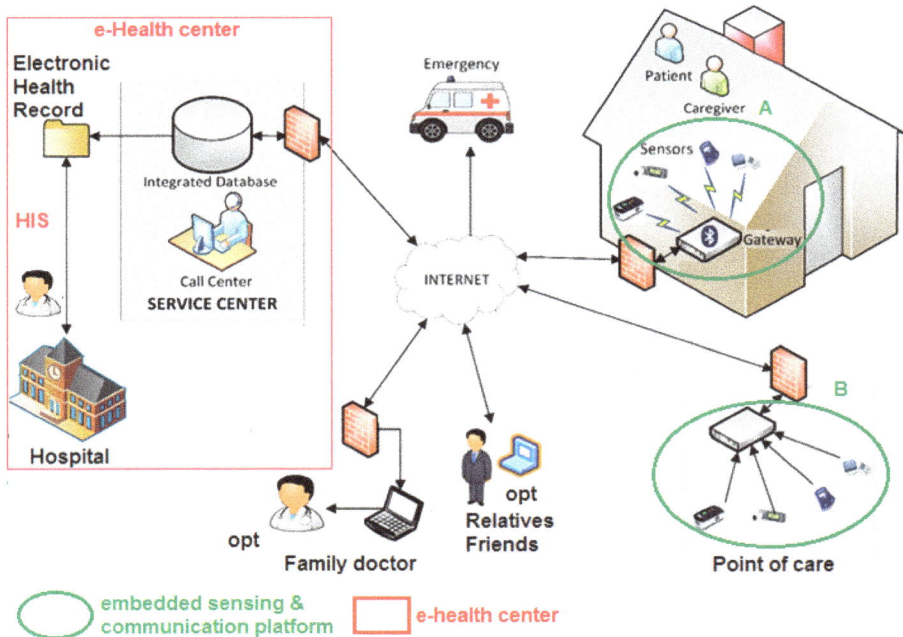

Figure 1. Distributed health care system for chronic illness monitoring.

3.1. Biomedical Sensing and Communication Platform

The reference architecture of the biomedical monitoring kit is reported in Figure 2. It includes: a set of wireless sensors, connected through a Bluetooth wireless technology to an embedded system (called gateway) in charge of biomedical sensor signal acquisition, local processing, and data storage, and three interfaces: one for the user, one for the sensors, and the other for the e-Health center.

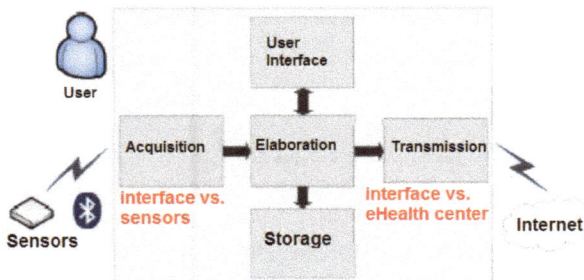

Figure 2. Main building blocks of the gateway.

The latter exploits the SOAP web service paradigm on standard wired or wireless communication technologies (e.g., WiFi, Mobile Broadband, Ethernet, etc.). To increase the system flexibility, besides the HL7 CDA standard data format, the proposed system also supports other protocols for client–server communication like JSON (JavaScript Object Notation) and XML (eXtensible Markup Language).

For the sensors interface, Bluetooth (BT) technology has been preferred to other 802.11x WLAN (Wireless Local Area Network) or 802.15x WPAN (Wireless Personal Area Network) technologies for its large diffusion, increasing the number of commercial sensors that can be selected and used. The gateway handles the communication via the dual-mode BT interface: the RFCOMM protocol (Serial Port Profile SPP) is used to handle legacy Bluetooth 2.0 sensors, whereas the profiles HDP (Health Device Profile) [32] or GATT (Generic Attribute Profile) are used to communicate with Bluetooth Low-Energy (BLE) devices. Relying on a dual-mode BT chipset, the gateway is able to handle both legacy BT and BLE devices, acquiring data from a single sensor per time. The number of connectable sensors is unlimited and the gateway is able to manage both master and slave sensors. Pin-based pairing procedure and eventual data encryption are available. The performance of BT, even in a low-power protocol version, is well suited for telemedicine applications. Indeed, the main features of BLE are: data rate up to 1 Mb/s; connection distance up to 100 m outdoor line-of-sight but at least meters indoor; supported security technology (128 bit Advanced Encryption Standard) and robustness techniques (Adaptive frequency hopping, Lazy Acknowledgement, 24-bit Cyclic Redundant Code, 32-bit Message Integrity Check), low communication latency of few ms, and low power consumption limited to a few tens of mW, since a current well below 15 mA is drained from a power supply of few Volts.

It is worth noting that the gateway can also implement local signal processing tasks and not only acquisition, communication, and user interface tasks. Supported signal processing functionalities are:

- Collection of the acquired data from the configured BT or BLE sensors to create statistics of the biomedical parameters acquired according to the specific plan.
- Graphical rendering for the visualization of the historical evolution of the biomedical parameters acquired according to the specific plan (see an example in Figure 3 related to the evolution of the SpO_2 parameter). The statistics and graphical rendering of the historical evolution of biomedical parameters are useful at the gateway side mainly in the "1:N" and in the "point-of-care" scenarios (where the remote acquisition of bio-signals is supervised by a professional caregiver). They are also made available at the remote server side (service center and HIS).

- Threshold-based analysis of the acquired data so that an early warning can be sent when one of the acquired parameters is above or below a specific threshold that can be changed dynamically and remotely by the doctor. The early warning can be used to force an immediate hospitalization in case the chronic illness enters into an acute phase.

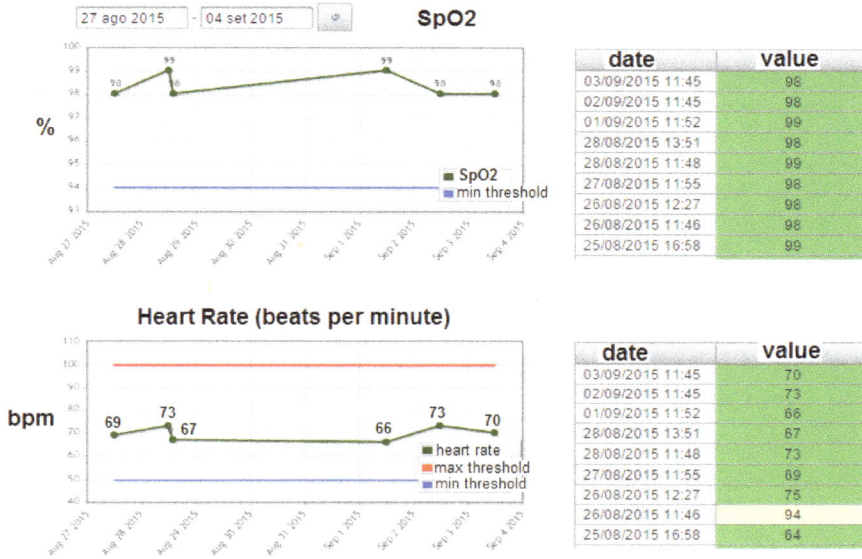

| 27 ago 2015 | - | 04 set 2015 | ⌗ | | SpO2 |

date	value
03/09/2015 11:45	98
02/09/2015 11:45	98
01/09/2015 11:52	99
28/08/2015 13:51	98
28/08/2015 11:48	99
27/08/2015 11:55	98
26/08/2015 12:27	98
26/08/2015 11:46	98
25/08/2015 16:58	99

Heart Rate (beats per minute)

date	value
03/09/2015 11:45	70
02/09/2015 11:45	73
01/09/2015 11:52	66
28/08/2015 13:51	67
28/08/2015 11:48	73
27/08/2015 11:55	69
26/08/2015 12:27	75
26/08/2015 11:46	94
25/08/2015 16:58	64

Figure 3. Graphical rendering of the historical evolution of SpO$_2$ and heart frequency.

From a hardware point of view, the gateway can be implemented on a general purpose system such as smartphones, tablets, or custom boards. The minimal requirements for the implementation are: 32 bit ARM Cortex processor, 1 GB RAM memory, at least 4 GB Flash Non-volatile storage, dual-mode Bluetooth chipset, network connectivity (e.g., Wifi, Ethernet, mobile broadband), and Android OS. On top of the hardware, a custom software layer has been developed using Java and the Android Software Development Kit (SDK). The strategic decision to use Java technology provides extreme flexibility concerning the configurability. Java also guarantees easy portability. The system is easily scalable in different processors and Printed Circuit Boards (PCBs). For example, one implemented configuration resulted in a compact size of 15 cm × 7 cm × 7 cm. Instead, in a configuration requiring a screen of 10 inches, the hardware is organized so that a size of about 22 cm × 14 cm × 1 cm is obtained. Solutions based on commercial tablets and smartphones have also been developed.

3.2. "1:1" Scenario

Changing the scenario ("1:1" or "1:N" or "point of care") means changing who is supervising the biomedical measurement activity: the patient or his/her relatives (non-professional users), or a nurse (professional user). With reference to the architecture of the monitoring kit in Figure 2 (which includes a set of wireless sensors connected through a BT or BLE technology to the embedded platform for data acquisition and communication), which is valid for all scenarios, the user interface and also the type of sensors to be used have to be specifically adapted to the different scenarios and the different users.

For example, a conventional 12-lead ECG biomedical instrumentation, providing complete and accurate measurements of heart activity, is not suited to the "1:1" scenario where a non-professional caregiver (the patient or his/her relatives) is supervising the measurement acquisition. If used by a non-professional caregiver, an ECG with lots of derivations will often have the electrodes placed in the wrong position or with a bad electrode–body contact, thus giving inaccurate results. This is why, in the following, for the different scenarios, different sets of building blocks must be used. Moreover, for some specific blocks, e.g., ECG, a new custom instrumentation has been designed (e.g., the single-lead ECG in Section 5.2).

In the "1:1" healthcare model, already described in Section 2, the gateway allows the user to view the individual care plan established by the family or specialist doctor, collect the measurements from medical sensors, through a BT or BLE connection, and send them to the e-Health Center (a service center and/or the HIS) through a wired or wireless Internet connection (see Figures 4 and 5). The patient also has the possibility to carry out a measurement not foreseen in the standard care plan. At a time when the activity is required, visual and audible signals inform the patient that a new measurement is required (see Figure 5). An animated image guides the patient on how to use the device to complete the task. The numbers near the arrows in Figures 5–7 highlight the temporal consecution of the different steps. After the measurement, the gateway notifies the patient if the activity was successful or s/he will need to repeat the operation. The user interface has been optimized for elderly people with a simplified graphic, using a touch-screen display of at least 10 inches with large buttons. The language can be customized for the specific nation where the system is used.

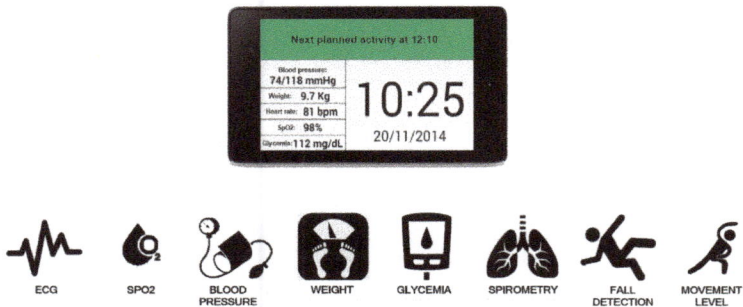

Figure 4. Gateway user interface in the 1:1 scenario.

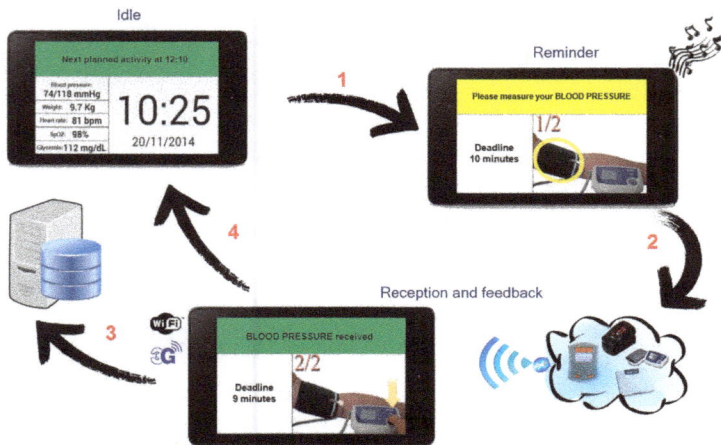

Figure 5. Evolution of the acquisition flow in multiple steps (from idle, to reminder, to reception and feedback, to communication towards the remote server, and again to idle).

3.3. "1:N" Scenario

In the "1:N" scenario, instead, the professional monitoring kit allows the nurse to take measurements of vital signs during home visits to different patients. As reported in Figure 6, first the gateway allows the nurse to select the specific patient (based on his/her health card or fiscal code, for example) and view the individual care plan for each patient. The user interface then guides the operator in the execution of the measures indicated in the plan. The system also allows the collection of unplanned measures and allows the repetition of the measurements made. Since the user is a professional one, more complex visualizations such as the ECG trace are available. The gateway can also work in offline mode, by sending all the measurements acquired at the end of the home visits. The professional operator selects the patient from the list of the clients and takes the measures foreseen in the individual care plan, making sure that the quality of the measurement is satisfactory. If it is not, s/he can proceed with a repetition of the measurement. The acquired measurements are sent to the e-Health Center, through the Internet connection. Figure 6 reports the complete flow.

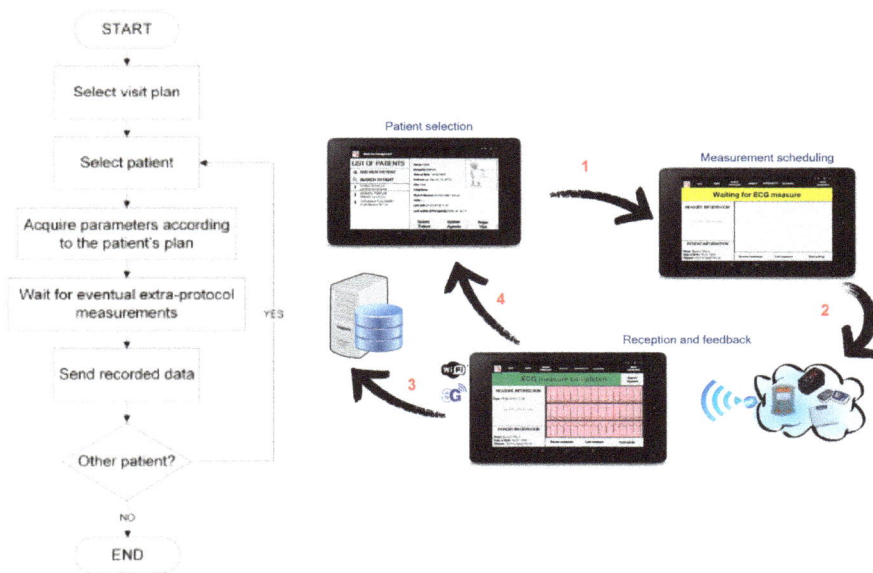

Figure 6. Evolution of the acquisition flow in case of the 1:N scenario; e.g., gateway for nurse.

3.4. "Point of Care" Scenario

In the "point of care" scenario, also called Totem mode (see Figure 7), the telemonitoring totem allows the measurement of the principal vital signs at dedicated facilities (pharmacies, residence for elderly people, etc.) and their transmission in the patient's electronic file through the Internet connection. The professional caregiver identifies the patient through his/her fiscal code or by scanning his/her health card and takes the measures provided for the individual care plan, making sure that the quality of the measurement is satisfactory. If it is not, s/he can proceed with a repetition of the measurement. The acquired measurements are sent to the e-Health Center, through the Internet connection. This scenario is similar to the "1:N" scenario, but here the patients are moving toward the health center. Moreover, the patient database is larger since a larger number of patients are followed. In all of the above cases the access by the user is done in secure mode through a login–password protocol.

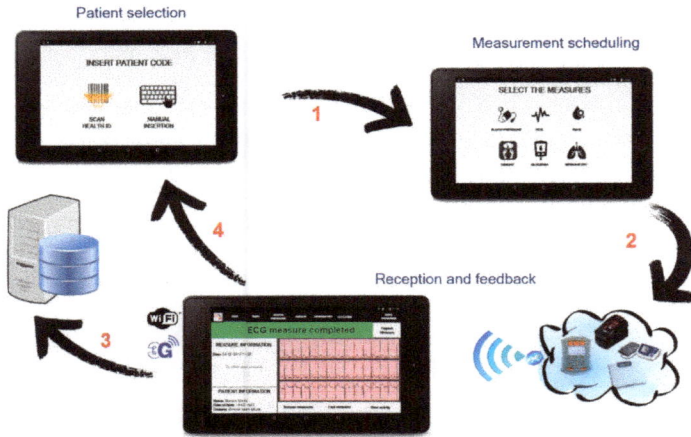

Figure 7. Evolution of the acquisition flow in the point-of-care scenario.

4. Home Monitoring Unit vs. e-Health Center Client–Server Communication

In the proposed health model the remote e-Health center in Figure 8 plays an essential role, since it integrates the services performed on the territory within the medical environment and the HIS. As reported in Figures 1 and 8, it includes a service center and the interface toward the HIS. The e-Health center is the central element of the overall modular system, in which specialized human operators and ICT resources allow for managing data flows and events.

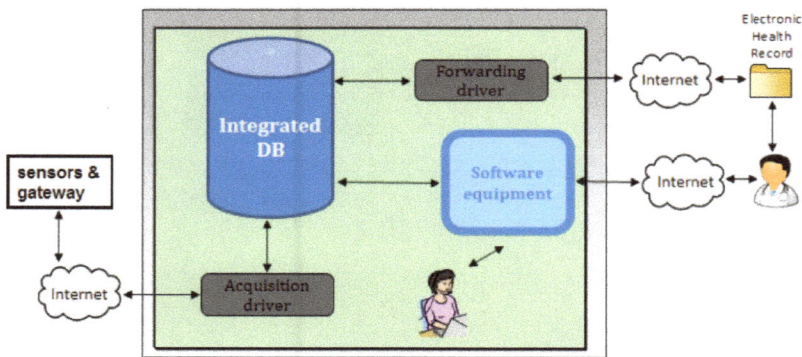

Figure 8. Block diagram of the remote e-Health center.

Concerning the presented telemedicine platform, the e-Health center is in charge of three important tasks. The first one is to manage the bidirectional communication with the gateway of the home monitoring module to receive data and send configurations. The second task is to provide the data processing capabilities and interfaces for the local service center operators and the other remote human operators of the system, respectively (i.e., users of the other modules). The last task is the synchronization of the clinical data with the electronic health record. Figure 8 shows the block diagram of the remote e-Health center including an intermediate node, the service center, and the connection with the HIS where the electronic health record of the patient is stored and where the specialist doctor can access and manage the telemedicine data and care plan.

Communications between the gateways, which are distributed throughout the area, and the acquisition block of the remote e-Health center take place through the public network. The acquisition driver implements the receiving endpoint, which is a SOAP (Simple Object Access Protocol) web service based on the HTTPS (Hyper Text Transfer Protocol, known as HTTP, over Secure Socket Layer, known as SSL) protocol able to manage both XML and HL7 CDA standard contents. Data coming from gateways are the results of the measurements performed by the caregivers. These are stored in the integrated database in order to be available for the other actors of the distributed telemedicine system. The remote e-Health center also manages the personalized monitoring protocol for each patient enrolled in the telemedicine program in terms of measurements to take and thresholds for alarm generation. The protocols of several patients can be loaded and updated remotely by calling from the gateway a dedicated configuration endpoint.

Dedicated software processes the received data by means of specific algorithms in order to find critical situations or dangerous alterations of vital signs. For example, those for CHF have already been published by the authors in [4]. It is possible to define personalized analysis profiles based on the needs of a specific patient. The processed information is stored into the integrated database and made available to the operators through different levels of the interface. The local interface is available only to service center operators and enables them to manage all the phases of telemedicine service provision. It allows for the enrollment and classification of new patients, the definition and updating of the treatment protocol for each patient, the establishment of personalized vital signs analysis profiles, and visualization and interaction with current and past measurements of vital signs. Moreover, it permits the operator to manage alarms or critical situations, eventually involving territorial emergency services or other appointed modules of the integrated system. Indeed, the remote interface allows professional operators (e.g., family doctors in Figure 1, etc.) to remotely access a subset of clinical information contained in the integrated database.

Another important role of the e-Health center is that of the relay agent for the clinical information received from the gateways with respect to the final destination, i.e., the electronic health record (see Figure 8). This process is completely transparent to the medical personnel and allows the information collected by the caregiver on the territory to be available in a few minutes in all offices that have access to the HIS. In this way, the electronic health record of each patient represents a deep anamnestic database and helps the medical personnel to improve the quality of treatment by developing therapy tailored to the specific patient's needs.

In such a system, in which critical and personal information are exchanged through the public Internet, the confidentiality, data integrity, and authenticity of the communicating parts are among the major requirements. In the presented platform we selected the HTTPS protocol to protect all the communications between the gateways and the e-Health center. This protocol provides the normal HTTP request–response mechanism typical of web-based or web service-oriented applications over an SSL or TSL (Transport Layer Security) encrypted end-to-end tunnel. The communicating parties firstly authenticate themselves using the X.509 certificate, then use their asymmetric keys to establish and exchange a symmetric session key that will be used to encrypt all the traffic. Message hashing ensures the data integrity over the session tunnel.

Concerning hardware and software implementation, the service center consists of a room where specialized operators (i.e., trained operators, nurses, or specialist doctors) interact with the ICT infrastructure through dedicated terminals of a Linux-based server machine that hosts the integrated database (Oracle), runs the communication drivers and the processing software for alarm generation, and finally provides the graphical user interfaces. All the equipment running on the service center server received the CE certification according to the 93/42/CEE directive following integrations for data generation, interpretation, and visualization. In fact, this software has been classified as a class IIa medical device. It is compliant with the standard ISO 62304—medical software life-cycle, the standard ISO 14971—Risk management in medical devices, the standard ISO 60601—alarm systems in medical devices, and ISO 62366—usability engineering in medical devices.

In this telemedicine platform, vital signs data collected by CE certified biomedical devices are provided and interpreted for diagnostic purposes only through CE certified software. In this way, the acquisition and transmission chain involves certified elements and medical devices only at the extremities, while the gateway and the other intermediate elements simply propagate the information without dealing with the content. Patients use the wireless medical devices assigned to them, and the data is sent automatically to the e-Health Center through a gateway. The data received by the gateway are managed in raw format and wrapped in XML before transmission. The e-Health Center parses the raw data received through a CE-marked data interpretation driver in order to allow their use for medical purposes (diagnosis, therapy, etc.). In this way, even in case the intermediate gateway is not CE-marked as medical devices, the whole chain maintains the certification, because critical data are only generated by and managed by CE-certified elements.

The e-Health Center is a multi-disease, multi-device, multi-parameter, multi-language, multi-tenant web platform for the management of patients and the remote monitoring of their vital parameters. An alarm is signaled every time a parameter is not received within the patient's schedule, and also if a parameter falls outside the ranges. Each patient has different ranges for red, yellow, and cyan alarms on each parameter. Specialized operators receive the alarms and handle them with appropriate protocols, which typically include contacting in a defined order one or more of: the patient, his/her caregivers, agreed neighbors and relatives, the family doctor, emergency services as ambulance, the fire brigade, etc. Depending on the established medical protocol, in the proposed system, the specialized operator in charge of alarm management can be the specialist doctor taking care of that specific patient, or a generic caregiver (i.e., trained operators, doctors of the hospital, or also a nurse). The operator performs further calls as needed and monitors the situation until resolved, recording in the e-Health Center all his/her activities and their outcomes. Patients can also have emergency ("panic") buttons to directly call operators for remote assistance. The server application for the e-Health center is composed of the following software components:

- Relational database: stores all the data and contains most application logic—including object-oriented PL/SQL data models, patient schedules, and alarm triggers. It is in charge of enforcing users' permissions.
- Java Enterprise Edition (JEE) web application, which implements and publishes the AJAX-based web 2.0 interface.
- Driver: receives the raw data sent by gateways, parses them, and inserts the parsed measurements into the database.
- Audit and Security System: monitoring component that detects and reports any malfunctioning. It also records the system activity.

The other main element of the architecture is the server gateway integration engine. This element is the link between data and the large set of heterogeneous management platforms on which the telemedicine services are based on. The technology used for the development of this part of the system is the JEE. From the functional point of view this module:

- Receives raw data embedded into XML tags from the client gateway.
- Transmits to the client gateway the agenda of the configured patient.
- Allows the complete management of patients.
- Transmits data to the server of the service centers with specific adapters.
- Receives agenda by external clinical data management tools.

5. Wireless Biomedical Sensors

5.1. Wireless Sensor Selection

As discussed in previous sections, the proposed system exploits wireless biomedical sensors with BT and BLE connectivity. The telemedicine market is still growing, so standards are not frozen. Moreover, qualification and CE certification for medical use of a new device entail significant development time and costs. Development of a new sensor makes also sense for devices with high added value, with a key difference vs. the state of the art, or with a high potential market. For this reason, the approach we followed is developing custom sensors in three cases:

- A single-lead ECG sensing device, patent-filed technology [33], which allows for self-monitoring of the heart in an easy way without the need to connect lots of electrodes in different parts of the body, but simply placing the two hands of the patient on top of a couple of electrodes. This sensor is further discussed in Section 5.2. This sensor, although simple and easy to use, can provide a graphical trace of the ECG and automatic measurement of heart rate and its statistics, thus being useful for arrhythmia monitoring.
- An integrated multi-channel Biomedical ASIC with a configurable sensor front-end [40,41], which allows multiple electrodes for multi-channel ECG or EEG or EMG measurements plus body temperature and blood pressure monitoring. The ASIC also supports automatic detection of pacemaker signals to avoid false alarm generation. This sensor is further discussed in Section 5.3.
- A motion sensor for correct detection of the patient's posture and possible falls. Indeed, the measurement of most biomedical parameters is influenced by posture. Therefore, the posture of the patient has to be acquired during remote monitoring to reduce the rate of false alarms or missed detection. As an additional service, motion sensors allow for the detection of patient falls and consequent alarm generation. Falls in elderly people are one of the main causes of accidents at home. This sensor is further discussed in Section 5.4.

The other sensors we selected are already qualified (i.e., CE certified) commercial wireless medical devices. Here the focus of our work has been first, together with medical staff from Fondazione Toscana Gabriele Monasterio, to define the set of measurements to take for each of supported chronic illness (CHF, diabetes, COPD) and the relevant requirements. Sensor specifications have been set in terms of dynamic range, acceptable noise and interference levels, signal bandwidth, sensitivity, and usability. Starting from this analysis, a set of BT and BLE sensors has been selected and integrated within the acquisition platform discussed in previous sections. Table 1 shows the main characteristics of the commercial sensors that have been used for the different illnesses monitored by the proposed biomedical platforms. Power consumption of commercial devices, as reported in their respective user manuals, is suitable for two or three months of use in a telemedicine service that requires one or two measurements per day. For example, the Cardioline Microtel Cardiette ECG device ensures at least 7 h of use with the same battery. This means that, assuming 2 min per measurement, the batteries need to be replaced about every 100 days.

Table 1. List of COTS biomedical sensors.

Sensor	Chronic Illness	Device Characteristics
Cardioline Microtel Cardiette	CHF	3/6/12 channels derivation; ECG continuous measurement; 0.05–150 Hz; sampling rate 500 Hz; pacemaker detection; BT 2.0 SPP
A&D weighting scale (UC-321PBT)	CHF	Range 0–200 kg, resolution 100 g; BT 2.0 SPP
A&D blood pressure (UA-767PBT)	CHF, COPD	Range: pressure 20–280 mmHg, pulse: 40–200 bpm; Accuracy: pressure $\pm$ 3 mmHg, pulse: $\pm$5%; BT 2.0 SPP
Nonin saturimeter (Onyx II 9560)	CHF, COPD	Range: SpO_2 0%–100%, pulse 20–250 bpm; Accuracy: $SpO_2 \pm 1$, pulse $\pm$ 3 bpm
Lifescan Glucometer (onetouch ultraeasy)	diabetes	Range: 20–600 md/dL; Accuracy: $\pm$5%; BT 2.0 SPP
MIR spirometer (Spirodoc)	COPD	Range: flux $\pm$ 16 L/s, Accuracy: flux $\pm$ 5%; BT 2.0 SPP

5.2. Single-Lead ECG Sensor

5.2.1. State of the Art Review and Specifications of Devices for Patient ECG Self-Measurements

Many modern ECG devices for telemonitoring use a wide variety of technologies and methods to record the electrocardiogram from the patient and send it to the service center. For example, systems like Intelsens V-Patch [34], Iansys Lifecare [35], or Lifewatch Lifestar ACT [36] make use of adhesive disposable electrodes attached to the chest in order to detect the ECG. Then, proprietary wireless protocols and gateways are used for the transmission of the ECG through an Internet connection. A different method of ECG acquisition is adopted by DOCOBO doc@home [37], which uses four dry metal electrodes in contact with the hands and transmits the recorded ECG through a wired telephone connection. Other devices like Card Guard PMP4 SelfCheck ECG [38] or SolutionMD ECG Mobile [39] use Bluetooth technology for sending data to a third party gateway. The device in [38] acquires the signal through two dry metal electrodes put directly on the chest or, as an alternative, using 10 wet adhesive electrodes connected to the device by a cable. The device in [39] instead uses a number of capacitive wearable electrodes embedded in clothing. In many of these solutions the ECG measurement device has a complex human–machine interface (HMI), presenting many different functions and showing much information.

The main objective of the proposed ECG device is to be ergonomic and easy to use in order to encourage patients to periodically record and send their own electrocardiograms following an assigned plan. To achieve this goal patients, mostly elderly, should be able to use and maintain the device in complete autonomy. Thus, it is important to ensure that a low number of simple operations are required to record and send the ECG. Moreover, the interaction with the device should be easy and immediate. The need to provide supplies of disposable specific materials, such as the electrodes, could represent a problem especially for elderly people, who often have poor mobility. To develop an ECG device satisfying the requirements described above, the main issue is represented by the electrical contacts with the patient. The most practiced solutions involve the use of adhesive electrodes placed on the chest, arms, and legs, connected to the ECG device with cables. A less frequent alternative is the use of dry or wet metal electrodes, located on the device, to be put directly in contact with the chest. Although these devices are able to acquire a number of ECG leads ranging from three to 12, these solutions are not optimal for our purpose. An affordable solution may be to reduce the number of leads to one and to find an easy way to establish and maintain the contact of the electrodes with the body. For example, the patient could record the first lead of his/her own ECG signal just by placing his/her hands on the device. Then the recorded ECG is sent to the gateway discussed in Section 3, through an automatic preconfigured Bluetooth connection. Once the ECG device is paired with the gateway, the user only has to make sure that both are switched on and wait for the ready-to-record signal. Another important aspect is the HMI, which should be as simple as possible in order to allow the device to be user-friendly. Since elderly patients usually do not have confidence with technology, an interface that presents more than one or two buttons and many functions and indicators may be a cause of rejection. The proposed solution involves a HMI with only one button for switching the device on/off, a few LED indicators (green for power and red for heart rate), and a simple display LCD (128 × 64 pixels).

5.2.2. Analysis of the Skin–Electrode Contact and the Shape of a Hand-Based Single-Lead ECG Device

To acquire a good-quality ECG in a comfortable, quick, and easy way for the patient, it is important to find the best configuration for contact with the hands and an ergonomic shape for the device. Thus, one of the first steps was a study about the position of the hands on the contacts that was carried out with the help of Lifepak 15, a commercial professional electrocardiograph. The Lifepak 15 is equipped with two metal paddles, used for defibrillation, with which it is also possible to detect an ECG lead. By placing the hands on the paddles in different configurations, it was possible to compare the different qualities of the signal detected. In a previous work [17] we tested several different configurations, of which the four most interesting configurations are reported here:

(1) paddles kept singularly in each hand with the palms in contact with the metal electrodes;
(2) paddles attached to each other on the insulated back and kept in contact with the skin through the pressure of the hands;
(3) paddles placed on a plane with the electrodes on the top, in contact with the fingers of each hand;
(4) paddles placed on a plane with the electrodes on the top, in contact with the proximal part of the palm of each hand.

Comparing the ECG obtained, the best quality signal is achieved in the case where the palms of the hands are placed on the electrodes. In the other cases, where the fingers are involved or where muscle tension is needed to keep contact, the ECG presents some artifacts due to the difficulty of avoiding small movements that cause variations in the pressure of the skin over the electrodes.

Once we assessed the best hand–electrode contact configuration, another step was to investigate the best shape for an ECG device that will be ergonomic and comfortable to use. Several different shapes and ways to handle them were proposed to tens of elderly people. From the feedback of the testers, the configuration that best allows one to keep stable contact with the electrodes is a parallelepiped having dimensions of about $30 \times 5 \times 3$ cm^3, with the electrodes placed on a plane with the long side parallel to the chest, on which the user lays the proximal part of the palm of the hand. Testers also reported that the greatest comfort is obtained when laying the hands on the shape instead of keeping it steady with one's hands. Furthermore, it is better to keep only the proximal part of the hand on the electrodes to avoid pressure on the wrists. This will cause an annoying feeling of pulsation during the measurement.

One of the most important aspects to be considered when using dry metal electrodes concerns the contact impedances of the electrodes with the skin, which represents the source impedances of the system [42]. Figure 9 shows a simplified model of the skin–electrode contact impedance that presents resistive and capacitive components.

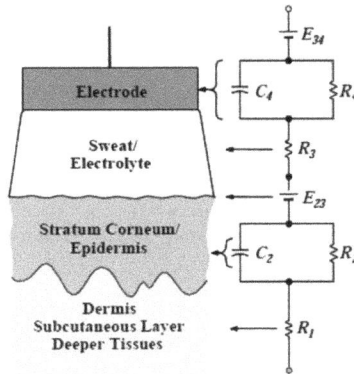

Figure 9. Skin–electrode contact electric model.

When a differential amplifier is used to capture the electrical cardiac signal between two electrodes, the values of the two source impedances have a significant influence on the quality of the output signal. If the source impedances are unbalanced, an amount of common mode signal transforms into a differential component at the input of the amplifier, worsening the Common Mode Rejection Ratio (CMRR) of the system. Since it is impossible to guarantee two identical contact impedances, especially if dry electrodes are used, it is important that the impedance values are as low as possible in order to minimize the absolute value of the difference. Moreover, the impedances are susceptible to variations in time due to changes in the pressure of contacts or in the local conductivity of the skin. This introduces low-frequency artifacts over the desired signal. Furthermore, the presence of a capacitive component of the electrode–skin interface could also introduce a phase distortion in the signal if the magnitude of the contact impedances is not negligible with respect to the value of the input differential impedance of the amplifier. Thus, it is important to study the properties of the contact impedances at the point of the body where the signal is detected.

To understand the order of magnitude of the source impedances, a simple frequency characterization of the impedance of the hand–electrode contacts was made. A steel electrode, obtained from pediatric defibrillation paddles, was fixed to each hand using an elastic bandage to keep a constant pressure. A voltage-to-current converter and voltage amplifier circuit was realized to impose a current between the two contact interfaces and to measure the voltage that occurs between the two electrodes. By comparing the output signal with the input measured at different frequency values with the aid of an oscilloscope, it was possible to obtain the sum of the two contact impedances. Assuming the two impedances are equal, the value of one of them is half of the observed value. From experimental measurements the maximum value of the impedance is about 10 kΩ at low frequencies (DC to 3 kHz), while it decreases at higher frequencies. Since the magnitude of the skin–electrode impedance is negligible with respect to the input impedance of a common instrumentation amplifier, the phase value of the former was considered irrelevant.

5.2.3. Architecture of a Single-Lead ECG Device

Figure 10 represents the block diagram of the single-lead ECG biomedical device. It includes two dry electrodes, an analog front-end, a microcontroller, a Bluetooth module, and a user interface. The analog front-end circuit used to detect the ECG signal from the two contacts and to condition the signal includes as input stage an instrumentation amplifier (IA). The IA stage converts the differential input signal to a single-ended signal. The gain of this stage is low to avoid saturations due to input offsets caused by the half-cell potential of the skin–electrode interfaces, usually ranging from 300 mV to 1 V when unbalanced source impedances are present. The feedback loop on the reference pin of the IA has been sized to implement a first-order high-pass filter that eliminates the output DC offset acting on the reference voltage of the IA. The output of the IA is then amplified by a gain stage that realizes a single pole low-pass filter and is finally supplied to the A/D converter of the microcontroller. Filters are sized so that the overall band of the analog front-end is in the range of 0.5 Hz to 50 Hz. The common mode unwanted signal at the input of the first stage is collected in the middle point of the gain resistance of the IA. Then it is amplified by an inverting amplifier and inserted again into the body through a third electrode in order to realize a negative feedback on the common mode disturbing signal. This feedback loop is historically called "Right Leg Drive" and allows us to substantially reduce the noise caused by the pairing of the patient with many sources of disturbing signals such as power lines.

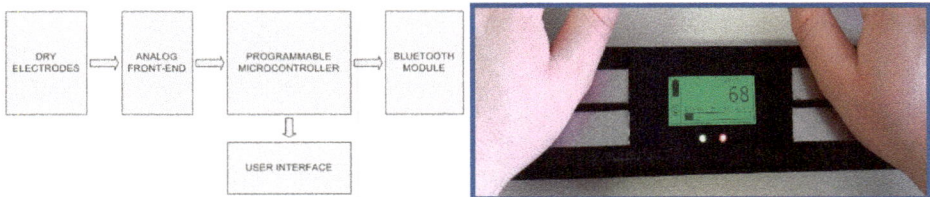

Figure 10. Architecture of the single-lead ECG biomedical device (**left**) and its implementation (**right**).

An eight-bit RISC microcontroller evaluation board, the Atmel XMEGA-A3BU Xplained, was used to realize the sampling, the digital elaboration, and the control of the system. The firmware performs a fine-grained filtering of the ECG signal in the digital domain, provides the elaborated samples to the Bluetooth 2.0 module with SPP profile, and manages the user interface. Moreover, an algorithm calculates the heart rate. The sampling rate is 500 sps and the filter applied is a highly selective FIR filter, built with 516 coefficients, having a passband ranging from 0.6 Hz to 37 Hz, and presenting a notch response at 50 Hz (i.e., the power line frequency in Europe; for other nations the filter response can be moved to 60 Hz). This filter achieves a clean signal in a frequency range defined "monitoring band" that is suitable for ECG monitoring. The lower band limit of the filter is as low as possible in order to allow the low-frequency components of the ECG signal to pass, but high enough to attenuate the oscillations caused by the hand–electrode contact impedance variations. The heart rate is calculated using the method of thresholding of the energy signal. In order to highlight the QRS complex, the ECG signal is filtered again with a 15-Hz low-pass FIR filter and the resulting samples are squared, obtaining a power signal. The power signal is then compared with a threshold that is proportional to the mean value of the power signal itself over a time window, which represents the local energy of the signal. When the threshold is exceeded, a QRS complex is detected and the time distance between near QRS complexes allows us to calculate the heart rate. The microcontroller uses an UART port to send to the Bluetooth module the filtered ECG samples to be forwarded to the gateway using the SPP profile. The module can be configured in two working modes by pressing one of two buttons while powering on the ECG device. In slave mode the module always waits for Bluetooth connection requests from the gateway, whereas in the master mode the module itself sends connection requests when a previously paired device is found. When a button is pressed the paired list is cleared and the device always waits for a pairing request. Pairing procedure requires a PIN. The master mode allows the user to keep the gateway always on and to establish an automatic connection just by powering on the ECG device. Once connected, the firmware of the microcontroller provides autonomy and transparency to send the ECG samples to the gateway. The microcontroller also drives the user interface, consisting of four LEDs indicating the status of the device and a LCD that shows the heart rate value and the progress of the acquisition.

For the overall power consumption let us consider a current consumption of about 100 mA, drained from a supply voltage of 3.3 V. The 4-AA battery pack ensures more than 15 h of continuous data streaming, which, considering its typical use in a telemedicine service (i.e., one measurement per day), means approximately six months of autonomy.

5.2.4. Testing of a Prototype Single-Lead ECG Device

The first prototype was realized and its performance in terms of user-friendliness and ECG quality was analyzed in an experimental trial with tens of testers, among them 12 elderly people with an average age higher than 60. While simplicity is important to ensure usability by all patients, ergonomics also has a relevant role in the quality of the ECG signal because it influences the stability of the position of the patient during the acquisition. After receiving a few instructions, all the testers were able to use the device without any help and reported as feedback that the position of the hands was comfortable and easy to keep for some minutes. Figures 10 and 11 show the hands of a tester on the prototype and the related ECG acquired. Another test campaign was done for comparing the quality of the ECG obtained using the prototype with the ECG obtained using the Lifepak 15 configured in the "monitor band" mode (i.e., 0.5 Hz–40 Hz). Figure 11 shows the result obtained, acquiring an ECG simultaneously on the same tester using both the prototype (by placing the hands on the electrodes) and the Lifepak 15 (by attaching the adhesive electrodes on the chest). The black trace on the graph paper in Figure 11 is the printed output of the Lifepak, whereas the blue trace is the prototype data using the new single-lead ECG device. The first two leads obtained at the same time with the two different devices are very similar and overlap almost perfectly. The minimal worsening of the signal quality, mainly due to the limits imposed by the skin–electrode contact set-up, is acceptable considering the significant improvements to the device usability.

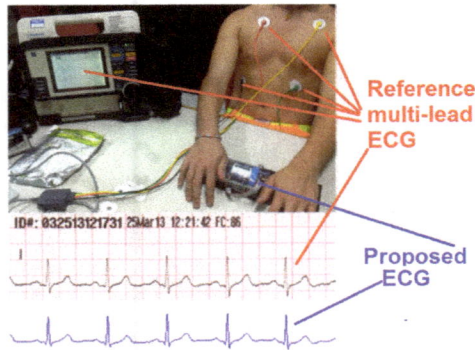

Figure 11. Comparison of the ECG acquisition with the new single-lead ECG biomedical device (blue trace) vs. a golden reference multi-lead ECG instrumentation (black trace). Electrodes placed on the chest are needed only for the reference ECG. The proposed ECG device just requires placing the hands, without any conductive gel, on top of the ECG device, where a couple of electrodes are placed.

Differently from [17], the newly developed single-lead ECG sensor can also be used as a stand-alone personal care device in connection with a smartphone or tablet acting as a storage and display unit (see Figure 12; the interface, in Italian in this specific example, can be regionalized according to the device language settings). Through the BT connection the acquired signal is transmitted to an Android OS-based terminal where a custom-developed app exploits the terminal processing and memory hardware resources to filter the signal in the digital domain, displays the ECG trace on the LCD screen, and allows us to build a repository of recorded ECG traces. Table 2 summarizes the main characteristics of the single-lead ECG device.

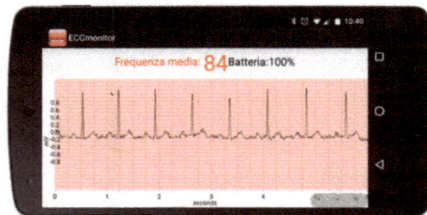

Figure 12. Single-lead ECG application running on an Android device.

Table 2. Characteristics of the single-lead ECG.

Dimensions (mm)	Sampling	Analog Band (Hz)	Digital Band (Hz)	Notch Filter (Hz)	ECG Trace	Processing	Outputs
300 × 70 × 40	12 bit 500 sps	0.5–50	0.6–37	50	First lead	Heart rate detection	LCD display 128 × 64 pixel Green and red LEDs BT 2.0 SPP 115 kbps

5.3. Multi-Channel Biomedical ASIC Sensor

5.3.1. General Architecture of the Biomedical ASIC

As discussed in Section 5.1, an ASIC has been designed and manufactured in AMS CMOS 2 Metal Layers 2 Poly technology, to ensure multi-sensor integration in the same wearable and portable medical device. The ASIC is assembled in a $14 \times 14 \times 1.4$ mm^3 TQFP (Thin Quad Flat Package) with 128 pins and used at 3.3 V. The current drained from the power supply is 10 mA in typical operating mode (i.e., 33 mW power consumption). Its operating temperature range is from 0 °C to 70 °C, which is fully compliant with use at home. It ensures multiparametric biomedical signal acquisition. The ASIC has been designed in the framework of the Health@Home European (EU) project in collaboration with CAEN spa, taking into account the typical constraints of biomedical signals monitoring like multi-lead ECG, blood pressure monitoring, and body temperature. With reference to the ASIC functional block in Figure 13, the main building blocks of the ASICs are described hereafter. A fully configurable multi-channel ECG block is present, with eight input differential channels (e.g., 16 input pins) for signal conditioning including amplification, filtering, and offset regulation for 3-5-12-leads ECG systems. There is also an adder generating the Central Terminal Point (CTP) for precordial leads, a right leg (RL) driver, and a SHIELD driver to reduce the Common Mode 50 Hz noise. A pacemaker detector section allows for pace pulses detection on a dedicated pin. All is in conformity with the IEC 60601-2-51 2003-02 standard. There is also one analog channel designed for decoupling and amplifying a blood pressure signal coming from an optional external pressure Sensor. Another input analog channel is used to process a temperature signal provided by an external temperature sensor. One programmable analog MUX allows for switching among the channels to be converted by the ADC. A 16-bit SAR (Successive Approximation Register) ADC is used to convert the voltages from the analog channels. The relevant Effective Number of Bits (ENOB) is 12. A serial peripheral interface (SPI) [41] is used to configure the chip settings and for data readout. A dedicated battery channel is present to monitor the battery status. On-chip diagnostic features, as in [43], are also integrated in the biomedical ASIC.

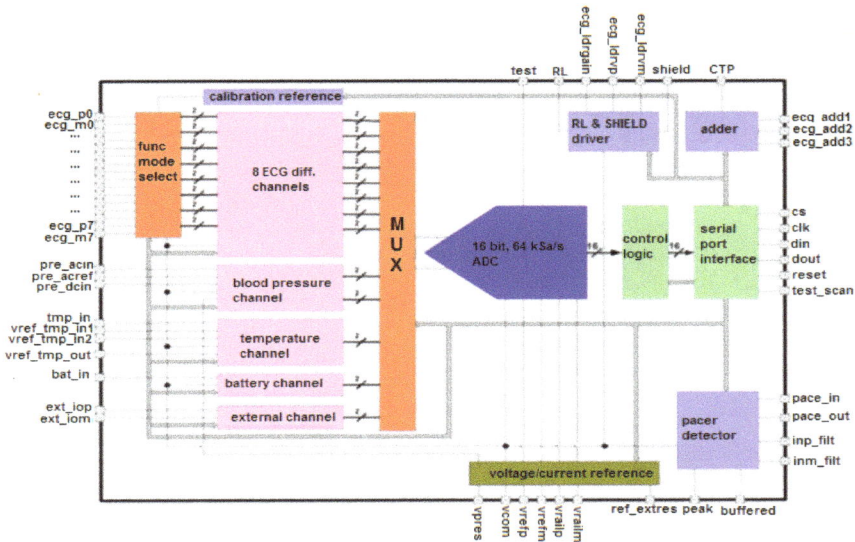

Figure 13. Multi-channel front-end of the biomedical ASIC.

5.3.2. Analysis of Specific ASIC Channels

Hereafter we describe the architecture of the specific channels for the monitoring of ECG, blood pressure, temperature, and device battery status.

Each single ECG channel is composed of an IA with high CMRR, a programmable gain amplifier (PGA), a BUFFER, and an offset regulator block. The IA has a high CMRR (100 dB typical; 92 dB minimum) in order to reduce environmental electric interferences, like the 50 Hz noise from the industrial network, always present in both electrodes that are connected with human body, and the ground. The power supply rejection ratio (PSRR) is 100 dB typical, 96 dB minimum. A first-order high-pass filter, obtained through an external capacitor, removes the low-frequency baseline wandering that is so common in ECG circuits (usually due to electrodes). An anti-aliasing filter is obtained with an external RC network. A PGA is used to provide four gains, 18, 24, 36, and 48, for standard ECG measurement. The IA has a gain of 15.6 set with an external resistor of 1.2 MΩ. Therefore, the total gain can be configured up to roughly 700. The third stage is a buffer section with fast settling that sends out the signal to the ADC. Each ECG channel has some switching MUXs placed in the input section, and is configurable via SPI command. The architecture of the ECG channel implements a baseline, therefore quickly rectifying any time variations in the signals due to artifacts that cause a temporary saturation of the IA. Offset regulator and gains regulator procedures are also available for each channel. If the ECG measurement is chosen, a cyclic procedure is activated, and the analog MUX connects this section to the ADC. Within 100 µs (worst case), all the channels are processed. Each ECG processing channel has also been characterized in terms of noise: the input referred noise measured in the range 0.1 Hz to 150 Hz is within 10 µVrms.

The Adder block is used to obtain the CTP signal reference to be used as inverting input for each channel that receives a pre-cordial signal. A driven RL circuit (RL Driver block) helps to set the common mode, and is safer than connecting the RL to voltage reference. The circuitry, able to drive in active mode the shield of the ECG cables (SHIELD Driver block), helps reduce the Common Mode 50 Hz noise.

A pacemaker detector block is also integrated in the ASIC to provide band pass filtering on the ECG signal, full wave rectification for detecting pacemaker pulses of either polarities, peak detection on the filtered and rectified signals, and discrimination relative to a programmable threshold level. Once a pacemaker pulse is detected, it provides on the dedicated output digital pin a pulse whose duration is configurable via SPI.

The Blood Pressure section includes two channels. One is an analog channel that converts a DC single-ended, ground-referred voltage to a differential voltage suitable to be converted by the on-chip ADC; it amplifies this signal. Input signal ranges between 0.5 V (0 mmHg) and 1.7 V (300 mmHg). The offset voltage for single-ended/differential conversion is provided by the internal references generator block. Another analog channel amplifies the AC component of the input pressure signal and translates the DC component around a reference voltage. Extraction of the AC component is assured by two external capacitors and one external resistor. If the BP measurement is chosen, a cyclic procedure is activated, and the MUX connects this section to the ADC.

The skin temperature measurement channel uses an NTC resistor as input sensor: the NTC resistor has a nonlinear voltage–temperature characteristic so an additional circuit is used to improve the linearity of the response in the range of the acquisition system from 32 °C to 42 °C, including body temperature.

The ADC structure used in this design is a SAR architecture, made by three main blocks: a sample and hold stage, a comparators stage, and a residue-multiplying DAC stage. ADC needs 16 clock cycles (1 MHz) to produce a 16-bit output code. The ADC has an INL of ±1 LSB and a DNL of 0.75 LSB (worst case). The offset and error gain of the ADC are below 0.5% and 1.5% of the full scale range, respectively.

Finally, the battery channel is used to monitor the battery status. Integrated bandgap circuits are used to internally generate reference levels from 1.1 V to 3.5 V.

In the experimental trials in Section 6, the biomedical ASIC has been configured so that the ECG is acquired involving three channels and the RL driver. The blood pressure channel and the temperature monitoring are also used in some configurations. Table 3 summarizes the characteristics of the biomedical ASIC.

Table 3. Characteristics of the biomedical ASIC.

Tech.	Temp. V Supply	ADC	Channels	ECG Channel	Gateway Connection
CMOS 0.8 µm	0 to 70 °C	16 b 64 kSa/s	8 ECG, 1 blood press	100 dB CMRR, 57 dB gain	BT 2.0 SPP
TQFP128 pin	3 V to 5 V	INL/DNL1/0.75 LSB	1 temp., 1 batt status	100 dB PSRR, 0.1 to 150 Hz	
$14 \times 14 \times 1.4$ mm^3		offset/gain error 0.5/1.5%	Pacer detect,	10 µVrms/150 Hz input noise	

5.4. Motion Sensor for Fall and Posture Detection

As discussed in Section 5.1, a custom mobility sensor to study body movement has been developed in the framework of the EU-Tuscany Region FESR project RIS (Research and Innovation in Healthcare Systems) in collaboration with TD Nuove Tecnologie S.p.a. In more detail, the newly developed sensor is able to extract the following parameters: fall detection, static detection (no movement is detected), step detection, and stride estimation. The sensor has been designed to be small enough and have a low enough weight that it can be worn, also for medium-term periods, without any impairment of normal activities for the patient. The sensor in Figure 14 has been implemented on a small PCB, equipped with System in Package (SiP) device containing a nine-axis MEMS Inertial Measurement Units (IMU). The sensor contains a 3-axial gyroscope plus a three-axis accelerometer and a three-axis digital compass. To fuse the information coming from the three sensors avoiding realignment or calibration issues, proper digital signal processing algorithms are implemented in real time on a 32-bit ARM Cortex-M processor. Two built-in algorithms run on the device: the first computing the step detection, stride estimation, and energy consumption; the second determining falls or the absence of movement. Communication with the gateway is based on a Bluetooth 4.0 LE chipset, integrated into the device. The motion sensor can be supplied at 3.3 V draining a current of few mA in normal operating mode. The onboard Li-Po rechargeable battery (3.7 V, 240 mAh) ensures about 80 h of continuous data streaming. Its size is comparable to the size of a 50 euro cent coin.

Figure 14. Newly developed motion sensor.

The algorithm implemented for step detection consists of four main stages. In the first stage, the magnitude of the acceleration *ai* for each sample *i*, captured by the accelerometer, is computed. In the second stage, the local acceleration variance is computed to remove gravity. The third stage uses two thresholds: the first (*T1*) is applied to detect the swing phase, whereas the second (*T2*) is applied to detect the stance phase (*B2i*) in a single step while walking. The fourth stage is detected in sample *i* when a swing phase ends and a stance phase starts. Estimating the Stride Length (SL) is necessary, at every detected step, in order to calculate the total forward movement of a person while walking. Here, SL depends on the person, his/her leg length and walking speed, and the nature of the movements during walking, etc. The algorithm proposed by Weinberg [20] assumes that SL is proportional to the bounce, or vertical movement, of the human hip. This hip bounce is estimated from the largest acceleration differences at each step. The algorithm implemented for SL estimation consists of the following steps. The first step computes the magnitude of accelerations a_i. The second step estimates the SL using the Weisberg expression in Equation (1) [19], where the maximum and minimum operations are applied over the filtered accelerations in a window of size $2w+1$ around the sample $i(p)$ corresponding to the p stance detection. In Equation (1) K is a constant that has to be selected experimentally or calibrated. If the length *SL*, estimated by the method above, and the frequency of the *step* is known, it is possible to derive the velocity of each step as the ratio of the two sizes.

$$SL = K \cdot \sqrt[4]{max_{j=(ip\pm w)}^{\tilde{a}_j} - min_{j=(ip\pm w)}^{\tilde{a}_j}} \tag{1}$$

If the length *SL*, estimated by the method above, and the frequency of the step is known, it is possible to derive the velocity of each step as the ratio of the two sizes. The algorithm implemented to determine a fall is based on the controls of the thresholds. A fall-like event is defined as an acceleration peak of magnitude greater than 3 g followed by a period of 2500 ms without further peaks exceeding the threshold. The accelerometer-sampling rate has been set at 50 Hz, a trade-off between resolution and power consumption. Threshold values around 3 g (ranging from 2.5 g to 3.5 g) have been widely used in other fall detection systems [22]. The value 3 g is small enough to avoid false negatives, since real falls are likely to present an acceleration pattern containing a peak that exceeds such a value. Several sensor placements have been already tested, e.g., the waist, trunk, leg, hip, and foot. From our test, although data from all locations provided similar levels of accuracy, the hip was the best single location for recording data for activity detection. It provides better accuracy than the other investigated placements [23]. This location is optimal for the implementation of more efficient algorithms, as it allows a cleaner signal from the IMU. However, the exact position and orientation of the platform on the hip are not important, because many algorithms only work with the magnitude of sensor readings. Table 4 summarizes the characteristics of the motion sensor.

Table 4. Characteristics of the motion sensor.

Sampling Rare	V Supply	Sensor	Digital Core	Processing	Output
50 sps	3.3 V	9D IMU (gyroscope, accelerometer, digital compass)	STM32	Fall event detection Posture detection	BLE 4.0

6. Experimental Trials

6.1. Experimental Trial Projects and Motivations

The proposed telemedicine platform, comprising remote monitoring kits (i.e., biomedical sensors and gateways), service center applications, and Electronic Health Records, has been developed and improved in the framework of different funded projects (ongoing or recently closed).

Health@Home EU Ambient Assisted Living project: remote monitoring of CHF patients recently dismissed from hospital through self-measurement of the main vital signs.

Domino project, funded by Arezzo's local health authority: remote monitoring of chronic patients (mainly cardiac) with periodic in-house visits performed by nurses and circulation of clinical data among all personnel involved in the patients' care (e.g., family doctors, specialists, etc.).

RIS (Research and Innovation in Healthcare Systems) EU-Tuscany Region FESR project: personalized and integrated remote monitoring of chronic patients, connecting in-hospital care and out-of-hospital follow-up based on the "1:1" scenario (self-measurements) or "1:N" scenario (measurements made by professional caregivers).

RACE (Research and Accuracy in Cardiology based on Evidence of clinical data) EU-Tuscany Region FESR project: long-term trial with health economic analysis in order to evaluate the impact of the telemedicine model on the economic burden of the PHS.

All these projects included a trial phase in the real-world telemedicine scenario with the involvement of patients and medical personnel. An exhaustive in-field demonstration phase is crucial to assess and improve the effectiveness of telemedicine platforms. The primary outcomes of our validations were, from a technical point of view, the usability of the system and its affordability, whereas from the medical point of view they were the quality of the clinical data and the ability to improve patient care.

6.2. Health@Home Experimental Trial Results

In the trial prepared for the Health@Home project, the platform was used according to the "1:1" scenario. The validation of the telemedicine platform was implemented involving 30 patients (average age of 62 years) affected by CHF coming from three different hospitals (Hospitales Universitarios Virgen del Rocio in Spain, Zdravstveni Dom Koper in Slovenia, and Fondazione Toscana Gabriele Monasterio in Italy), under the supervision of two doctors specializing in cardiology from each hospital. The minimum monitoring period was three months. Inclusion criteria included diagnosis of heart failure, New York Heart Association (NYHA) classes III and IV, at least one hospitalization for acute heart failure in the previous six months, and agreement to take part in the study. Acute coronary syndrome within three months before the enrollment was the only exclusion criterion. Patients were enrolled in the study at time of discharge from the hospital where they were admitted for acute heart failure or during a routine ambulatory visit. At enrollment time they received a monitoring kit with digital scale, blood pressure monitor, oximeter, ECG device, telemedicine gateway, and brief training on how to use the system. Both a technical and a medical contact were available for patients during the trial.

In order to validate the platform, specialist doctors were asked to check the information that arrived to HIS, evaluating the quality and coherence of the data collected and the relevance of the alarms automatically generated by the platform. Moreover, a specific questionnaire was developed to gather feedback from patients, caregivers, family, and specialist doctors about the end-user usability. The robustness of data storage and data transmission was also evaluated.

The results show a very limited number of activity misses (<3%), mostly in the first days of monitoring, also confirming the property of such a telemedicine system to improve the therapy compliance. Additionally, the number of false positive alarms is less than 5%. No connectivity, storage, and transmission problems, including data loss, occurred. All end-users reported valid impressions of the platform and a good satisfaction level in the final questionnaire. Table 5 and Figure 15 show the scores reached in each macro-parameter for medical staff and patients, respectively. Medical personnel reported that the use of this platform does not impinge on their regular activity, while it represents a valid means of controlling at a distance the progress of their patients thanks to the high quality of acquired signals and alarm detection capability. All specialist doctors are definitively in favor of the adoption of the platform. In addition, 89% of patients report a very high satisfaction level, highlighting the friendliness of the solution and the ease of following the daily therapy.

Table 5. Aggregated feedback from specialist doctors.

Macro-Parameter	Score
Simply decision and increase effectiveness of diagnosis and treatment of patient based on better evidence	9,125/10
In general terms, easy to use with clear and understandable interactions	9,5/10
Flexibility of the system and compatibility with other systems already in use	9,75/10
Quality of the provided signal	9,1/10
Sensibility of the alarm detection function	9,15/10
In favor of the adoption of the H@H system	9,2/10

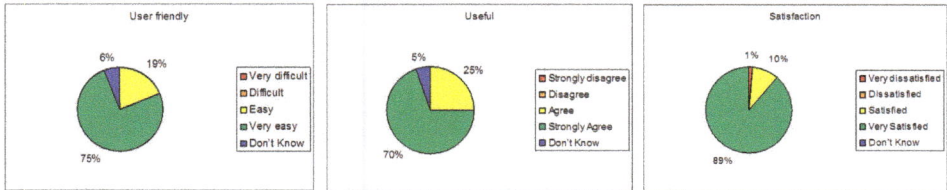

Figure 15. Aggregated feedback from patients.

6.3. Domino Experimental Trial Results

The trial carried out in the Domino project exploited the potential of the presented platform for the provisioning of a telemedicine service according to the "1:N" scenario. In particular, under the supervision of the Local Health Authority of the city of Arezzo in Italy (ASL8), a group of 50 consenting chronic patients mainly affected by cardio-respiratory diseases (e.g., CHF, COPD) and already following the Chronic Care Model procedures was monitored for a period of six months. According to the "1:N" scenario, the proposed telemedicine equipment was used by two nurses allocated by the ASL8 to this trial during the domiciliary visits. In this study, a telemedicine kit with a gateway and a set of biomedical sensors (i.e., 12-lead ECG device, oximeter digital scale, blood pressure monitor, glucometer, and spirometer) was provided to each nurse in order to collect vital signs during the domiciliary visits scheduled according to the personalized healthcare plan of each patient. This ICT-assisted method allowed an almost real-time distribution of the acquired vital signs to all people interested in the patient's care thanks to the automatic transmission of data directly after the acquisition from the gateway to the online platform.

In order to evaluate the platform in term of usability, impact with respect to the traditional method, and capacity to improve the management of the chronic patients, medical personnel involved in the pilot were interviewed at the end of the period. Additionally, the robustness of data storage and data transmission was evaluated.

From a technical point of view, no data loss occurred during the trial. In some cases real-time transmission of the collected data was not possible due to missing availability of mobile broadband connectivity. The main benefits of the "1: N" solution, highlighted by the medical staff, are:

- Synchronization of the vital signs of the patients as soon as they are acquired in the e-Health center, through an automatic procedure that minimizes the possibility of errors due to manual insertion of values.
- Clinical information sharing among the family doctor, the specialists, and the rest of the caregivers, without any time or distance barriers.
- Reduction of the duration of the domiciliary visits, and a better scheduling of the work flows.

Table 6 summarizes the interviews with the nurses who used the telemedicine kit.

Table 6. Nurses interviewed in Domino project.

"The patient selection process and the measurement acquisition process are quickly and easy to use"
"The medical devices provided with the gateway cover acquisition of all the requested vital signs"
"The telemedicine kit, composed of a gateway and medical devices, represents an important improvement and optimization of the domiciliary visit with respect to the traditional model"

6.4. RIS Experimental Trial Results

In the framework of the RIS project, the complete telemedicine platform described in previous sections was extensively tested. The rationale was to test the ability of the platform to provide effective telemedicine services involving in an integrated and coordinated fashion the hospitals, the territorial services (i.e., family doctors, local Health Authorities, etc.) the clinical specialists, and in general all the caregivers and stakeholder of the healthcare system. For these reasons, all paradigms described in Section 3 were used in the pilot: self-measurements performed by the patients, domiciliary visits assigned to nurses and health corners in public places (e.g., pharmacies, medical centers, residences for elderly people, etc.). Moreover, the platform enhanced the role of the service center as a central point of the architecture for the circulation, management, and sharing of clinical information. Operators received specific training in how to manage the data coming from the telemedicine systems and how to handle alarms. A group of 10 chronic patients, with CHF as the main complaint but also affected by some comorbidities such as diabetes, was monitored for at least one month according to a structured and personalized healthcare program that included self-measurements and periodic (i.e., weekly) in-house visits.

At the end of the trial, the overall platform was evaluated through questionnaires and direct interviews. The metrics established for the evaluation of the system belong to two main categories: objective and subjective. The first are related to items that are unequivocally measurable; the latter depend on the personal experience during the demonstration and the individual's feeling about the system. The quality of the vital signs, self-measured by the patients or collected by the nurses allocated to domiciliary visits, was confirmed by specialist doctors, who also assessed the capability of the platform to improve the treatment of chronic patients in terms of therapy compliance, clinical outcomes, alarm situation handling, and better allocation of medical resources. The platform, as already assessed in the previous pilots, confirmed its robustness and flexibility, avoiding data loss and ensuring secure circulation of clinical information.

7. Conclusions

This paper has presented the implementation and experimental verification of a remote monitoring system including the whole value chain from the top (health care model) down to the technical implementations of sensors, data acquisition, processing, a communication platform (gateway), and integration with a service center and HIS. The proposed system is scalable in different telemedicine scenarios involving different roles for all the actors in a health system (patients, their family, nurses, doctors, institutions, local health corners, call center operators): to this end the "1:1," "1:N," and "point of care" scenarios are presented and discussed. A mixture of commercially available sensors and new custom ones are presented and used. The new custom-designed sensors range from a single-lead ECG for easy self-measurements at home, to a multi-channel biomedical ASIC for acquisition of multi-channel bio signals (e.g., ECG, EEG, EMG), to a new motion sensor for patient posture estimation and fall detection. Specific focus has been placed on aspects such as the user interface and easy use of the device for non-professional users. From a communication point of view, BT and BLE wireless PAN links are used between sensors and gateway, whereas wireless LAN or wired technologies are used between the gateway and the remote server. All data can be transferred though several protocols, including HL7 CDA. Experimental trials in real-world telemedicine applications assess the proposed system in terms of easy usability for patients, family members, specialist doctors, and caregivers; in terms of scalability in different scenarios; and in terms of suitability for the implementation of needed care plans.

Since the number and type of patients involved in the reported experimental trials (see Sections 6.2–6.4) are still not enough to claim a complete survey and reliable statistics, there is ongoing experimental activity. The aim of the ongoing activity is also to validate the efficacy, in terms of economic advantages and impact on the conventional organization of people and infrastructures, of the telemedicine scenarios for the PHS. The psychological acceptance of the telemedicine model from a large part of the population should also be validated. To this end, the telemedicine system can also be supported by a remote videophone service, exploiting compression technologies we already developed in [44–46]—thus the patient can also "see" the doctor. Moreover, ongoing activity is focused on implementing a reliable, automatic check of the quality of the acquired biomedical measurements. Currently, an automatic check is done by the system but with a coarse grain (e.g., if acquired data or their difference with previous acquisitions are above or under some specific thresholds), whereas a fine-grain analysis of the quality of the acquired signal is left to the subjective analysis of the professional caregiver. Finally, in the current implementation the alarms generated by the gateway must be first validated by the caregiver before an emergency plan is activated and the patient is re-hospitalized. This increases the latency of reaction to an alarm, which can be critical particularly if the patient's home is far from a hospital (e.g., mountainous zones and/or regions with low population density). In the future, when reliable automatic diagnosis of the quality of the measurements is reached and the system has been tested on a large population of patients with reliable statistics, automatic activation of the emergency plan can be instituted.

Acknowledgments: The authors gratefully acknowledge discussions with L. Benini from the University of Pisa, C. Passino from the Fondazione Toscana Gabriele Monasterio, and P. Barba from CAEN S.p.a.

Author Contributions: S.S. and L.F. defined the specifications and conceived the architecture of the telemedicine platform, whereas M.D. and A.C. managed its implementation (hardware and software) and testing.

Conflicts of Interest: The authors declare no conflict of interest.

References

1. Go, S.A.; Mozaffarian, D.; Roger, V.L.; Benjamin, E.J.; Berry, J.D.; Borden, W.B.; Bravata, D.M.; Dai, S.; Ford, E.S.; Fox, C.S.; et al. Heart disease and stroke statistics 2013 update: A report from the American Heart Association Statistics Committee and Stroke Statistics Subcommittee. *Circulation* **2013**, *127*, 6–245. [CrossRef] [PubMed]

2. Berry, C.; Murdoch, D.R.; McMurray, J.J. Economics of chronic heart failure. *Eur. J. Heart Fail.* **2001**, *3*, 283–291. [CrossRef]

3. Masella, C.; Zanaboni, P.; Borghi, G.; Castelli, A.; Marzegalli, M.; Tridico, C. Introduction of a telemonitoring service for patients affected by chronic heart failure. In Proceedings of the 11th International Conference on e-Health Networking, Applications and Services, Sydney, Australia, 16–18 December 2009; pp. 138–145.

4. Fanucci, L.; Saponara, S.; Bacchillone, T.; Donati, M.; Barba, P.; Sanchex-Tato, I.; Carmona, C. Sensing Devices and Sensor Signal Processing for Remote Monitoring of Vital Signs in CHF Patients. *IEEE Trans. Instrum. Meas.* **2013**, *62*, 553–569. [CrossRef]

5. Parissis, J.; Athanasakis, K.; Farmakis, D.; Boubouchairopoulou, N.; Mareti, C.; Bistola, V.; Ikonomidis, I.; Kyriopoulos, J.; Filippatos, G.; Lekakis, J. Determinants of the direct cost of heart failure hospitalization in a public tertiary hospital. *Int. J. Cardiol.* **2015**, *180*, 46–49. [CrossRef] [PubMed]

6. McMurray, J.J.V.; Adamopoulos, S.; Anker, S.D.; Auricchio, A.; Böhm, M.; Dickstein, K.; Falk, V.; Filippatos, G.; Fonseca, C.; Gomez-Sanchez, M.A.; et al. ESC guidelines for diagnosis and treatment of acute and chronic heart failure 2012. *Eur. Heart J.* **2012**, *14*, 803–869. [CrossRef]

7. Go, A.S.; Mozaffarian, D.; Roger, V.L.; Benjamin, E.J.; Berry, J.D.; Blaha, M.J.; Dai, S.; Ford, E.S.; Fox, C.S.; Franco, S.; et al. Heart disease and stroke statistics, 2014 Update. *Circulation* **2014**, *129*, 28–292. [CrossRef] [PubMed]

8. Zannad, F.; Agrinier, N.; Alla, F. Heart failure burden and therapy. *Europace* **2009**, *11*, 1–9. [CrossRef] [PubMed]

9. Gardner, R.S.; McDonagh, T.A. Chronic heart failure: Epidemiology, investigation and management. *Medicine* **2014**, *42*, 562–567. [CrossRef]

10. Ambrosy, A.P.; Ambrosy, A.P.; Fonarow, G.C.; Butler, J.; Chioncel, O.; Greene, S.J.; Vaduganathan, M.; Nodari, S.; Lam, C.S.; Sato, N.; et al. The global health and economic burden of hospitalization for heart failure. *J. Am. Cardiol.* **2014**, *63*, 1123–1133. [CrossRef] [PubMed]

11. Lee, W.C.; Chavez, Y.E.; Baker, T.; Luce, B.R. Economic burden of heart failure: A summary of recent literature. *Heart Lung* **2004**, *33*, 362–371. [CrossRef] [PubMed]

12. Fergenbaum, J.; Bermingham, S.; Krahn, M.; Alter, D.; Demers, C. Care in the home for the management of chronic heart failure. *Cardiovasc. Nurs. J.* **2015**, *30*, 44–51. [CrossRef] [PubMed]

13. Ekeland, A.G.; Bowes, A.; Flottorp, S. Effectiveness of telemedicine: A systematic review of reviews. *Int. J. Med. Inform.* **2010**, *79*, 736–771. [CrossRef] [PubMed]

14. Finet, P.; Le Bouquin Jeannès, R.; Damerond, O.; Gibaudb, B. Review of current telemedicine applications for chronic diseases. *IRBM* **2015**, *36*, 133–157. [CrossRef]

15. Kligfield, P.; Gettes, L.S.; Bailey, J.J.; Childers, R.; Deal, B.J.; Hancock, E.W.; van Herpen, G.; Kors, J.A.; Macfarlane, P.; Mirvis, D.M.; et al. Recommendations for the standardization and interpretation of electrocardiogram. *Circulation* **2007**, *155*, 1306–1324.

16. Lopez Sendon, J. The heart failure epidemic. *Medicographia* **2011**, *33*, 363–368.

17. Benini, A.; Donati, M.; Iacopetti, F.; Fanucci, L. User-friendly single-lead ECG device for home telemonitoring application. In Proceedings of the 2014 8th International Symposium on Medical Information and Communication Technology (ISMICT), Firenze, Italy, 2–4 April 2014.

18. Donati, M.; Bacchillone, T.; Fanucci, L.; Saponara, S.; Cstalli, F. Operating protocol and networking issues of a telemedicine platform integrating from wireless home sensors to the hospital information system. *Hindawi J. Comput. Netw. Commun.* **2013**, *2013*, 1–12. [CrossRef]

19. Li, J.-F.; Wang, Q.-H.; Liu, X.-M.; Cao, S.; Liu, F.-L. Pedestrian Dead Reckoning System Integrating Low-Cost MEMS Inertial Sensors and GPS Receiver. *J. Eng. Sci. Technol. Rev.* **2014**, *7*, 197–203.

20. Weinberg, H. *Using the ADXL202 in Pedometer and Personal Navigation Applications*; AD AN-602 App. Note; Analog Devices: Norwood, MA, USA, 2002.

21. Abbate, S.; Avvenuti, M.; Bonatesta, F.; Cola, G.; Corsini, P.; Vecchio, A. A smartphone-based fall detection system. *Pervasive Mobile Comp.* **2012**, *8*, 883–899. [CrossRef]

22. Bourke, A.K.; O'Brein, J.V.; Lyons, G.M. Evaluation of a threshold-based tri-axial accelerometer fall detection algorithm. *Gait Posture* **2007**, *26*, 194–199. [CrossRef] [PubMed]

23. Cleland, I.; Kikhia, B.; Nugent, C.; Boytsov, A.; Hallberg, J.; Synnes, K.; McClean, S.; Finlay, D. Optimal placement of accelerometers for the detection of everyday activities. *Sensors* **2013**, *13*, 9183–9200. [CrossRef] [PubMed]

24. World Health Organization. *mHealth: New Horizons for Health through Mobile Technologies: Second Global Survey oneHealth*; WHO: Geneva, Switzerland, 2011; p. 1.

25. Bacchillone, T.; Donati, M.; Saponara, S.; Fanucci, L. A flexible home gateway system for telecare of patients affected by chronic heart failure. In Proceedings of the 2011 5th International Symposium on Medical Information and Communication Technology, Montreux, Switzerland, 27–30 March 2011; pp. 1–5.

26. Zhou, F.; Yang, H.-I.; Reyes Álamo, J.M.; Wong, J.S.; Chang, C.K. Mobile personal health care system for patients with diabetes. In *Aging Friendly Technology for Health and Independence*; Lecture Notes in Computer Science; Springer: London, UK, 2011; Volume 6159, pp. 94–101.

27. Yadav, J.; Rani, A.; Singh, V.; Murari, B.M. Near-infrared LED based non-invasive blood glucose sensor. In Proceedings of the 2014 International Conference on Signal Processing and Integrated Networks (SPIN), Noida India, 20–21 February 2014; pp. 591–594.

28. Gan, K.B.; Yahyavi, E.S.; Ismail, M.S. Contactless respiration rate measurement using optical method and empirical mode decomposition. *Technol. Health Care* **2016**, 1–8, preprint.

29. Daw, W.; Kingshott, R.; Scott, A.; Saatchi, R.; Elphick, H. Development of the 'BreathEasy' contactless portable respiratory rate monitor (CPRM). *Eur. Respir. J.* **2015**, *46*, PA1583. [CrossRef]

30. Fang, D.; Hu, J.; Wei, X.; Shao, H.; Luo, Y. A Smart Phone Healthcare Monitoring System for Oxygen Saturation and Heart Rate. In Proceedings of the IEEE International Conference on Cyber-Enabled Distributed Computing and Knowledge Discovery, Shanghai, China, 13–15 October 2014; pp. 245–247.

31. Lutze, R.; Waldhor, K.A. Smartwatch Software Architecture for Health Hazard Handling for Elderly People. In Proceedings of the IEEE International Conference on Healthcare Informatics, Dallas, TX, USA, 21–23 October 2015; pp. 356–361.

32. Stankevich, E.; Paramonov, I. Using Bluetooth on Android Platform for mHealth Development. In Proceedings of the 10th Conference of FRUCT Association, Tampere, Finland, 7–11 November 2011; pp. 140–145.

33. Fanucci, L.; Benini, A.; Donati, M.; Iacopetti, F. Self Electrocardiogram Acquisition Device. Patent N. WO2015083036 A1, 11 June 2015.

34. V-Patch Medical System. Available online: http://www.vpatchmedical.com (accessed on 18 June 2016).

35. Isansys Lifecare. Available online: http://www.isansys.com (accessed on 18 June 2016).

36. LifeWatch. Available online: http://www.lifewatch.com/ACT (accessed on 18th June 2016).

37. Docobo. Available online: http://www.docobo.co.uk (accessed on 18 June 2016).

38. Augustyniak, P.; Tadeusiewicz, R. Telemedical Solutions in Cardiac Diagnostics. In *Chapter 2 in Ubiquitous Cardiology: Emerging Wireless Telemedical Applications, Medical Information Science Reference*; IRMA book: Hershey, PA, USA, 2009; pp. 72–109.

39. SolutionMD. Available online: http://www.solutionmd.com (accessed on 18 June 2016).

40. Saponara, S.; Petri, E.; Fanucci, L.; Terreni, P. Sensor modeling, low-complexity fusion algorithms, and mixed-signal IC prototyping for gas measures in low-emission vehicles. *IEEE Trans. Instrum. Meas.* **2011**, *60*, 372–384. [CrossRef]

41. Bacciarelli, L.; Lucia, G.; Saponara, S.; Fanucci, L.; Forliti, M. Design, testing and prototyping of a software programmable I2C/SPI IP on AMBA bus. In Proceedings of the IEEE PRIME, Otranto, Italy, 12–15 June 2006; pp. 373–376.

42. Baba, A.; Burke, M.J. Electrical characterisation of dry electrodes for ECG recordings. In Proceedings of the 12th WSEAS International Conference on Circuits, Heraklion, Greece, 22–24 July 2008.

43. Costantino, N.; Serventi, R.; Tinfena, F.; D'Abramo, P.; Chassard, P.; Tisserand, P.; Saponara, S.; Fanucci, L. Design and test of an HV-CMOS intelligent power switch with integrated protections and self-diagnostic for harsh automotive applications. *IEEE Trans. Ind. Electron.* **2011**, *58*, 2715–2727. [CrossRef]

44. Saponara, S.; Martina, M.; Casula, M.; Fanucci, L.; Masera, G. Motion estimation and CABAC VLSI co-processors for real-time high-quality H.264/AVC video coding. *Microprocess. Microsyst.* **2010**, *34*, 316–328. [CrossRef]

45. Fanucci, L.; Saleti, R.; Saponara, S. Parametrized and reusable VLSI macro cells for the low-power realization of 2-D discrete-cosine-transform. *Microelectron. J.* **2001**, *32*, 1035–1045. [CrossRef]

46. Chimienti, A.; Fanucci, L.; Locatelli, R.; Saponara, S. VLSI architecture for a low-power video codec system. *Microelectron. J.* **2002**, *33*, 417–427. [CrossRef]

![electronics logo] *electronics*

MDPI

Article

A Multi-Modal Sensing Glove for Human Manual-Interaction Studies

Matteo Bianchi [1,2,*], Robert Haschke [3], Gereon Büscher [3], Simone Ciotti [1,2], Nicola Carbonaro [1] and Alessandro Tognetti [1,4,*]

[1] Research Center "E. Piaggio", University of Pisa, Largo Lucio Lazzarino 1, Pisa, Italy; simone.ciotti@ing.unipi.it (S.C.); nicola.carbonaro@centropiaggio.unipi.it (N.C.)

[2] Advanced Robotics Department, Istituto Italiano di Tecnologia, via Morego 30, 16163 Genova, Italy

[3] Cluster of Excellence Cognitive Interaction Technology, Bielefeld University, Inspiration 1, 33619 Bielefeld, Germany; rhaschke@techfak.uni-bielefeld.de (R.H.); gbuescher@techfak.uni-bielefeld.de (G.B.)

[4] Information Engineering Department, University of Pisa, via G. Caruso 16, 56122 Pisa, Italy

[*] Correspondence: matteo.bianchi@centropiaggio.unipi.it (M.B.); a.tognetti@centropiaggio.unipi.it (A.T.); Tel.: +39-50-2217050 (M.B.); +39-50-754790 (A.T.)

Academic Editors: Enzo Pasquale Scilingo and Gaetano Valenza
Received: 27 May 2016; Accepted: 15 July 2016; Published: 20 July 2016

Abstract: We present an integrated sensing glove that combines two of the most visionary wearable sensing technologies to provide both hand posture sensing and tactile pressure sensing in a unique, lightweight, and stretchable device. Namely, hand posture reconstruction employs Knitted Piezoresistive Fabrics that allows us to measure bending. From only five of these sensors (one for each finger) the full hand pose of a 19 degrees of freedom (DOF) hand model is reconstructed leveraging optimal sensor placement and estimation techniques. To this end, we exploit a-priori information of synergistic coordination patterns in grasping tasks. Tactile sensing employs a piezoresistive fabric allowing us to measure normal forces in more than 50 taxels spread over the palmar surface of the glove. We describe both sensing technologies, report on the software integration of both modalities, and describe a preliminary evaluation experiment analyzing hand postures and force patterns during grasping. Results of the reconstruction are promising and encourage us to push further our approach with potential applications in neuroscience, virtual reality, robotics and tele-operation.

Keywords: hand pose sensing; tactile/force sensing; wearable sensing; optimal design; human hand synergies

1. Introduction

The human hand is the principal means of interaction with the external world and has a crucial role in many tasks related to common daily life activities. Consequently, numerous neuro-scientific works have focussed their attention on the quantitative analysis of hand kinematics and kinetics to deepen our understanding of the motor control mechanisms underlying the remarkable manipulation skills of human hands and thus paving the way to develop more efficient human-machine interfaces and robotic hands. Many application fields can benefit from these research results, e.g., virtual reality/video-games [1,2], rehabilitation [3,4] and remote manipulation/tele-operation [5,6].

Existing hand pose reconstruction (HPR) systems are visual-based or glove-based devices that track and estimate the hand kinematics, see [7–9] for a detailed review. Visual-based HPR systems, such as the ones described in [1,2,6], are quite accurate, inexpensive and unobtrusive, but they are not suitable for ambulatory monitoring during daily life activities. In contrast, glove-based HPR devices, relying on off-the-shelf flex sensors [3,10], dielectric elastomer stretch sensors [11], optical fibers [12], or inertial measurement units [4], are intrinsically ambulatory, but they are often less usable than visual-based systems due to obtrusive wiring and rigid sensor technology that does not adapt to the dynamically changing hand shape of the users. Indeed, the human body, and the human hand in particular, have a high number of degrees of freedom (DOFs) that act on a continuously compliant structure. For this reason, stretchability and adaptability of the sensing technology are mandatory requirements for efficient and ecological monitoring of the human hand.

In addition to hand pose, tactile sensing and force control play a crucial role for successful manipulation. In an early experiment it was shown, that subjects have severe difficulties in maintaining stable grasps when their sense of touch was eliminated by local anesthesia [13]. Similarly, the lack of tactile feedback in today's robots restricts their use to highly structured environments where contact with unknown objects and humans has to be avoided by external security measures. To endow future service robots with modern tactile sensors, e.g., [14–17], and thus to enable their operation also in unstructured and unknown environments, calls for new tactile-based control approaches [18–20]. To develop those, in turn requires a deeper understanding of the interplay between kinematic and kinetic hand control in human manipulation, i.e., particularly also considering interaction forces, which will open new insights into the overall motor control processes underlying manual intelligence [21].

Several works have dealt with the development of flexible tactile sensors suitable for ambulatory hand monitoring. A common technology employs flexible printed circuit boards (PCBs) [22–24], which can be bent in one direction at a time. Cutting the film carrier, surfaces with two-dimensional curvature can be covered as well [14,25]. However, stretchable material can much better adapt to the variable human hand shape. For example, methods using conductive rubber with interwoven wiring were reported [26]. Using electrical impedance tomography, it is possible to get rid of the wiring and only use electrodes along the circumference of the conductive rubber sensor to reconstruct applied force patterns [27]. However, this method is prone to ghosting and mirroring (detecting spurious tactile locations). Optical sensors, exploiting intensity modulation based on reflection [28] or strain in optical fibers [29], require cumbersome wiring to/from light sources and thus are too bulky for unobtrusive usage. Very high spatial resolution was achieved with a sprayed-on silicone elastomer [30], but this method was restricted to single-time usage. All these tactile sensing technologies present some drawbacks, mainly related to poor robustness, scarce adaptability, or complex electronics design.

To the best of our knowledge, none of the existing monitoring systems can measure at the same time the hand pose and tactile/kinetic information in an ambulatory and unobtrusive fashion. The aim of the current work is to report about the development and preliminary assessment of a multi-modal sensing glove able to provide both kinematic and kinetic information.

The kinematic part of our prototype—the kinaesthetic glove of Section 2.1—relies on our previous achievements on the development of textile-based, flexible and stretchable electro-goniometers, described in [31–33]. Similar sensors were used to estimate the pose of a continuum soft robot [34]. Textile goniometers, besides being stretchable and well adaptable to the different body structures, demonstrated a reliable performance in angular measurement (errors below five degrees [33]). On the other hand, they are quite complex bi-layer devices that need at least six connecting wires per DOF, thus limiting their use in multi-DOF monitoring (such as the human hand). For this reason—to maintain the wearability of our prototype—we combined textile goniometers with synergy-based optimal design and reconstruction techniques, we have described in [35,36]. In particular, we have exploited human hand synergy information, i.e., inter-joint covariation patterns, to design an under-sensing glove able to reconstruct the full hand posture (of a 19-DOF hand model) from only five goniometric sensors. The sensors are placed on the hand joints according to a synergy-based optimal design and the complete hand kinematics is reconstructed leveraging synergistic information again, following the theoretical findings we reported in [36].

The kinetic part of our prototype—the tactile sensing glove described in Section 2.2—is based on our previous developments of fabric-based, flexible and stretchable tactile sensors [37,38]. The sensor is composed of multilayer, conductive and piezo-resistive fabric that allows for multi-taxel measurements. The fabric substrate makes the sensor soft, light, and conformable to natural shapes. The force measurement range is $[0.1 - 30]$ N, thus covering the range of grasping forces observed in human daily-life activities [39]. The multi-taxel sensor design mimics the distribution of mechano-receptors in the human hand, achieving spatial resolutions of less than 1 cm in the fingertips and less than 3 cm in the palm.

In the current work, we integrated the kinaesthetic and tactile sensing gloves to develop a unique hand monitoring interface. In particular, we have conceived grasping experiments where the kinaesthetic glove was worn underneath the tactile one, while both the kinematic and kinetic signals were acquired and interpreted. A dedicated software developed in ROS enables to synchronize data acquisition and implements result visualization through a realistic dedicated hand model. Preliminary experiments performed with one participant show a good level of reliability. In this manner, we also demonstrate the feasibility of the here proposed approach, which encourages us to further proceed towards a more effective hardware integration. At the same time, a thorough quantitative validation and exploration of potential application fields are envisioned.

2. Materials and Methods

2.1. Kinaesthetic Sensing

The kinaesthetic sensing glove reconstructs the hand poses associated to daily life grasping activities. The key design objectives were: (i) hand pose reconstruction according to a 19 degrees of freedom kinematic model of the human hand; (ii) real time performance; and (iii) wearability, which is a key aspect to enable the glove employment during daily life activities (e.g., rehabilitation, robotic tele-operation, entertainment, etc.).

2.1.1. Textile-Based Goniometers

We employed "e-textile" goniometers based on knitted piezoresistive fabrics (KPFs). As described in our previous studies [31,32], KPF goniometers are made of two identical piezoresistive layers that are coupled through an electrically insulating textile. As shown in Figure 1a, we fold a single KPF on the insulating layer to obtain the goniometer. The theoretical working principle assumes the flexion angle (θ), defined as the angle between the tangent planes at the sensor extremities (Figure 1b), to be proportional to the difference of the electrical resistance (ΔR) between the two layers.

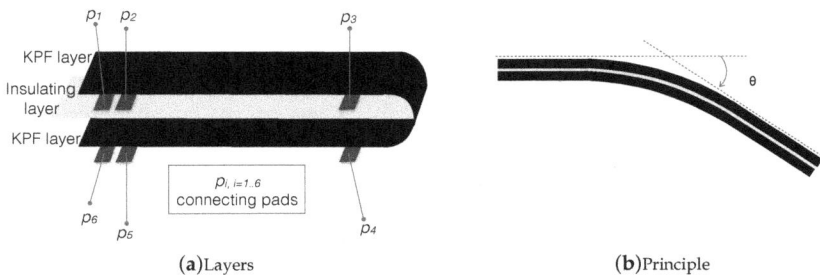

(a) Layers   **(b)**Principle

Figure 1. Textile goniometer made of two knitted piezoresistive fabric (KPF) layers. (**a**) The two conductive KPF layers (in black) are coupled through the electrically insulating stratum (in light gray). The device has six connecting pads for power supply and signal acquisition (p_i, $i = 1, ..., 6$, in dark grey); (**b**) The difference of electrical resistance (ΔR) between the two sensing layers is proportional to the flexion angle (θ), defined as the angle between the planes tangent to the sensor extremities.

As demonstrated in [31], the relationship $\theta(\Delta R)$ can be approximated using the following linear function:

$$\theta = \frac{\Delta R - \Delta R_o}{s}, \tag{1}$$

where s and ΔR_o are the angular sensitivity and offset, respectively, which can be determined in the calibration phase by measuring the sensor output in two known angular positions (0°: flat hand, palm down, 90°: closed hand (fist)). Note, that due to the differential measurement ($\Delta R = R_1 - R_2$), the sensor is agnostic to pressure and stretching, but only measures bending.

As shown in Figure 2, the goniometer can be regarded as the series of six strain-variable electrical resistances (three resistances for each layer). The resistances R_1 and R_2 represent the active layers of the goniometer and their difference $\Delta R = R_1 - R_2$ depends on the flexion angle (θ) through Equation (1). We acquire R_1 and R_2 by means of a 4-wires measuring method (to reduce the influence of the contact resistances at the pads). We supply a constant current I between the pads p_1 and p_6 and we amplify the voltage across consecutive signal pads ($V_1 = V_{p_2} - V_{p_3} = R_1 I$ and $V_2 = V_{p_4} - V_{p_5} = R_2 I$) through the instrumentation amplifiers INS_1 and INS_2. We chose instrumentation amplifiers to avoid current loss through $p_2 - p_5$ thanks to the high input impedance. We set both amplifications to the same quantity (A), thus obtaining: $V_{o1} = A R_1 I$ and $V_{o2} = A R_2 I$. Finally, the differential amplifier ($DIFF$) amplifies the difference between V_{o1} and V_{o2} to obtain the final output (V_{out}) that is proportional to ΔR. Note that the proportionality constant depends on well known values: I, A and the gain of $DIFF$. $DIFF$ can be also used to perform hardware compensation of the sensor offset.

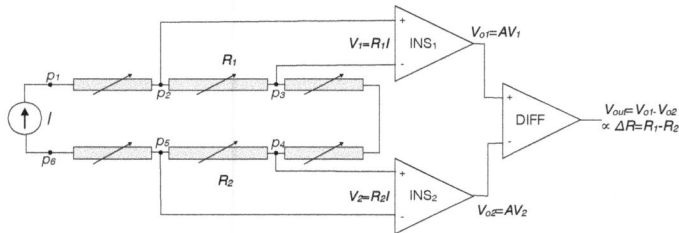

Figure 2. Electrical model of the KPF goniometer and schematic of the front-end of the acquisition electronics. The light grey boxes are strain-variable electrical resistances. A 4-wires resistance measuring method is employed: (i) a constant current (I) is supplied (pads p_1 and p_6); (ii) the voltages across the two sensing layers are measured ($p_1 - p_2$ and $p_4 - p_5$) through the two instrumentation amplifiers (INS_1 and INS_2), and (iii) the difference between the measured voltages—proportional to ΔR—is performed through the differential amplifier ($DIFF$).

In an earlier sensing glove prototype [33], we integrated three KPF goniometers to measure the flexion-extension of thumb (trapezius-metacarpal joint), index (metacarpal-phalangeal joint) and medium (metacarpal-phalangeal joint) fingers. The KPF goniometers exhibit errors below 5 degrees during natural hand opening/closing movements. Despite these promising results, that can be considered a consistent step forward in human motion detection through e-textiles, KPF goniometers cannot be easily employed in multi-DOF measurements due to the high number of connecting wires per DOF (6 pads per goniometer, see Figure 1a).

2.1.2. Glove Design

As a compromise between wearability/comfort and reconstruction performance, we have designed and engineered the kinaesthetic sensing glove, which uses only five KPF sensors optimally placed over the hand to measure five joints according to the optimal design guidelines described in [36]. We chose a five sensor design as a good trade-off between the retrieved kinematic information and the wearability of the prototype. The hand pose reconstruction can then be performed exploiting synergistic information to estimate the full hand kinematics according to a 19 DOF model. The 19 DOF kinematic model (reported in Figure 3) is partially derived from the study described in [40], augmenting the original 15 DOF model with the distal joints of the fingers. Their joint values are computed from proximal joints (P) using the relationship $\theta_D = \frac{2}{3} \cdot \theta_P$ [41].

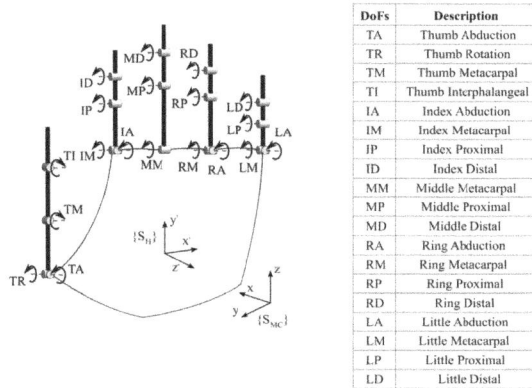

DoFs	Description
TA	Thumb Abduction
TR	Thumb Rotation
TM	Thumb Metacarpal
TI	Thumb Interphalangeal
IA	Index Abduction
IM	Index Metacarpal
IP	Index Proximal
ID	Index Distal
MM	Middle Metacarpal
MP	Middle Proximal
MD	Middle Distal
RA	Ring Abduction
RM	Ring Metacarpal
RP	Ring Proximal
RD	Ring Distal
LA	Little Abduction
LM	Little Metacarpal
LP	Little Proximal
LD	Little Distal

Figure 3. The 19 degrees of freedom (DOF) kinematic model of the human hand.

The main idea of the synergy-based approach is to exploit joint angle correlations observed in human hand poses during everyday tasks. Performing a principal component analysis (PCA) on those hand poses, it is possible to explain a huge fraction of hand motions using only a few eigen vectors (synergies), which effectively reduces the number of independently controlled degrees of freedom in the human hand. This prior knowledge on how humans most frequently use their hands can be exploited as a prior to reconstruct the most likely hand pose from only a few noisy measurements provided by any HPR device [35,42,43]. In this manner, undersensing, i.e., the usage of a number of sensors ($m = 5$) smaller than the number of DOFs ($n = 19$), becomes feasible, thus ensuring full hand reconstruction from a reduced amount of sensing elements. This aspect is particularly important to increase the wearability of the prototype.

At the same time, hand synergies can be also used to determine the sensor selection that maximizes the knowledge on the actual posture given a limited number of sensors [36,44]. For further details, the interested reader should refer to the mentioned references. However, for the sake of completeness, we will summarize the main equations in the following section.

2.1.3. Synergy-Based Hand Pose Reconstruction

Let us consider an n-DoF hand model to be reconstructed from m sensors. Assuming a linear relationship between joint variables $x \in \mathbb{R}^n$ and measurements $y \in \mathbb{R}^m$ we obtain the model:

$$y = Hx + \eta \tag{2}$$

where $H \in \mathbb{R}^{m \times n}$ is a full-row-rank matrix and $\eta \in \mathbb{R}^m$ denotes Gaussian measurement noise, with zero mean and covariance R. From a kinematic point of view, hand synergies can be defined in terms of inter-joint covariation patterns, which were observed both in free hand motion [40] and object manipulation [45]. Collecting a large number N of hand postures in a matrix $X \in \mathbb{R}^{n \times N}$, the synergy information can be summarized in the covariance matrix

$$C_o = \frac{1}{N-1} \sum_{i=1}^{N} (x_i - \bar{x})(x_i - \bar{x})^T \quad \in \mathbb{R}^{n \times n}, \tag{3}$$

where x_i and $\bar{x} \in \mathbb{R}^n$ are individual hand posture vectors and the mean hand posture vector respectively. According to [42], the hand pose reconstruction can be obtained through the minimum variance estimation (MVE) technique as:

$$\hat{x}(y) = \bar{x} - C_o H^T (H C_o H^T + R)^{-1} (H \bar{x} - y) \tag{4}$$

where the matrix $C_{map} = C_o - C_o H^T (H C_o H^T + R)^{-1} H C_o$ is the "a-posteriori" covariance matrix, that can be used as a measure of how much information an observable variable y_i carries about a joint variable x_j. In [36,44], we explored the role of the measurement matrix H on the estimation procedure and obtained as a result the optimal placement of sensors. Theoretical results were used to devise design guidelines for the kinematic sensing glove, as discussed in the next subsection.

2.1.4. Sensor Layout

Starting from the hypothesis that each goniometer measures a single hand DOF, the optimal design problem is reduced to the choice of the m DOFs (or joints) that ensure the best reconstruction performance. Assuming negligible measurement noise, the theoretical solution described in [36] with five measures (which represent a good trade-off between effectiveness and wearability) proposes to place the sensors at the following joints: TA, MM, RP, LA and LM. Then we engineered the kinaesthetic glove by sewing KPF goniometers on a Lycra glove, positioning them over the chosen joints, as shown in Figure 4. The KPF goniometers were specifically built for the specific application and, in particular, their length was specifically tailored to take into account the subject's anthropometric variability (avoiding cross-talk).

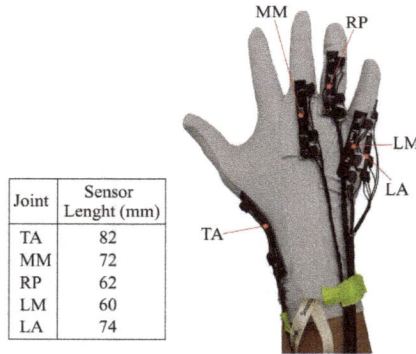

Joint	Sensor Lenght (mm)
TA	82
MM	72
RP	62
LM	60
LA	74

Figure 4. The kinaesthetic glove. According to the optimal design criteria, the five KPF goniometers are placed on TA, MM, RP, LA and LM joints of the hand.

For the acquisition phase, we designed and developed a dedicated acquisition unit. The analog front-end has five channels, one for each goniometer, which replicate the circuit reported in Figure 2. We filtered each channel with a low-pass (anti-aliasing) filter with 10 Hz cut-off frequency. An "Arduino Micro" board was then employed to digitally convert (10 Sa/s) and stream the data to a PC for further processing.

2.2. Tactile Sensing

The tactile sensing glove attempts to acquire contact forces during interaction with objects like everyday grasping and manipulation. Key design objectives during the development of the sensorized glove were: (i) high sensitivity to allow for detection of small first-touch contact forces around 0.1 N;

(ii) coverage of a large range of forces up to 30 N; and (iii) a high degree of wearability and robustness. Particularly, wearability is a very important aspect in order to minimize interference with human motion execution and tactile sensing thus maintaining a manipulation experience that is as natural as possible. Accordingly, we looked for a thin and stretchable fabrics solution, and –after evaluating numerous combinations of conductive fabrics– we decided for a design comprising four layers of fabrics as shown in Figure 5.

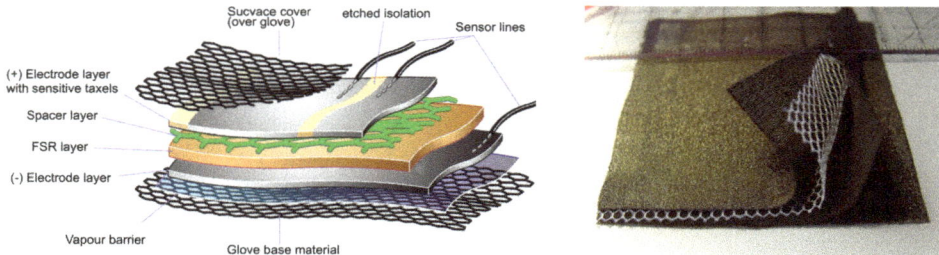

Figure 5. The flexible tactile sensing glove is manufactured from several layers of conductive and piezo-resistive fabrics.

The sensor exploits the piezo-resistive effect, i.e., mechanical pressure applied to the material induces a change of its electrical resistance that can be easily measured. The piezo-resistive material employed here is a highly stretchable knitted fabric (72% nylon, 28% spandex) manufactured by Eeonyx. The individual fibers within the fabric are coated on a nano-scale with inherently-conductive polymers. The material is available at different resistances, determined by the thickness of the applied coating. During experimental testing, we found a material with a surface resistivity of 70–80 k$\Omega/\square$ to be most suitable for our application.

By placing the piezo-resistive fabric between two highly conductive materials, we can measure the change in the resistance between the two outer layers when pressure is applied to the compound. These outer layers constitute the low impedance electrodes that transport current into and out of the sensor with minimal losses. A low impedance is achieved by plating nylon knitted fabric (78% polyamide, 22% elastomer) with pure silver particles.

Wrinkles in the fabrics can generate spurious contact observations. Thus, in order to reduce this effect, we integrated an additional meshed layer between the piezo-resistive layer and an electrode layer, which keeps them insulated as long as too small forces are applied. Obviously, the sensor's first-touch sensitivity, i.e., the smallest measurable force threshold, depends on the thickness of the meshed layer and on the size of its mesh openings: Smaller openings and thicker layers result in a higher force threshold, because more force is required to establish contact between conductive layers. We evaluated meshes with openings in the range of 0.2 to 5 mm and found that a 0.23 mm thick fabric with a honeycomb structure (Figure 5) and openings of ca. 2 mm balances best between high first-touch sensitivity and suppression of spurious contacts. In experiments we measured an initial force threshold of 0.1 N using a 3 mm^2 probe tip.

The more contacts between the conductive layers are established, the more the resistivity decreases. Hence, above the initial force threshold, the sensor can be considered as a parallel circuit of force-sensitive resistors (Figure 6). Hence, as already pointed out in [46], two major factors contribute to the force sensitivity of a piezo-resistive sensor: (i) the increase of contact area between conductive layers and (ii) the piezo-resistive effect in the middle layer.

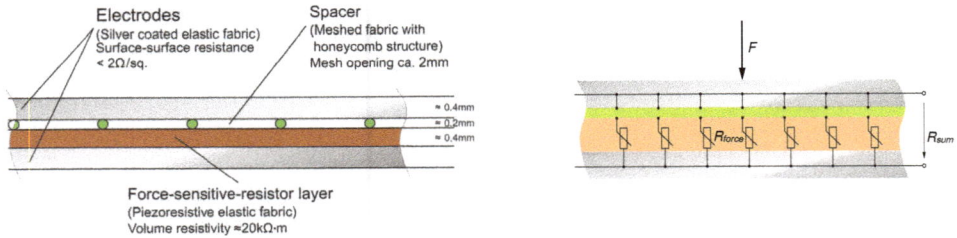

Figure 6. Schematic representation of the fabric-based tactile sensor: Two highly conductive electrode layers (silver) enclose a piezo-resistive layer (orange), measuring changes in its resistivity. The mesh layer, pictured in green, guarantees high resistance (GΩ range) during no-contact idle state. When contacts are established between the electrode and piezo-resistive layer, the sensor acts as parallel circuit of force sensitive resistors.

The high resistance of the sensor in its idle state has the additional benefit of minimizing the current flow through the sensor and thus minimizing the energy loss. This ensures a longer runtime for battery-powered portable systems and also significantly reduces the heat produced by the sensor.

2.2.1. Tactile Sensor Characteristics

To determine the sensor characteristics and to evaluate various fabric materials, we used a measurement bench with a calibrated, strain gauge sensor mounted on a vertical linear axis to measure ground truth forces. A motion performed along the linear axis is transformed into force changes via a coil-spring.

We evaluated the sensor characteristics by first increasing the force from 0 to 10 N and subsequently decreasing back again to 0 N, always measuring the resulting voltage between the electrodes. Between individual measurements, the linear axis was displaced by an amount of 0.1 mm, decreasing or increasing the applied force in a non-linear fashion. Before each measurement we waited for 300 ms for the mechanics to settle, thus reducing transient effects. Each measurement sweep was repeated 5 times to evaluate the repeatability of measurements. The resulting characteristic curves for various sensor materials are shown in Figure 7.

All curves exhibit an hysteresis effect, i.e., measuring smaller voltages (or higher force) during unloading compared to the loading phase for the same applied forces (up to 14% of the measurement range). This hysteresis effect is common to all piezo-resistive materials due to an increasing intertwinement of the material and memorized compression due to the applied load. However, as can be observed from the graphs, the sensor's repeatability is very high, having a standard deviation of less than 0.4% of the measurement range.

We have chosen a fabrics that nicely covers the whole measurement range of 0–5 V within the typical force range of 0–30 N observed in typical human manual interaction [39]. The nonlinear, saturating sensor characteristics is beneficial for measuring forces across several orders of magnitude: While the sensor provides high force resolution for small forces (< 10 N), it can also measure forces up to 60 N. This is in accordance to the human sense of touch, which exhibits a power-law dependency of resolution as well [47].

2.2.2. Taxel Layout and Glove Design

Employing the fabrics-based tactile sensor, we produced a multi-taxel tactile-sensing glove to be worn by humans allowing us to record force interaction patterns during manipulation. To this end, different taxel areas need to be isolated from each other. We employed two different methods to do so: First, individual fabric patches are sewed onto a very thin and breathable support glove. As the seams will disturb the tactile experience of the human operator, we tried to minimize the number of seams. Hence, as a second method to create isolated taxels, we used etching: Employing a FeCl$_3$ acid, the conductive silver coating of the electrode layer is removed along thin paths between individual taxels, thus electrically isolating them from each other. The resulting glove with its individual taxels is shown in Figure 8.

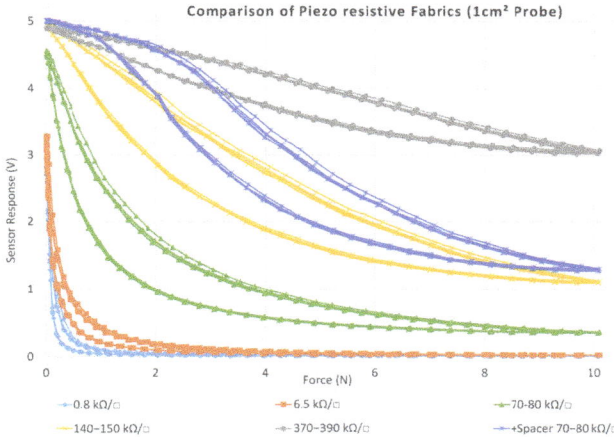

Figure 7. Tactile sensor characteristics was measured using a 1 cm^2 flat probe tip. Curves of like colors represent a measurement sweep of increasing and decreasing force in the range of 0–10 N for a specific material. All curves exhibit hysteresis that is common to piezo-resistive sensors. We have chosen the material that best covers the whole measurement range of 0–5 V, namely the fabrics with a surface resistance of 70–80 kΩ/□ without the spacer layer (Ω/□ indicates the unit of surface resistivity).

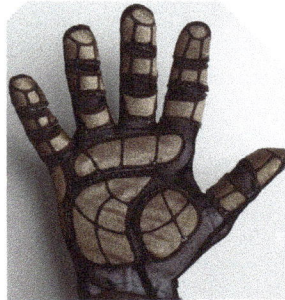

Figure 8. Tactile-sensing glove composed from various sensing patches sewed onto a very thin and breathable base glove. Individual sensing patches are further subdivided into taxels by etching away the conductive silver coating. The corresponding insulating gaps are clearly visible as dark lines in contrast to the bright taxels regions.

In order to reduce sensor degradation due to sweat and other moisture (which corrodes the silver coating), the sensing layers are augmented with an additional vapor barrier underneath the lower electrode layer as shown in Figure 5. The individual taxels are connected to the acquisition electronics, which is located in a wrist band, using Teflon coated wires interwoven into the electrode layers. Using a voltage divider circuit and an ADC for each single taxel, a PIC18 micro-controller on the acquisition board collects the sensor data of all taxels and transmits them via USB to the host PC.

The taxel distribution across the palmar side of the glove was chosen to mimick the density distribution of mechano-receptors in the human hand [48]; however, on a much coarser scale. Each finger tip comprises four individual taxels, measuring contact forces at the tip, the central fingertip area, and on the sides. Similarly, the other finger segments are covered by 2–3 taxels. Their sensor area ranges from 34 to 130 mm^2 corresponding to a spatial resolution of 7.2 to 9.6 mm.

For the palm we have chosen much larger taxel areas (195 to 488 mm^2) corresponding to a spatial resolution of 12.4 to 29.7 mm. All sensor patches are designed to minimize wrinkling within individual taxels as this would induce spurious measurement peaks. Consequently, each finger segment is covered by an individual fabrics patch, which is further subdivided by etching into individual taxels. The sensor patches within the palm were separated along typical folds of the human skin. The final tactile-sensing glove comprises 54 individual taxels as shown in Figure 5.

2.3. Experiment Approach

For the experiments, we placed the two types of glove one over the other. A dedicated software was developed based on ROS to enable modular and synchronous data acquisition as well as visualization. More specifically, the data acquisition and processing modules of the two sensing gloves published their time-stamped measurements (joint angle and normal force values) to specific ROS topics, which were integrated by rviz for visualization. For pose visualization we developed a rigid-link 3D model of a human hand, which is available as URDF. Tactile contact forces are rendered on top of this hand model using mesh markers matching the shape of individual taxels and mapping force magnitudes with a color-map from black over green to red, i.e., from zero over medium to maximum force. To specify the tactile sensor configuration, i.e., location and shape of taxels, we augmented URDF with appropriate information. The corresponding software is available on on our repository [49].

For the preliminary experiments, we asked a right handed male participant (28) to grasp the following objects, which are representative of human grasp workspace as we did in [43,50]: hammer, credit card, pen and ball. What is observable is the high kinematic coherence with human biomechanics: at the same time, contact force can be visualized thus providing information on which parts of human hand are the most involved for grasp. Note that the effectiveness of the integration of synergistic information for hand posture reconstruction via Conductive Elastomer (CE) and KPF sensors was already validated in terms of estimation error and pose classification in [35] and in [50], respectively. For further details, the interested reader can refer to these references. In Figure 9 we report the original grasps performed by the subject wearing the integrated glove as well as their reconstructions in a side-by-side manner. As already mentioned, the reconstructions are defined in $\mathbb{R}^{19}$, completing the measurements of the five KPF sensors with synergistic information. For the sake of visualization, in Figure 9 the grasped object is also shown.

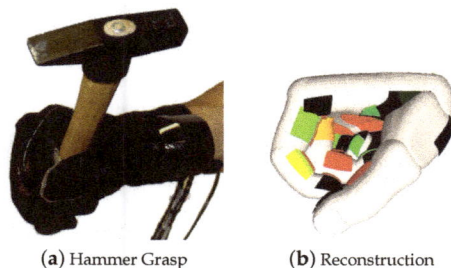

(**a**) Hammer Grasp   (**b**) Reconstruction

Figure 9. *Cont.*

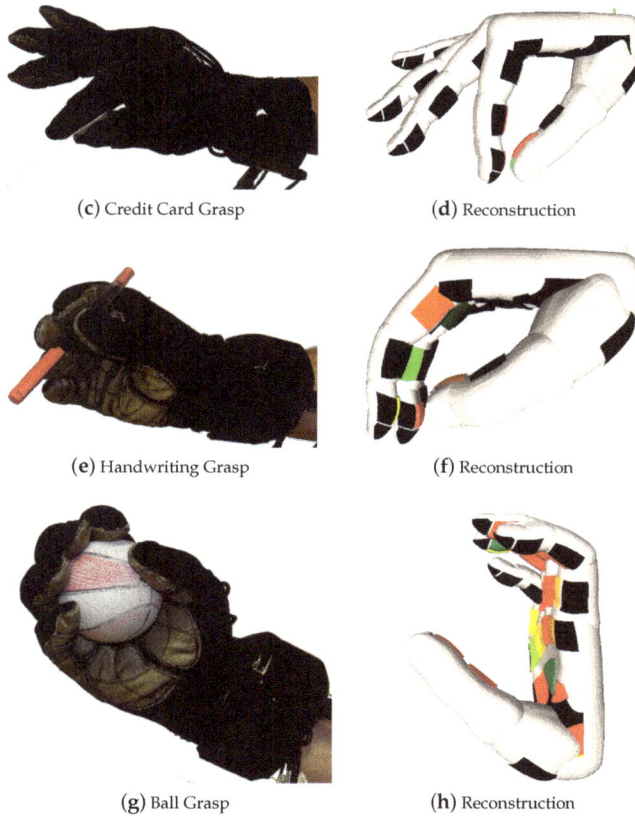

(**c**) Credit Card Grasp

(**d**) Reconstruction

(**e**) Handwriting Grasp

(**f**) Reconstruction

(**g**) Ball Grasp

(**h**) Reconstruction

Figure 9. Side-by-side comparison of real and reconstructed grasps.

3. Discussion and Conclusions

In this paper, we have reported on the (software) integration between a tactile glove and an HPR under-sensed system based on KPF technology. The latter uses synergistic information, i.e., hand joint covariation patterns observed in grasping tasks, to complete the estimation of the hand pose from only five measures to a set of 19 joint angles. Outcomes of such an integrated sensing can be displayed using a visualization tool, which enables to show both postural and tactile information of the user's hand. Results are very promising and show a high level of consistency with human hand bio-mechanics, under a qualitative point of view. This encourages us to push further our wearable approach that could have numerous applications in different fields, such as neuroscience. For example, the system could be used to investigate how the central nervous system copes the redundancy problem by examining how humans control grasping of hand-held objects in unconstrained grasping tasks, enabling the analysis of anticipatory control in both the position and force/tactile domains [51,52]. Another potential application field could be in rehabilitation for the assessment of hand recovery in post-stroke patients. Indeed, the integrated measurement of both hand posture and force/tactile information on the patient's hand, both jointly recorded in an unobtrusive and ecological fashion, is important for the evaluation of the effectiveness of therapeutic outcomes. Our integrated glove could be also employed to map user's hand kinematics and tactile data onto a slave robotic hand controlled in tele-operation tasks. Finally other potential application fields could be in virtual reality and entertainment. Future works will be devoted to develop our integration also under a hardware point of view, in order to get a more and more unobtrusive sensing system. At the same time, a more thorough quantitative evaluation of our techniques will be performed. Finally, we will also consider the usage of other wearable sensing systems, which could provide additional information on the interaction between the human hand and the external environment (e.g., [53], as well as the integration of the here proposed approach with feedback mechanisms, see e.g., [54,55] to be used in tele-operation applications for the remote control of robotic hands).

Acknowledgments: This work is supported in part by the European Research Council under the Advanced Grant "SoftHands: A Theory of Soft Synergies for a New Generation of Artificial Hands" (No. ERC-291166), by the EU H2020 projects "SoftPro: Synergy-based Open-source Foundations and Technologies for Prosthetics and RehabilitatiOn" (No. 688857) and "SOMA: Soft Manipulation" (No. 64559) and by the EU FP7 project (No. 601165) "WEARable HAPtics for Humans and Robots (WEARHAP)".

Author Contributions: M.B., R.H. and A.T. conceived and designed the experiments. G.B. and R.H. developed and preliminary tested the tactile glove. N.C. and A.T. developed and preliminary tested the kinaesthetic glove and the acquisition electronics. M.B. and S.C. performed the optimal design and reconstruction. R.H. designed and implemented the visualization tool. S.C. performed the integration and conducted the preliminary experiments. M.B., R.H. and A.T. drafted and supervised the manuscript writing phase.

Conflicts of Interest: The authors declare no conflict of interest.

Abbreviations

The following abbreviations are used in this manuscript:

DOF   degree of freedom
HPR   hand pose reconstruction
KPF   knitted piezoresistive fabric
PCA   principal component analysis

References

1. Lu, G.; Shark, L.K.; Hall, G.; Zeshan, U. Immersive manipulation of virtual objects through glove-based hand gesture interaction. *Virtual Real.* **2012**, *16*, 243–252.
2. Hürst, W.; Van Wezel, C. Gesture-based interaction via finger tracking for mobile augmented reality. *Multimed. Tools Appl.* **2013**, *62*, 233–258.
3. Borghetti, M.; Sardini, E.; Serpelloni, M. Sensorized glove for measuring hand finger flexion for rehabilitation purposes. *IEEE Trans. Instrum. Meas.* **2013**, *62*, 3308–3314.
4. Kortier, H.G.; Sluiter, V.I.; Roetenberg, D.; Veltink, P.H. Assessment of hand kinematics using inertial and magnetic sensors. *J. Neuroeng. Rehabil.* **2014**, *11*, doi:10.1186/1743-0003-11-70.
5. Liarokapis, M.V.; Artemiadis, P.K.; Kyriakopoulos, K.J. Telemanipulation with the DLR/HIT II robot hand using a dataglove and a low cost force feedback device. In Proceedings of the 2013 21st Mediterranean Conference on Control Automation (MED), Platanias-Chania, Greece, 25–28 June 2013; pp. 431–436.
6. Kim, Y.; Leonard, S.; Shademan, A.; Krieger, A.; Kim, P.C. Kinect technology for hand tracking control of surgical robots: Technical and surgical skill comparison to current robotic masters. *Surg. Endosc.* **2014**, *28*, 1993–2000.
7. Erol, A.; Bebis, G.; Nicolescu, M.; Boyle, R.D.; Twombly, X. Vision-based hand pose estimation: A review. *Comput. Vis. Image Underst.* **2007**, *108*, 52–73.
8. Suarez, J.; Murphy, R.R. Hand gesture recognition with depth images: A review. In Proceedings of the 21st IEEE International Symposium on Robot and Human Interactive Communication, Paris, France, 9–13 September 2012; pp. 411–417.
9. Dipietro, L.; Sabatini, A.; Dario, P. A Survey of Glove-Based Systems and Their Applications. *IEEE Trans. Syst. Man Cybern. Part C Appl. Rev.* **2008**, *38*, 461–482.
10. Cyberglove. Available online: https://www.cyberglovesystems.com (accessed on 23 May 2016).
11. O'Brien, B.; Gisby, T.; Anderson, I.A. Stretch sensors for human body motion. In Proceedings of SPIE, Electroactive Polymer Actuators and Devices (EAPAD), San Diego, CA, USA, 9 March 2014; Volume 905618.

12. Sareh, S.; Noh, Y.; Li, M.; Ranzani, T.; Liu, H.; Althoefer, K. Macrobend optical sensing for pose measurement in soft robot arms. *Smart Mater. Struct.* **2015**, *24*, 125024, doi:10.1088/0964-1726/24/12/125024.

13. Westling, G.; Johansson, R.S. Factors influencing the force control during precision grip. *Exp. Brain Res.* **1984**, *53*, 277–284.

14. Cannata, G.; Maggiali, M.; Metta, G.; Sandini, G. An embedded artificial skin for humanoid robots. In Proceedings of the 2008 IEEE International Conference on Multisensor Fusion and Integration for Intelligent Systems (MFI 2008), Seoul, Korea, 20–22 August 2008; pp. 434–438.

15. Fishel, J.A. Design and Use of a Biomimetic Tactile Microvibration Sensor with Human-Like Sensitivity and Its Application in Texture Discrimination Using Bayesian Exploration. Ph.D. Thesis, University of Southern California, Los Angeles, CA, USA, August 2012.

16. Kõiva, R.; Zenker, M.; Schürmann, C.; Haschke, R.; Ritter, H. A highly sensitive 3D-shaped tactile sensor. In Proceedings of the 2013 IEEE/ASME International Conference on Advanced Intelligent Mechatronics, Wollongong, NSW, Australia, 9–12 July 2013; pp. 1084–1089.

17. Kappassov, Z.; Corrales, J.A.; Perdereau, V. Tactile Sensing in Dexterous Robot Hands—Review. *Robot. Auton. Syst.* **2015**, *74*, 195–220.

18. Li, Q.; Haschke, R.; Ritter, H. A Visuo-Tactile Control Framework for Manipulation and Exploration of Unknown Objects. In Proceedings of the 2015 IEEE-RAS 15th International Conference on Humanoid Robots (Humanoids), Seoul, Korea, 3–5 November 2015.

19. Dang, H.; Weisz, J.; Allen, P. Blind grasping: Stable robotic grasping using tactile feedback and hand kinematics. In Proceedings of the 2011 IEEE International Conference on Robotics and Automation (ICRA), Shanghai, China, 9–13 May 2011; pp. 5917–5922.

20. Ward-Cherrier, B.; Cramphorn, L.; Lepora, N.F. Tactile Manipulation With a TacThumb Integrated on the Open-Hand M2 Gripper. *IEEE Robot. Autom. Lett.* **2016**, *1*, 169–175.

21. Maycock, J.; Dornbusch, D.; Elbrechter, C.; Haschke, R.; Schack, T.; Ritter, H. Approaching Manual Intelligence. *KI Künstliche Intell.* **2010**, *24*, 287–294.

22. Kerpa, O.; Weiss, K.; Wörn, H. Development of a flexible tactile sensor system for a humanoid robot. In Proceedings of the 2003 IEEE/RSJ International Conference on Intelligent Robots and Systems (IROS 2003), Karlsruhe, Germany, 27–31 October 2003.

23. Lowe, M.; King, A.; Lovett, E.; Papakostas, T. Flexible tactile sensor technology: Bringing haptics to life. *Sens. Rev.* **2004**, *24*, 33–36.

24. Kim, K.; Lee, K.R.; Kim, W.H.; Park, K.B.; Kim, T.H.; Kim, J.S.; Pak, J.J. Polymer-based flexible tactile sensor up to 32×32 arrays integrated with interconnection terminals. *Sens. Actuators A Phys.* **2009**, *156*, 284–291.

25. Ohmura, Y.; Kuniyoshi, Y.; Nagakubo, A. Conformable and scalable tactile sensor skin for curved surfaces. In Proceedings of the 2006 IEEE International Conference on Robotics and Automation, Orlando, FL, USA, 15–19 May 2006.

26. Shimojo, M.; Namiki, A.; Ishikawa, M.; Makino, R.; Mabuchi, K. A tactile sensor sheet using pressure conductive rubber with electrical-wires stitched method. *IEEE Sens. J.* **2004**, *4*, 589–596.

27. Alirezaei, H.; Nagakubo, A.; Kuniyoshi, Y. A highly stretchable tactile distribution sensor for smooth surfaced humanoids. In Proceedings of the 7th IEEE-RAS International Conference on Humanoid Robots, Pittsburgh, PA, USA, 29 November–1 December 2007; pp. 167–173.

28. OptoForce Ltd. Optical Force Sensors. Available online: http://optoforce.com (accessed on 19 July 2016).

29. Sareh, S.; Jiang, A.; Faragasso, A.; Noh, Y.; Nanayakkara, T.; Dasgupta, P.; Seneviratne, L.D.; Wurdemann, H.A.; Althoefer, K. Bio-inspired tactile sensor sleeve for surgical soft manipulators. In Proceedings of the 2014 IEEE International Conference on Robotics and Automation (ICRA 2014), Hong Kong, China, 31 May–7 June 2014; pp. 1454–1459.

30. Sagisaka, T.; Ohmura, Y.; Kuniyoshi, Y.; Nagakubo, A.; Ozaki, K. High-density conformable tactile sensing glove. In Proceedings of the 2011 11th IEEE-RAS International Conference on Humanoid Robots (Humanoids), Bled, Slovenia, 26–28 October 2011; pp. 537–542.

31. Tognetti, A.; Lorussi, F.; Dalle Mura, G.; Carbonaro, N.; Pacelli, M.; Paradiso, R.; De Rossi, D. New generation of wearable goniometers for motion capture systems. *J. Neuroeng. Rehabil.* **2014**, *11*, doi:10.1186/1743-0003-11-56.

32. Dalle Mura, G.; Lorussi, F.; Tognetti, A.; Anania, G.; Carbonaro, N.; Pacelli, M.; Paradiso, R.; De Rossi, D. Piezoresistive goniometer network for sensing gloves. In Proceedings of the XIII Mediterranean Conference on Medical and Biological Engineering and Computing 2013, Seville, Spain, 25–28 September 2013; Volume 41, pp. 1547–1550.

33. Carbonaro, N.; Dalle Mura, G.; Lorussi, F.; Paradiso, R.; De Rossi, D.; Tognetti, A. Exploiting wearable goniometer technology for motion sensing gloves. *IEEE J. Biomed. Health Inf.* **2014**, *18*, 1788–1795.

34. Cianchetti, M.; Renda, F.; Licofonte, A.; Laschi, C. Sensorization of continuum soft robots for reconstructing their spatial configuration. In Proceedings of the 2012 4th IEEE RAS & EMBS International Conference on Biomedical Robotics and Biomechatronics (BioRob), Rome, Italy, 24–27 June 2012 ; pp. 634–639.

35. Bianchi, M.; Salaris, P.; Bicchi, A. Synergy-based Hand Pose Sensing: Reconstruction Enhancement. *Int. J. Robot. Res.* **2013**, *32*, 396–406.

36. Bianchi, M.; Salaris, P.; Bicchi, A. Synergy-based hand pose sensing: Optimal glove design. *Int. J. Robot. Res.* **2013**, *32*, 407–424.

37. Büscher, G.; Kõiva, R.; Schürmann, C.; Haschke, R.; Ritter, H. Tactile dataglove with fabric-based sensors. In Proceedings of the 12th IEEE-RAS International Conference on Humanoid Robots (Humanoids 2012), Osaka, Japan, 29 November–1 December 2012; pp. 204–209.

38. Büscher, G.; Kõiva, R.; Schürmann, C.; Haschke, R.; Ritter, H. Flexible and stretchable fabric-based tactile sensor. *Robot. Auton. Syst.* **2015**, *63*, 244–252.

39. Kõiva, R.; Hilsenbeck, B.; Castellini, C. FFLS: An accurate linear device for measuring synergistic finger contractions. In Proceedings of the Annual International Conference of the IEEE Engineering in Medicine & Biology Society (EMBC 2012), San Diego, CA, USA, 28 August–1 September 2012.

40. Santello, M.; Flanders, M.; Soechting, J.F. Postural hand synergies for tool use. *J. Neurosci.* **1998**, *18*, 10105–10115.

41. Lin, J.; Wu, Y.; Huang, T.S. Modeling the constraints of human hand motion. In Proceedings of the 2000 Workshop on Human Motion, Los Alamitos, CA, USA, 7–8 December 2000; pp. 121–126.

42. Bianchi, M.; Salaris, P.; Turco, A.; Carbonaro, N.; Bicchi, A. On the use of postural synergies to improve human hand pose reconstruction. In Proceedings of the 2012 IEEE Haptics Symposium (HAPTICS), Vancouver, BC, Canada, 4–7 March 2012; pp. 91–98.

43. Bianchi, M.; Carbonaro, N.; Battaglia, E.; Lorussi, F.; Bicchi, A.; De Rossi, D.; Tognetti, A. Exploiting hand kinematic synergies and wearable under-sensing for hand functional grasp recognition. In Proceedings of the 2014 EAI 4th International Conference on Wireless Mobile Communication and Healthcare (Mobihealth), Athens, Greece, 3–5 November 2014; pp. 168–171.

44. Bianchi, M.; Salaris, P.; Bicchi, A. Synergy-based optimal design of hand pose sensing. In Proceedings of the 2012 IEEE/RSJ International Conference on Intelligent Robots and Systems, Vilamoura, Portugal, 7–12 October 2012; pp. 3929–3935.

45. Santello, M.; Flanders, M.; Soechting, J.F. Patterns of hand motion during grasping and the influence of sensory guidance. *J. Neurosci.* **2002**, *22*, 1426–1435.

46. Weiß, K.; Wörn, H. The working principle of resistive tactile sensor cells. In Proceedings of the IEEE International Conference Mechatronics and Automation (ICMA), Karlsruhe, Germany, 29 July–1 August 2005; Volume 1, pp. 471–476.

47. Stevens, S.S. On the psychophysical law. *Psychol. Rev.* **1957**, *64*, 153–181.

48. Vallbo, Å.B.; Johansson, R.S. Properties of cutaneous mechanoreceptors in the human hand related to touch sensation. *Hum. Neurobiol.* **1984**, *3*, 3–14.

49. Software to handle tactile sensors in ROS. Available online: http://github.com/ubi-agni/tactile_toolbox (accessed on 19 July 2016).

50. Ciotti, S.; Battaglia, E.; Carbonaro, N.; Bicchi, A.; Tognetti, A.; Bianchi, M. A Synergy-Based Optimally Designed Sensing Glove for Functional Grasp Recognition. *Sensors* **2016**, *16*, 811, doi:10.3390/s16060811.

51. Naceri, A.; Santello, M.; Moscatelli, A.; Ernst, M.O. Digit Position and Force Synergies During Unconstrained Grasping. In *Human and Robot Hands: Sensorimotor Synergies to Bridge the Gap Between Neuroscience and Robotics*; Bianchi, M., Moscatelli, A., Eds.; Springer International Publishing: Cham, Switzerland, 2016; pp. 29–40.

52. Fu, Q.; Zhang, W.; Santello, M. Anticipatory planning and control of grasp positions and forces for dexterous two-digit manipulation. *J. Neurosci.* **2010**, *30*, 9117–9126.

53. Battaglia, E.; Bianchi, M.; Altobelli, A.; Grioli, G.; Catalano, M.G.; Serio, A.; Santello, M.; Bicchi, A. ThimbleSense: A Fingertip-Wearable Tactile Sensor for Grasp Analysis. *IEEE Trans. Haptics* **2016**, *9*, 121–133.

54. Casini, S.; Morvidoni, M.; Bianchi, M.; Catalano, M.; Grioli, G.; Bicchi, A. Design and realization of the CUFF-clenching upper-limb force feedback wearable device for distributed mechano-tactile stimulation of normal and tangential skin forces. In Proceedings of the 2015 IEEE/RSJ International Conference on Intelligent Robots and Systems (IROS), Hamburg, Germany, 28 September–2 October 2015; pp. 1186–1193.

55. Bianchi, M.; Battaglia, E.; Poggiani, M.; Ciotti, S.; Bicchi, A. A Wearable Fabric-based display for haptic multi-cue delivery. In Proceedings of the 2016 IEEE Haptics Symposium (HAPTICS), Philadelphia, PA , USA, 8–11 April 2016; pp. 277–283.

electronics

MDPI

Article

Analysis of In-to-Out Wireless Body Area Network Systems: Towards QoS-Aware Health Internet of Things Applications

Yangzhe Liao [1,*], Mark S. Leeson [1], Matthew D. Higgins [2] and Chenyao Bai [1]

[1]   School of Engineering, University of Warwick, Coventry CV4 7AL, UK;
      mark.leeson@warwick.ac.uk (M.S.L.); chenyao.bai@warwick.ac.uk (C.B.)
[2]   WMG, University of Warwick, Coventry CV4 7AL, UK; m.higgins@warwick.ac.uk
*   Correspondence: yangzhe.liao@warwick.ac.uk; Tel.: +44-759-647-6435

Academic Editors: Enzo Pasquale Scilingo and Gaetano Valenza
Received: 6 April 2016; Accepted: 6 July 2016; Published: 13 July 2016

Abstract: In this paper, an analytical and accurate in-to-out (I2O) human body path loss (PL) model at 2.45 GHz is derived based on a 3D heterogeneous human body model under safety constraints. The bit error rate (BER) performance for this channel using multiple efficient modulation schemes is investigated and the link budget is analyzed based on a predetermined satisfactory BER of 10^{-3}. In addition, an incremental relay-based cooperative quality of service-aware (QoS-aware) routing protocol for the proposed I2O WBAN is presented and compared with an existing scheme. Linear programming QoS metric expressions are derived and employed to maximize the network lifetime, throughput, minimizing delay. Results show that binary phase-shift keying (BPSK) outperforms other modulation techniques for the proposed I2O WBAN systems, enabling the support of a 30 Mbps data transmission rate up to 1.6 m and affording more reliable communication links when the transmitter power is increased. Moreover, the proposed incremental cooperative routing protocol outperforms the existing two-relay technique in terms of energy efficiency. Open issues and on-going research within the I2O WBAN area are presented and discussed as an inspiration towards developments in health IoT applications.

Keywords: health IoT; in-to-out body channel; WBAN; QoS-aware; BER; routing protocol

1. Introduction

The health Internet of Things (IoT) is one of the most promising approaches in improving the quality of human life. This is through healthcare monitoring and remote telemedicine support systems, which are able to deliver real-time data collection, transmission and visualization via the Internet [1,2]. In our previous work, we have presented a flexible quality of service (QoS) target-specific smart healthcare system [3]. This consisted of a smart gateway, a collection of sensor nodes and wireless communication links that can continuously acquire, process and transmit human vital signs to a remote medical server. Such a system makes remote patient health status monitoring by doctors and nurses a feasible proposition [1,3]. Moreover, the large volume of data collected makes it possible for researchers to develop new healthcare products and provide effective health education to people via the internet [1,4]. A wireless body area network (WBAN) is a networking technology that offers the prospect of the early detection of abnormal health conditions, real-time healthcare monitoring and remote telemedicine support systems. The Institute of Electrical and Electronic Engineers (IEEE) 802.15.6 Task Group has established the first international wireless communication standardisation of WBANs that optimises power consumption and provides safety guidance for medical and non-medical applications operating inside, on, or around the human body [4]. To date, the existing research work has mainly focused on on-body and off-body communication systems rather than in-body or I2O body communication systems [1,4,5].

We previously proposed a biological implant WBAN communication channel [5], where the communication link was specific for the human cephalic region and covered a limited communication range. Thus in this paper, the proposed I2O body WBAN systems enable transmission of real-time medical data from the in-body to the on-body area, which can enhance several practical health IoT scenarios [1,4]. Figure 1 shows the system in total, with the focus of this paper being the I2O tranmission link shown on the frontal thorax of the human outline.

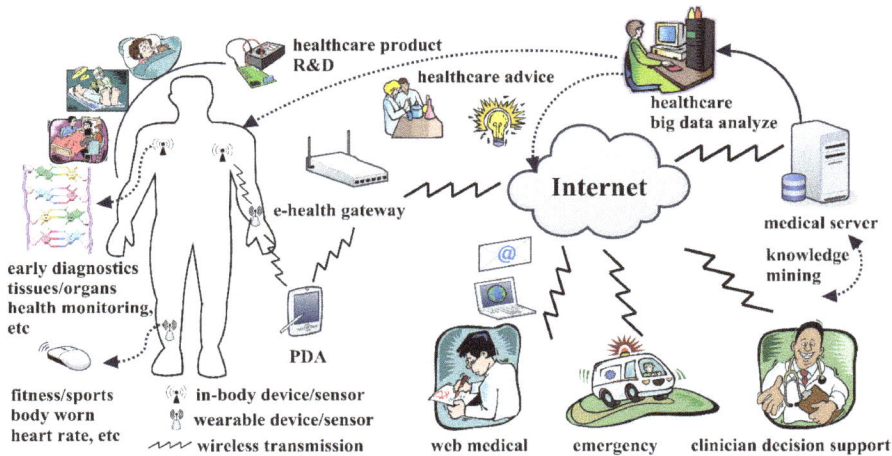

Figure 1. The proposed typical structure of the health IoT system.

WBANs differ from traditional wireless communication systems in terms of propagation medium, transmission power restrictions and human tissue/organ safety requirements [5]. Over 65% of the human intra-body region is composed of water. Moreover, organ-tissue communications and drug transportation is via the blood [6]. This makes radio frequency (RF) signals remarkably attenuated when transmitting data through tissues/organs, even at relatively low frequencies such as the medical implant communication service (MICS) 402–405 MHz band proposed by the IEEE [5]. The 2.45 GHz industrial, scientific and medical (ISM) radio band is investigated in this article, as this brings advantages for WBAN systems. These are primarily that: (a) various compact small-sized antennas are available for implantation; (b) satisfactory performance is obtained with low energy consumption; (c) there is significant bandwidth to support high data transmission rates [4]. Increasing transmission power is one option to account for the energy attenuation but may result in damage to human tissues or organs due to heat absorption [4,7]. Thus, the transmission power must comply with the Federal Communications Commission (FCC) regulations and the specific absorption rate (SAR) must be lower than that laid down by IEEE and the International Commission on Non-Ionizing Radiation Protection (ICNIRP) [8,9].

The concept of telemedicine has changed from 'medicine practiced at a distance' to 'personalised ubiquitous healthcare on the move' under the health IoT [1]. Concerning hospital or home healthcare monitoring scenarios, wireless implanted healthcare monitoring devices would significantly improve the comfort and mobility for patients when compared with wired connected medical devices. RF can cover longer operating distances and enable interactive implanted sensors and devices to communicate with on-body devices wirelessly [1,4,5]. Furthermore, due to the technical constraints of batteries, improving the energy efficiency of RF modules is a key factor in the I2O body channel and one that can be realised by using small-sized implantable antennas [10–13].

Kurup et al. proposed the first in-body path loss (PL) model for homogeneous human muscle tissue at 2.45 GHz in 2009 [14]. Later they reported an extended PL model, which includes the conductivity and permittivity of the human tissues in [15]. However, those PL models and other studies are antenna-specific and only applicable to the virtual family models provided by either of the SEMCAD X or FEKO software packages [4,5,14,15]. In this paper, the PL model is obtained by using software from Computer Simulation Technology (CST) [16] and the recently proposed 3D heterogeneous human body model developed by Kurup et al. [7]. The PL result is believed to be more accurate because the CST software takes the loss tangent parameters of human tissues and organs into account in the simulations and the deviation value is smaller than that seen in [11–15,17]. Furthermore, it is vital to choose suitable digital modulation techniques to overcome the strong power attenuation over the shadow fading channel caused by the intra-body environment [18]. The relationship between a large range of data rates from 250 kbps to 30 Mbps and the operating communication distance for safe transmission powers is discussed and investigated when using the selected modulation schemes.

Routing protocols have been reported to discover and analyze the most energy efficient route [19,20]. Relay based protocol solutions have also been studied that minimize the energy consumption of the in-body sensor nodes by reducing the length of the transmission distance [21]. A two relay based WBAN routing protocol has been investigated by Deepak and Babu [22] and the results show that the proposed routing scheme outperforms direct communication and single relay methods in terms of transmission reliability and energy efficiency. Cooperative communication techniques have gained significant attention as an effective strategy to improve energy efficiency and spatial diversity in wireless fading channels [21]. However, the incremental relaying routing protocol strategy has not been investigated in I2O WBANs.

In this paper, an accurate statistical I2O body path loss model that describes the signal propagation between the transmitter (Tx) and receiver (Rx) antennas is obtained by using CST electromagnetic solvers, based on a heterogeneous and innovative 3D virtual human body model at 2.45 GHz. The SAR is assessed to ensure adherence to the authorities' safety requirements. The BER performance for the I2O shadow fading channel employing the binary phase shift keying (BPSK), quadrature PSK (QPSK), 16 quadrature amplitude modulation (16QAM) and 16PSK modulation schemes is obtained. The threshold signal-to-noise ratio (SNR) of the four modulation methods is derived when using an acceptable BER performance of 10^{-3}. Here, our proposed incremental relaying routing protocol is analyzed and compared with the existing two-relay WBAN routing scheme. A series of QoS metrics such as network lifetime, throughput, average energy consumption, residual energy and propagation delay are also investigated.

The rest of the paper is organized as follows: In Section 2, we briefly introduce a typical I2O WBAN system; details of our proposed I2O WBAN simulation setup, path loss, channel modeling and link budget are given in Section 3; comparison of the proposed incremental relaying protocol and the existing two-relay protocol, including the analysis of mathematical expressions of related QoS metrics, communication flow and QoS metrics performance is shown in Section 4; Section 5 analyzes numerous emerging I2O WBAN topics and health IoT requirements; finally, Section 6 presents the conclusions.

2. System Model

Health IoT communication systems cover short-range telemetry communication channels such as from in-body to on-body. The on-body device can also collect and transmit high data rates and real-time medical information to doctors and nurses via a gateway [1,3]. Figure 2a–c illustrates the elements of the CST model, whilst the three primary components of the health IoT WBAN system are shown in Figure 2d with further details given below:

- *Implant device/sensor*: A biological compatible and miniaturized size implant device that is located inside the human body, either in the tissue/organ region (deep region) or under the skin (near surface) [10].

- *On-body device/sensor*: An on-body (or wearable) device that can be located on either surface or up to 20 mm away from skin [2].
- *Gateway*: Typically, this has no direct connection to an implanted device or sensor. A smartphone or other personal data device is needed to enable to the collection, processing and transmission of data to doctors and nurses via the internet [3].

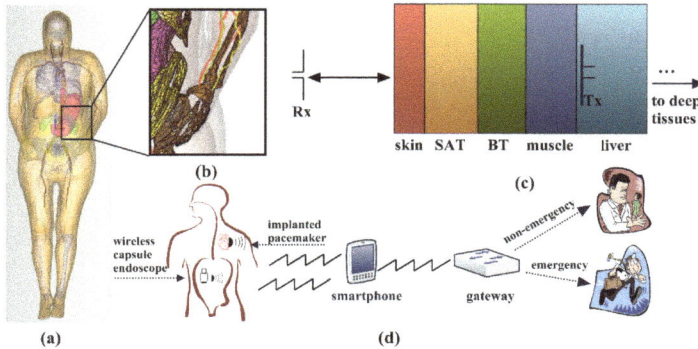

Figure 2. (**a**) The computational 3D human body model in CST; (**b**) the cross section of the human frontal thorax; (**c**) equivalent human frontal thorax model; (**d**) typical healthcare applications.

3. Analysis of I2O WBAN Systems

In this section, the proposed communication system is considered in likely operating scenarios, such as in hospitals or homes, where a patient with an in-body implant device/sensor transmits data to a wearable device placed at the maximum distance of the on-body region. A communication link is established for an I2O body scenario between the Tx located in the liver tissue and the Rx placed 2 cm from the human body surface. The BER performance for the I2O shadow fading channel employing BPSK, QPSK, 16PSK and 16QAM modulation schemes is obtained. The relationship between the data transmission rates, transmitting power and the achievable communication distance is discussed under the link budget analysis.

3.1. Configuration and Human Safety Analysis

Implantable biomedical antenna design in WBAN systems is affected by posture, body mass index, aging and so on. Detailed information about biocompatible in-body antenna design can be found in the review by Movassaghi et al. [4]. Here, we focus our attention mainly on I2O body communication systems analysis. An efficient and multi-layered heterogeneous human body model (Figure 3) proposed in [23] for WBANs, at the frequency of 2.45 GHz, is investigated in this paper. The configuration consists of a layer of air followed by layers equivalent to the frontal thorax of an adult, making it possible to divide the human body into numerous areas of a multi-layered model. We employ dipole antennas because they are well-studied antennas in free space and have a simple structure [5,15,17]. In agreement with [23], we place the Tx antenna of length 3.9 cm in the liver region and the Rx antenna is a free space, half wavelength dipole with a length of 6.12 cm, located 2 cm from the human body surface. Both the Rx and Tx antennas are made of perfect electric conductors (PEC) and are directionally aligned, with a thickness of 2 mm. The simulations use a current source and the simulation methods are the same for all the cases. The equivalent 3D human body model contains multiple layers that are built on dry skin, subcutaneous adipose tissue (SAT), breast tissue (BT), muscle tissue and liver; whose thicknesses are 2 mm, 5 mm, 1 mm, 10 mm and 10 mm with dielectric properties as shown in Table 1. Since it is difficult for both manufacturers and researchers to investigate their systems on an actual human body, the proposed human body model offers a viable alternative to investigate the performance of I2O body WBAN systems.

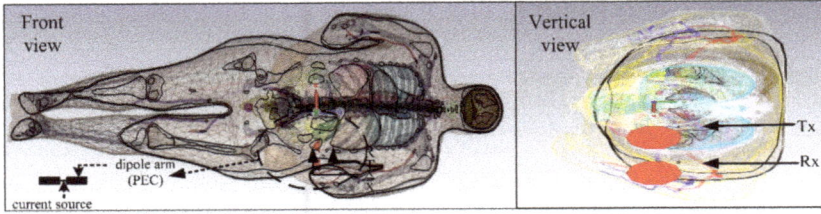

Figure 3. The 3D human body model and dipole antennas. (**left**) front view; (**right**) vertical view.

Table 1. Conductivity (ε_r), Relative Permittivity σ and Loss Tangent tanδ [24].

Parameter	Skin	SAT	BT	Muscle	Liver
ε_r	38	10.8	5.15	52.7	43
σ [S/m]	1.46	0.27	0.14	1.74	1.69
tanδ	0.28262	0.14524	0.19535	0.24194	0.28751

As reported in [5,7], signal propagation in the human body leads to high attenuation, which will result in heating of tissues and organs and to an increase in the temperature of the human body. Biological effects and health risks may occur by exposure to RF electromagnetic fields. The IEEE standard and the ICNIRP safety guidance specify that the averaged SAR over 10 g of tissue should be no more than 1.6 W per kg and 2 W per kg, respectively [8,9]. With an input power of 1 W provided to the implanted sensor or device, the finite-difference time domain (FDTD) approach provided by CST was used in association with the 3D human model investigated to calculate the SAR [25]. Results demonstrate that the SAR of the human body model is far lower than both the regulations. The maximum 1 g and 10 g SAR values calculated tissue are given in Table 2 for the skin minimal distance (skin region) and maximum distance (liver region).

Table 2. Typical Locations and Values of 1 g and 10 g SAR for I2O Body Model.

Distance	Maximum SAR (1 g)	Maximum SAR (10 g)
5 mm	36.8 mW·kg^{-1}	17.4 W·kg^{-1}
20 mm	31.5 W·kg^{-1}	19.3 W·kg^{-1}

3.2. Path Loss Model

The human body is a natural lossy environment, which therefore leads to high attenuation for signal transmission. The PL model of homogeneous human tissue/organ has been described in [14]. For the heterogeneous human body, the Tx antenna moves through the different layers (tissues) and the PL is obtained. When the antenna is placed in a specific layer, the surrounding layers differ from that containing the antenna, leading to deviations between the simulated and calculated PL values. A semi-empirical PL formula in dB between two implant devices can be expressed as [5,14,15,25]:

$$PL_{dB}(d) = PL_{dB}\left(d_{ref}\right) + 10n\, log_{10}\left(\frac{d}{d_{ref}}\right) + S_{dB},\ d \geqslant d_{ref} \qquad (1)$$

where $PL_{dB}\left(d_{ref}\right)$ is the PL value at the reference distance d_{ref} (5 mm in this paper). S_{dB} is the shadow fading parameter expressed in decibels (dB), which follows a normal distribution $S \sim N(0, \sigma_s)$ where σ_s is the standard deviation, since the logarithm of a lognormal distribution is normally distributed. Moreover, σ_s reflects the degree of the shadow fading strength [25]. The probability density function of the shadow fading effect S_{dB} can be expressed as:

$$P_{S_{dB}} = \frac{1}{\sigma\sqrt{2\pi}} exp \left(-\frac{S_{dB}^2}{2\sigma_{log}^2} \right) \tag{2}$$

The variable d is the separation distance between the Tx and Rx; n is the path loss exponent which depends on the propagation media. MATLAB® least square fit computation has been implemented to yield a fitted PL as seen in Figure 4. The simulation results are summarized in Table 3. Closer agreement than many proposed PL models [15,16] is obtained between the derived PL model and the simulations using CST with an average deviation of 2.93 dB.

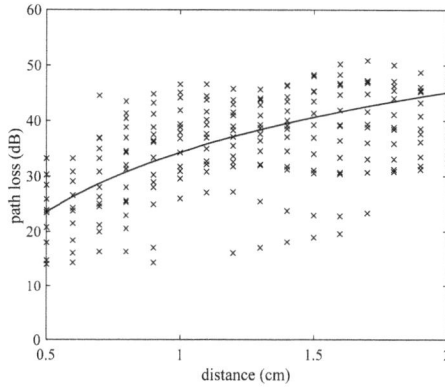

Figure 4. Path loss model versus distance between Tx and Rx antennas.

Table 3. Simulation Results of the Path Loss Model.

Parameter	Value (Unit)	Description
n	3.6	PL exponent
σ_{dB}	2.93	average deviation
d_{ref}	0.5 cm	reference distance
$PL_{dB}\,(ref)$	23.49 dB	PL at the reference distance

3.3. I2O Channel Model

Energy efficient modulation schemes are beneficial in reducing the hardware structure, lowering the noise interference and prolonging the communication network lifetime, due to the shortage of implanted device battery capacity. Moreover, low error probabilities are required in the modulation methods to ensure the reliability of the communication transmission. In [18], M-ary PSK (M-PSK) is reported to achieve significant energy saving and high system outage probability. In addition, 16QAM is the most spectral efficient modulation technique, which can meet higher data rates than QPSK and 8PSK when operating under the same bandwidth [10,26,27]. An I2O communication channel shares the property with all fading RF channels that the instantaneous SNR, γ, at the receiver side, can be regarded as a random variable due to the fading effect. Since this means that the bit errors in demodulation are therefore not fixed since the SNR varies randomly, we investigate the expected bit error rate (BER) for channel performance evaluation, where the expectation is with respect to the probability distribution of the SNR.

The average BER of the I2O shadow fading communication channel can be evaluated by the following integral [5,28]:

$$P_e(\overline{\gamma}) = \int_0^\infty P_{b,AWGN}(\gamma) P_0(\gamma) d\gamma \tag{3}$$

where $P_{b,AWGN}(\gamma)$ represents the BER performance of the additive white Gaussian noise (AWGN) channel for an SNR, γ, and assuming a mean SNR, $\bar{\gamma}$. The function $P_0(\gamma)$ denotes the probability density function of the γ, which is lognormal distributed with the same standard deviation σ_s as the BER performance for the I2O shadow fading channel when employing the four selected modulation techniques. The received power can be obtained simply from the transmitted power P_t and PL via:

$$P_r = \frac{P_t}{PL} \tag{4}$$

The received energy per bit E_b can be related to the data rate R_b and the received power and it then can be expressed as:

$$\frac{E_b}{N_o} = \frac{P_r}{N_0 R_b} = \frac{P_t}{R_b N_0 PL} \tag{5}$$

where N_0 is the noise power. Expressing (6) in dB:

$$\ln\left(\frac{E_b}{N_o}\right) = \ln\left(\frac{P_t}{R_b N_0 PL}\right) = \ln\left(\frac{P_t}{R_b N_0}\right) - \ln PL \tag{6}$$

The first term in the Expression (6) for $\ln(E_b/N_o)$ is fixed for a given data rate R_b and transmitter power P_t. Thus, since the PL model presented earlier follows a lognormal distribution, $\ln(E_b/N_o)$ is normally distributed, i.e., E_b/N_0 follows a lognormal distribution:

$$P_0(\gamma) = \frac{1}{\sqrt{2\pi}\sigma\gamma}e^{-\frac{(\ln\gamma-\mu)^2}{2\sigma^2}} \tag{7}$$

where $\mu = \ln\bar{\gamma} - \frac{1}{2}\left(\frac{\ln 10}{10}\right)^2\sigma_{dB}^2$, and $\sigma = \frac{\ln 10}{10}\sigma_{dB}$.

Detailed information on the four selected modulation techniques can be found in [5,18,26,27]. We use coherent BPSK modulation as a concrete example, where [5]:

$$P_{b,AWGN}(\gamma) = \frac{1}{2}\text{erfc}(\sqrt{\gamma}) \tag{8}$$

where erfc(.) is the complementary error function and so (3) becomes:

$$P_e(\bar{\gamma}_m) = \sum_{n=1}^{N}\frac{1}{2}\text{erfc}(\sqrt{\gamma})\frac{1}{\sqrt{2\pi}\sigma\gamma_n}e^{-\frac{(\ln\gamma_n-\ln\bar{\gamma}_m+\frac{1}{2}(\frac{\ln 10}{10})^2\sigma_{dB}^2)}{2\sigma^2}}(\gamma_n - \gamma_{n-1}) \tag{9}$$

The average BER performance of the I2O fading channel under BPSK is then obtained by numerical evaluation of (9) [5,28]. The same method can be applied to other proposed modulation schemes. The average BER performance of the four selected modulation methods is shown in Figure 5. At the receiver, AWGN is the dominant noise source [5] so we consider only thermal noise with one-sided power spectral density (PSD) [5,29–31]:

$$N_0 = k\left[T_I + (N_F - 1)T_O\right] \tag{10}$$

where k and N_F are the Boltzmann constant and receiver noise factor, respectively; T_I and T_O are the noise temperatures at the receiver and transmitter. The mean temperature of liver tissue is around the 306 K and the ambient temperature of human skin is 310 K [24,32]. N_F is the noise figure and can be defined as:

$$N_{F,dB} = 10\log_{10}(N_F) \tag{11}$$

where $N_F = 1 + T_I/T_O$. The received SNR in dB can be expressed as:

$$SNR_{dB} = P_{r,dBW} - 10\log_{10}(R_b) - N_{F,dB} \tag{12}$$

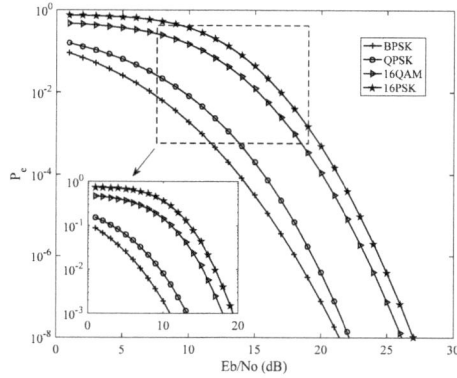

Figure 5. BER performance comparison between BPSK, QPSK, 16QAM and 16PSK.

3.4. Link Budget Analysis

The I2O human body environment is lossy producing and high attenuation, so it is important to analyze the link budget when designing wireless communication systems for several scenarios. Two significant elements are: (i) an energy consumption calculation for a required communication link quality; (ii) the effective communication distance estimation when employing a certain transmitting power.

Based on the European Research Council (ERC) regulations, the maximum input power is 25 µW [19]. In this paper, typical transmitting power values of 1, 10 and 25 µW are selected for further research and discussions. Given the high QoS requirements of healthcare communication systems, similar to other I2O body communication systems, a predetermined BER threshold of 10^{-3} was selected to ensure the communication performance is acceptable [5,28–31]. According to Figure 5, the threshold SNR_{thr} values for BPSK, QPSK, 16QAM and 16PSK are approximately 11 dB, 13 dB and 15.5 dB and 18 dB, respectively. The parameters used in the link budget simulations are summarized in Table 4.

Table 4. Parameters for the Link Budget Simulations.

Simulation Parameter	Value
Frequency band (GHz)	2.45
Tx output power (µW)	1, 10, 25
Antenna gain (dBi)	0
Coding gain (dB)	0
Ambient temperature (K)	310
Liver tissue temperature (K)	306
Boltzmann constant (JK^{-1})	1.38×10^{-23}
BER (predetermined)	10^{-3}
SNR (threshold) (dB)	11 (BPSK), 13 (QPSK) 15.5 (16QAM), 18 (16PSK)
Selected data rate (Mbps)	0.25, 5, 30
Selected distance (m)	2

One valuable system parameter that can effectively evaluate the reliability of the communication system using the threshold BER performance is the system margin, M_s. A communication channel with a negative link margin has insufficient power to transmit data and thus, is essential to offer adequate link to ensure that the communication system is reliable [33]. M_s is given by determining the SNR above the threshold level (SNR_{thr}) as:

$$M_s = SNR_{dB} - SNR_{thr} > 0 \qquad (13)$$

Figures 6–9 show the dependence of system margin versus communication link distance, with several levels of data rates and multiple modulation techniques utilizing transmitter powers of 1 µW, 10 µW and 25 µW. For health IoT applications, such as hospital real-time healthcare monitoring services, where the receiving sensor or device is normally placed 1 meter away from patients, all the above-mentioned methods could support satisfactory wireless data transmission [1,2,23]. Due to the constraints on the battery energy supply of the implantable device or sensor, trade-offs between transmitting power and communication channel quality should be taken into account [11,12]. Results show that, as one might expect, higher transmitting power can achieve longer communication distances for a certain data rate. The communication system can also cover longer distances by using lower transmission data rates when compared with higher data rates. For conditions when under the same data rates and transmitting powers, the BPSK modulation scheme can achieve more reliable transmission than the other investigated modulation schemes. Furthermore, we suggest a fade system margin of a few dBs that accounts for energy losses caused by the antenna orientations and body movements to ensure the reliability of the data transmission, when designing real-time healthcare wireless monitoring systems.

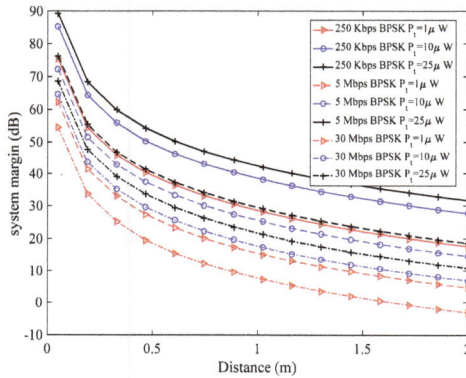

Figure 6. Link margin versus distance using BPSK at multiple levels of data rates and powers.

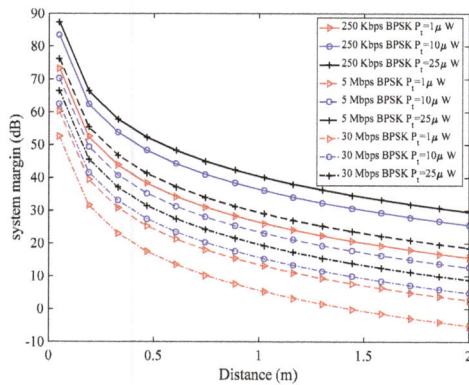

Figure 7. Link margin versus distance employing QPSK at multiple levels of data rates and powers.

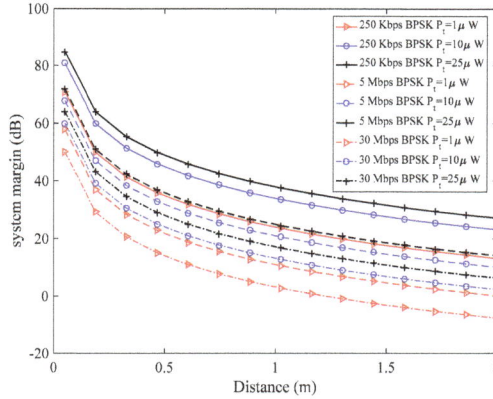

Figure 8. Link margin versus distance for 16PSK at multiple levels of data rates and powers.

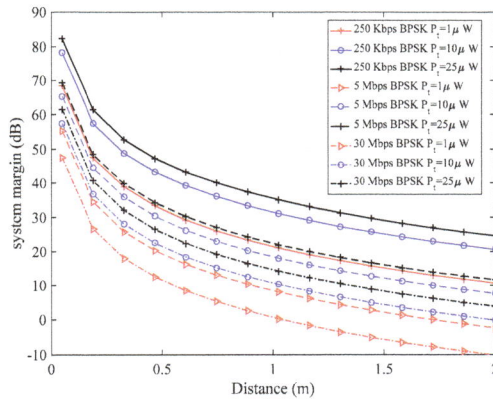

Figure 9. Link margin versus distance for 16QAM at multiple levels of data rates and powers.

4. Relay Based QoS-Aware Routing Protocol for I2O WBAN

4.1. Motivation

One of the major challenges in I2O WBANs is maximizing the WBAN lifetime [21]. To date, there are several routing protocols that have been reported in literature such as single or two-relay WBAN techniques [22]. However, those routing protocols are not very energy efficient and less likely to support long-term healthcare monitoring tasks. In [23], the authors stated that the energy consumption of implants is directly related to the transmission distance and therefore, energy efficient routing protocols are an effective approach in minimizing the overall length of communication paths. Moreover, by deploying an incremental relaying strategy, the complexity and energy consumption are transferred from the implant device to the on-body relay, which is a device that can be easily replaced and recharged, in contrast to the in-body sensor nodes [10,34]. QoS requirements in WBANs vary between applications [35]. A practical approach is to focus on data transmission models that are used in different applications and map the requirements of these onto a set of QoS metrics [35]. Figure 10 demonstrates the multiple relay-based routing protocols proposed in [21,22]; Figure 10a, single relay based scenario; Figure 10b–d, two-relay based selective routing techniques.

In this section, mathematical formulas for important QoS metrics in terms of network lifetime, network throughput and end to end delay are given, along with the related constraint functions. Details of our proposed incremental relaying strategy is given and compared with the two-relay protocol. The results show that the incremental routing protocol outperforms the two-relay based protocol in terms of network lifetime, throughput. However, the latter protocol could support high traffic load conditions when compared with the former.

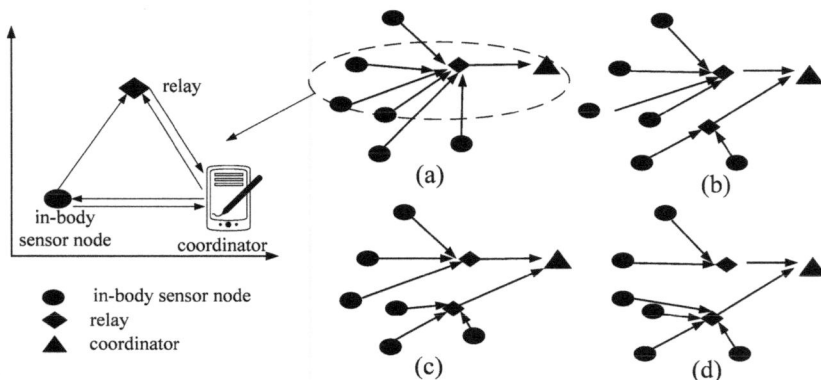

Figure 10. Demonstration of the relay-based routing protocol; (**a**) single relay based scenario; (**b**)–(**d**) two-relay based selective routing protocols.

4.2. Radio Model

The analysis of WBAN system energy consumption is given by extending our previous work on flexible QoS WBANs [3]. Assuming the transmission packet length is defined as k, the minimal transmission energy consumption of sensor nodes can be expressed as:

$$E_{Tx\_min}(d,k) = kE_{Tx_{elec}} + kE_{amp}dn \tag{14}$$

where E_{Tx_min} and E_{Tx_elec} mean the minimal required energy for data transmission from a sensor node to a relay and the essential energy consumption to activate the electronic circuit, respectively.

The distance between the transmitting sensor node and the receiver side is represented by d, and n is the path loss exponent. Similarly, the minimal energy consumption for the reception process can be regarded as $E_{Rx,min} = kE_{elec}$. The minimal total energy consumption for a sensor node E_{Total} can be expressed as:

$$E_{Total}(k,d) = E_{Tx\_min}(d,k) + kE_{elec} \tag{15}$$

The PL parameters in Table 3 are utilized when the in-body sensor nodes transfer data to the relay nodes. Two commercially available WBAN transceivers, the Nordic nRF2401A and Chipcon CC2420 are utilized and the related radio parameters have been summarized in Table 5 [36].

Table 5. Radio parameters of nRF2401A and CC2420.

Parameter (Unit)	nRF2401A	CC2420
Tx current (mA)	10.5	17.4
Rx current (mA)	18	19.7
Voltage (V)	1.9	2.1
E_{Tx_elec} (nJ/bit)	16.7	96.9
E_{Rx_elec} (nJ/bit)	36.1	172.8
E_{amp} (nJ/bit/m^2)	1.97	2.71

4.3. QoS Metric Modeling

4.3.1. Network Lifetime

The stability period and total network lifetime are defined as the lifespan of the network until the first in-body sensor node is energy depleted, including the time duration of the network, until all in-body sensor nodes are energy depleted, respectively [36]. Assuming the number of in-body sensor nodes is N and each node is initially equipped with energy E_0. The main object of I2O WBAN is to maximize the network lifetime T, which can be formulated via linear programing as [21]:

$$Objective{:}\ Max\ T = \sum_{r} t_r \tag{16}$$

where r and t_r denotes the current round and summation of rounds before all in-body sensor nodes energy deplete, respectively. The energy consumption per bit consists of sensing, processing and transmitting energy for an in-body sensor node represented by E_{sen}, E_{pro} and E_{trans} [21]. The remaining energy of the I2O WBAN network after each round can be defined as network residual energy E_i. In terms of constraints, the I2O WBAN is subject to:

$$t_r \geqslant \frac{E_i}{\sum_i k\left(E_{sen}^i + E_{pro}^i + E_{trans}^i + nE_{amp}^i d_{SR}\right)},\ \forall i \in N \tag{17}$$

$$E_o \geqslant E_i,\ \forall i \in N \tag{18}$$

$$E_i \to 0,\ \forall i \in N \tag{19}$$

$$\sum_i f_{SR} > \sum_r f_{RC},\ \forall i \in N \tag{20}$$

where f_{RC} and f_{SR} represent the data flow directions from relay R to the coordinator C and from in-body sensor node S to the corresponding relay R, respectively. The Constraint (17) illustrates the network energy consumption per round. Constraints in (18) and (19) are the energy requirements, the network residual energy E_i reduces after each round and is finally exhausted. Constraint (20) demonstrates that the data flow should be transmitted from node S to the coordinator via a corresponding relay R. Moreover, violation of (20) leads to heavy traffic conditions resulting in transmission delay and packet dropping.

4.3.2. Network Throughput

The network throughput represents the total number of successfully received information packets at the coordinator. It is important to maximize the number of successfully transmitted packets at the coordinator C, because all information is critical in I2O WBANs. The optimization expression for maximizing the number of successfully received packets P_s can be formulated as:

$$Objective:\ Max \sum_r P_s,\ \forall r \in T \tag{21}$$

Subject to:

$$P_{SR} > P_{RC},\ \forall S \in N,\ \forall R \in N \tag{22}$$

$$E_i \geqslant E_{Tx\_min} \tag{23}$$

$$P_{link} \geqslant P_{min} \tag{24}$$

The objective function (21) aims to maximize the number of successfully received packets P_s during the network lifetime T. Constraint (22) demonstrates that data packets may drop when data transmission occurs from R to C. Constraint (23) points out that no data information transmission is possible when the remaining energy E_i is lower than the minimal required transmission energy

E_{Tx_min} as mentioned in (14). Constraint (24) states that the probability of a transmission link P_{link} should be no less than the minimal predetermined required value P_{min}.

4.3.3. Delay

As analyzed in (17) and (20), maximizing the network lifetime will increase the delay. In addition, the I2O WBAN links suffer from high energy attention leading to transmission link instability causing higher data transmission delay. Propagation delay is an important factor in dealing with high data rate transmission scenarios. The mathematical model of the end to end delay can be expressed as:

$$Objective: Min\ \tau_{SC} = \tau_S + \tau_{RC} \tag{25}$$

where τ_{SC} is the delay for the in-body node S transmits to the coordinator C. τ_S and τ_{RC} represent nodal delay at S and delay for data transmission between R and C, respectively.

Subject to:

$$\tau_S \geq \tau_S^{Tx} + \tau_S^{queue} + \tau_S^{Proc} + \tau_S^{CC},\ \forall S \in N \tag{26}$$

$$x \geq N \geq 0,\ \forall x \in Z^+ \tag{27}$$

$$P_{SR} \geq P_S \tag{28}$$

$$\gamma_S^{dep} \geq \gamma_S^{arr} \tag{29}$$

$$BER_i \geq BER_{pre} \tag{30}$$

$$Min\ d_{SR} \rightarrow d_{min} \tag{31}$$

Constraint (26) illustrates that the nodal delay τ_S consists of the propagation delay τ_S^{Tx}, queuing delay τ_S^{queue}, data processing delay τ_S^{Proc} and channel capture delay τ_S^{CC} [21]. Constraint (27) provides the upper and lower bounds for N. Considering a dense I2O WBAN network, (number of N is very large), the channel access delay τ_S^{CC} will increase due to more sensor nodes contending the channel access. Constraint (28) regulates the number of data packets that need optimizing due to the limited packet handing capacity at the receiver. Similarly, Constraint (29) demonstrates that the packet arrive rate γ_S^{arr} should be less than the packet transmit rate γ_S^{dep} and this can reduce the queuing delay τ_S^{queue}. Constraint (30) points out that BER should be higher than a predetermined BER threshold, otherwise the rate of dropped packets will increase τ_S^{Proc}. Constraint (31) illustrates that minimizing the transmission distance is an effective method to decrease the propagation delay τ_S^{Tx}, which can be defined as:

$$\tau_S^{Tx} = d(S)/c \tag{32}$$

where $d(S)$ is the distance between the in-body sensor node to relay in each round and c is the speed of electromagnetic wave, respectively. Since there is no agreement or standard for superframe structures in I2O WBAN devices, we follow [37] and estimate the delay as stated in (32).

4.4. Proposed Protocols

The overall flow chart for the protocol is shown in Figure 11 and we now give details of the various operations within it.

4.4.1. Initialization Phase

All in-body sensor nodes and relays are assigned a unique ID. The data transmission time division multiple access (TDMA) schedules for sensors and relays are based on those IDs. The first round starts once the network initialization and configuration have finished. The coordinator checks the energy status of all in-body sensor nodes.

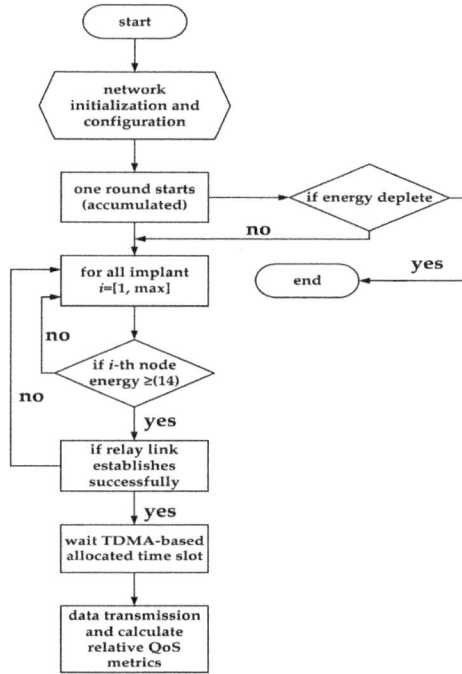

Figure 11. The information flow for the proposed protocol.

4.4.2. Routing Phase

If the in-body sensor node residual energy is greater than (14), then the coordinator will check the distance between a relay and the node. The coordinator assigns TDMA time slots to the in-body sensor node and a nearby relay, which is selected based on the cost function similarly to [36]:

$$C\left(S\right) = d\left(S\right)/R\left(S\right) \tag{33}$$

Relay nodes are available for all in-body sensor nodes. For the two-relay based protocol, when two possible relays have the same cost function values, the one with a smaller value of (15) is selected.

4.4.3. Transmission Phase

In the data transmission scheduling stage, relays assign TDMA based allocated time slots to the in-body sensor nodes. A communication link between the selected in-body sensor node and the nearby relay is established. The selected in-body sensor node transfers the sensed data during the allocated time slot to the nearby relay node. The relay node receives the data from the in-body sensor node and forwards them to the coordinator during the allocated time slots. The process will continue with the number of rounds accumulating until the energy of all the in-body sensor nodes is depleted.

4.5. Performance Evaluation and Results

The topology is introduced as follows, a coordinator is located in the center of the human body and six in-body sensor nodes are also positioned within the body. The coordinates of the coordinator and the in-body sensor nodes are summarized in Table 6. All in-body nodes have the same initial energy of 0.5 Joules. The number of relays is limited to two with coordinates (1.65, 0.75) and (0.9, 1.65). As presented in (24), a probabilistic approach (random uniform model) with a probability of packet loss of 0.3 is utilized in all simulation cases in agreement with [36]. The packet size is set as 2000 bits, which is defined as the maximum payload based on the IEEE 802.15.6 standard [22]. Simulation parameters are summarized in Table 7.

Table 6. The Coordinates of In-Body Nodes and Coordinator.

Node ID	X-Coordinate	Y-Coordinate
1	0.2	1.6
2	0.4	0.4
3	0.3	0.1
4	0.6	0.35
5	0.7	1.5
6	0.9	1.65
coordinator	0.45	0.85

Table 7. Simulation Parameters.

Simulation Parameter (Unit)	Value
Number of in-body nodes	6
Network Initial energy (Joule)	3
Payload size (bits)	2000
Electromagnetic wave speed (m/s)	3×10^{8}
Packet loss probability	0.3

Figure 12 illustrates the comparison of the stable period and total network lifetime for the proposed protocols. It can be seen that the incremental relay-based routing protocol achieves a longer stability period and total network lifetime than the two-relay based protocols with two different transceivers. This is because the second relay node only receives and forwards the data from the in-body sensor nodes when the first relay fails and therefore minimizes the network energy consumption. The stability periods of the incremental relay-based protocol are circa 4000 rounds using the using nRF2401A and circa 1500 rounds with the CC2420, whereas the two-relay based delivers 3400 rounds and 1200 rounds, respectively.

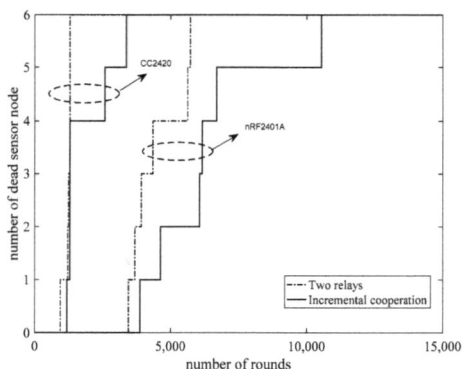

Figure 12. The relationship berween the number of dead nodes and network lifetime.

Figure 13 presents the comparison of network residual energy of two protocols for the two transceivers considered. The nRF2410A-based incremental relay-based protocol has the highest network residual energy and thus leads to the longest network lifetime. Figure 14 demonstrates that the average energy consumption per round of the incremental protocol is 0.27 mJ and 0.57 mJ when deploying nRF2401A and CC2420, which achieves nearly 52% and 45% less energy consumption when compared to the two-relay based protocol. Reasons for the above results are as follows: (a) relay nodes forward data even if when it is not necessary, resulting in channel resouce waste; (b) an alternative data transmission link only starts to work when the first link fails.

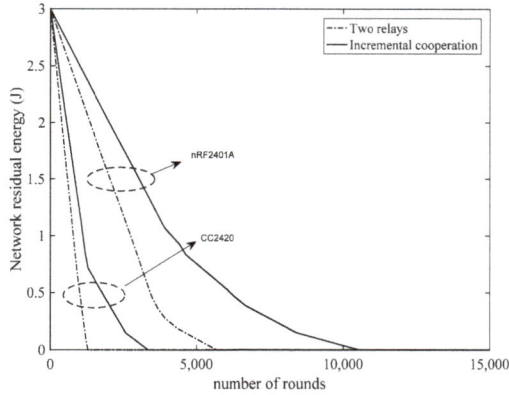

Figure 13. The residual energy versus the network lifetime.

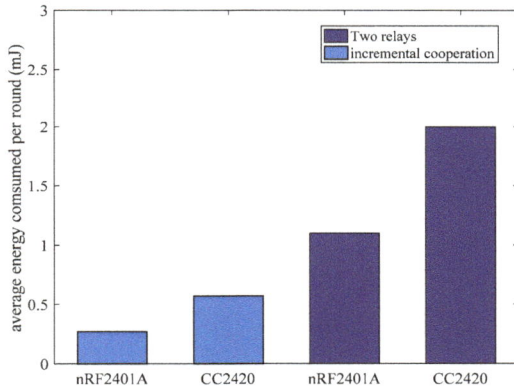

Figure 14. Average energy consumption per round.

Data loss is usually caused by transmission attenuation and shadow fading as presented in Section 3 [21]. Figure 15 shows the number of successfully transmitted information packets in the network and Figure 16, the network delay. It can be seen that the two-relay protocol can promise higher data transmission rates than the incremental relay-based protocol when deploying the same transceiver. The number of total transmission packets depends on the number of alive, in-body sensor nodes and the total network lifetime. The shorter lifetime of the two-relay routing technique is the major factor in its decreased throughput. The propagation delay depends on the overall I2O WBAN communication distance. Moreover, it can be seen from Figures 15 and 16 that more information packets generated by the system leads to higher propagation delays in the network.

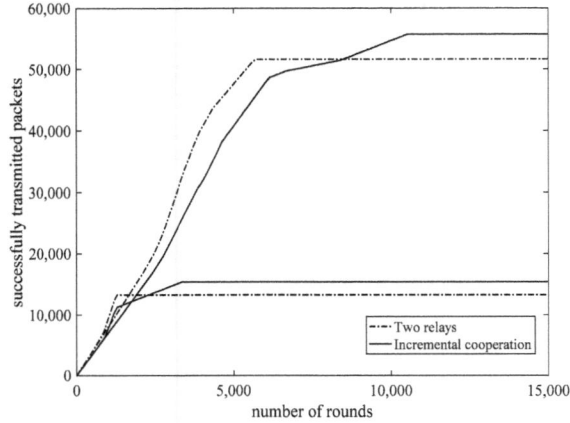

Figure 15. The successfully transmitted packets versus the network lifetime.

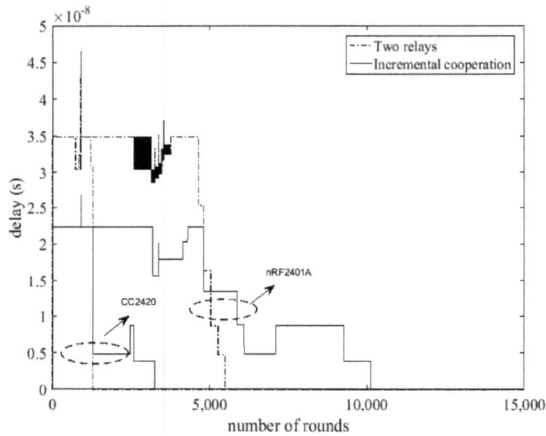

Figure 16. The propagation delay versus the network lifetime.

For related QoS metrics, the results show that the nRF2401A incremental technique can achieve a longer stability period and network lifetime, minimizes average network energy consumption and delivers the highest number of successfully transmitted packets. It is therefore a promising technique in the support of long-time healthcare services. For high data rate transmission or heavy traffic load requirement applications, the two-relay based approach would achieve better performance. Moreover, we compare the attributes of the other state of art existing routing techniques with our proposed protocol as summarized in Table 8. Further study will be required in order to optimise energy efficient routing protocols and to ensure data packet IP compatibility, both of which are considered essential for the ongoing health of IoT applications.

Table 8. Comparison of the Proposed Protocol With Other State of Art Techniques [21,34,37].

Protocol	Features	Weaknesses	Performance
Incremental relaying (this paper)	• Cooperative routing • Considering multiple QoS metrics • Cost function to select route • Probabilistic approach	• Not suitable for heavy load traffic	• Extend network lifetime • Higher throughput • Minimize energy consumption • Reduce propagation delay
Energy-Balanced Rate Assignment and Routing protocol (EBRAR)	• Routing strategy based on the residual energy of sensors • Intelligent transmission routes • Balance payload on sensors	• low bandwidth utilization • High drop packets • High delay	• Minimum energy consumption • Decrease packet loss • Fault resistance
Enhanced cooperative critical data transmission in emergency in static WBAN (EInCo–CEStat)	• Incremental relaying • Three relays • Cooperation communication	• Decrease stability period • Low energy efficiency	• Very high throughput • Improve packet error rate
Link-Aware and Energy Efficient protocol for WBANs (LAEEBA)	• Minimize path loss • Cost function to select the routine • Support emergency data transmission	• Throughput is not satisfactory • Only consider on-body communication channels	• Limited stability period • Limited network lifetime • Energy efficiency
Relay based routing protocol for in-body sensor networks	• Energy efficient • Improve faults tolerant performance	• Path loss model is not accurate • Limited throughput	• Prolong the network lifetime • Energy efficient

5. Discussion and Open Research Issues

In this section, we report and analyze the QoS mechanism, target-specific QoS requirements and the most relevant QoS factors for service systems. Some emergency issues and future research topics in I2O WBAN system are also listed. The requirements and potential architecture of the health of IoT based I2O WBANs is discussed.

5.1. QoS in I2O WBANs

5.1.1. Candidate Radio Technologies in I2O WBANs

An I2O WBAN is a type of wireless body area sensor network (WBASN) and centers on the human intra-body region using low power consumption [3,9]. There are several potential communication techniques in this area such as Bluetooth, WiFi, Zigbee and near field communication (NFC) [4]. Moreover, wireless coexistence communication technologies in I2O WBAN, such as LP-WiFi, are also promising candidates for future research [38].

5.1.2. QoS Metrics for I2O WBANs

Despite significant developments in wireless communication technology, QoS handling for different healthcare applications in I2O WBANs remains a challenging issue [39,40]. The general QoS metrics from the I2O WBAN network perspective involve network throughput, reliability transmission, energy efficiency, network lifetime and so on. It should be noticed that applications may request target-specific QoS support by specifying their requirements in terms of one or more of the QoS metrics. In general, an I2O WBAN is required to analyze the application requirements and deploy various QoS mechanisms [35]. Table 9 illustrates the QoS mechanism solutions based on requirements [4,41].

Table 9. QoS Mechanism Solutions Based on Requirements.

QoS Mechanism	Reliability	Real-Time Transmission	Energy Efficiency	Adaptability
Data collision	-	√	√	√
Data compression	-	-	√	√
Error control coding	√	√	√	√
Power control	√	√	√	√
Targeted ability	-	-	-	√

5.1.3. QoS Requirements for I2O WBANs

Optimization of communication systems that can realise the target-specific health IoT QoS requirements involves multiple factors [1]. As mentioned above, for transplanted organs, the battery lifetime of the implanted devices would be the crucial factor, with the data transmission rate not the primary concern. For diabetic patient implanted devices, a reliable transmission channel for glucose data transmission would be the vital issue and thus may require significant energy consumption to improve channel quality (e.g., a higher predetermined BER may be required). Surgical operations, such as wireless capsule endoscopes and biomedical image processing need relatively high data transfer speeds (nearly 10 Mbps) and our proposed I2O communication systems are able to reliably support these up to a few meters. The key factors of the selected in-body and on-body WBAN applications discussed in this paper are summarized in Table 10 [8,9,42].

Table 11 highlights the QoS requirements of WBAN applications [39]; Table 12 demonstrates detailed QoS parameters of WBAN service requirements in the application, transport, network, media access control (MAC) and physical layers [4,39]. Thus, I2O WBAN and system service QoS issue metrics and requirements in relation to each layer, are summarised and analysed in Tables 9–12.

Table 10. The Key Factors of Selected In-Body and On-Body WBAN Applications [6,7,20].

Application	Sensor	Energy Consumption	BER	Operating Distance	Lifetime	Data Rate
On-body applications	ECG	Low	Low	High	>1 week	>3 Kbps
	Blood pressure	Low	Low	Low	Very long	<10 Kbps
In-body applications	ICD	Moderate	High	Low	>40 h	Few Kbps
	Organ monitoring	Low	Moderate	Moderate	7–10 days	>100 Kbps
	Glucose	High	Moderate	Low	>1 week	Few Kbps
	Capsule endoscope	High	High	Moderate	>24 h	~10 Mbps
	Image processing	High	High	Low	>12 h	~10 Mbps

Table 11. The QoS Requirement of WBAN Applications [20,21].

QoS Requirement	WBAN
Data rate	WBAN communication systems should cover bit rates from few Kbps to ~30 Mbps
Tolerance	Stand ~3 s when sensor nodes either added or removed
Maximum number of sensor nodes	≤256
Mobility	Capable to reliable transmission when people moving Data should not loss even if capacity is reduced Anti-interference when people moving
Latency	Latency <125 ms for medical applications, Latency <250 ms for non-medical service Jitter <50 ms for all applications
Coexistence	In-body and on-body sensor nodes should able to work together

Table 12. Important QoS Parameters of WBAN Service Systems [4,21].

Layer	QoS Issues	QoS Metric	QoS Requirements
Application	• Total system lifetime • Data fusion ability • Error tolerant	• Communication distance • Resource allocation • Available working time • Fault awareness	• Maximum system lifetime • Transmission reliability • Security
Transport	• Link reliability • Latency • Packet loss • Transmission corruption	• Delay • Jitter (delay variance) • Buffer • Error packet ratio • Energy efficient strategy	• Minimum transmission latency • Minimum energy consumption • Decrease packet loss • Suitable data rates awareness • Fault resistance
Network	• Path latency • Routing • Mobility capability • Faults tolerant	• Channel latency • Traffic strategy • System robustness	• Minimise latency • Mobility support • Routing fault tolerance • Energy control (lower overhead)
MAC	• Throughput • Delay • Packets delivery • Collision management • Faults tolerant • Power management	• Data rate • Throughput • Collision probability • interference immunity • Bandwidth utilisation • link fault tolerant	• Improve reliability • Error control (channel coding) • Minimise data collision • Decrease interference
Physical	• Topology • Physical standards (such as interface) • Bandwidth management • Tx(s), Rx(s) faults tolerant	• Capacity • Data rate • SNR • Power attenuation	• Efficient topology • Data rate • Priority information guarantee • Channel allocation strategy

5.2. Emerging I2O WBAN Issues

5.2.1. I2O WBAN Packet Design

To incorporate the diverse QoS requirements of heterogeneous communication networks into a packet format, superframe structures are promising technologies that allow I2O devices with a particular traffic type to transmit during the period that is best suited in meeting the corresponding QoS [42]. Here, we assume that this aspect is in existence, whilst acknowledging that there is further work to be done in order to bring it into routine practice.

5.2.2. I2O WBAN Interface Design

The in-body environment makes I2O WBAN technologies more complex and challenging when compared with other communication networks. I2O WBAN systems can be regarded as a 'shared bus', where different kinds of entities that can generate data and transfer the data provided by other entities [43–46]. Recently, software-based results related to WBAN entity interfaces design have been presented [44]. However, the in-body WBAN interface design should consider both network configuration and network management and there is to date no agreed international standard for the I2O interface technologies [43,44]. Different layer requirements for in-body device (logical management) entities interface, are reported in [40]. One promising solution reported in [44], is a radio-based I2O system that enables data transmission between the in-body device(s) to the e-health gateway via an external programming device. For on-body WBAN interface technologies, smartphones and other on-body sensor nodes (wearable devices) with data collection, data buffering, information transmission, user authentication, computational and communication capabilities have been proposed to work together with the in-body interface [3,47].

5.2.3. I2O WBAN Models Validation

Owing to technical constraints and legal provisions, practical human I2O radio channel experiments are not possible [48,49]. Alternative approaches to validate the I2O WBAN communication systems involve advanced computational electromagnetics and biological phantoms [5,49]. In the future, multi-disciplinary collaboration with clinicians to measure radio channel in animals is one promising technique that can derive statistical I2O body models that could overcome the limitations of performing measurements in humans.

5.3. Analysis of I2O WBAN Based Health IoT

An effective approach for the interconnection of WBAN systems is to use the Internet Protocol (IP) [35]. WBAN information packets can be processed and translated into IP datagrams by a gateway or a smartphone on various available platforms, such as the Advanced Health and Disaster Aid Network (AID-N) and the Microsystems Platform for Mobile Services and Applications (MIMOSA) [20]. Smartphones equipped with multiple network interfaces could enable the user to interact with the linked WBAN and forward data to physicians in any location [35,50].

To date, the core of health IoT solutions is low power, wireless personal area networks (6LoWPAN) and Internet Protocol for Smart Objects (IPSO), which are predicted by the Internet Engineering Task Force (IETF) and aim to manage WBAN devices Internet connectivity issues through IP version 6 (IPv6) [35]. It is of great significance that the I2O WBAN infrastructure supports access to the health IoT. Meanwhile, more effort should be spent on I2O WBANs low power consumption routing protocol design in order to minimize the energy consumption and improve multiple QoS metric performance. An I2O WBAN sensor address configuration process should consider uniqueness, low energy consumption and offer address reclamation. Once a WBAN system collects the physical parameters of a human body, all sensor nodes in the WBAN must be configured with a unique address [48]. In recent years, the IEEE and other authorities have standardized numerous protocols to support WBANs [20]. However at present, the proposed standardized protocols are ineligible for an I2O WBAN [20,35,51]. For example, the IEEE 802.15.3 standard is designed for high data speed, wireless personal area networks (WPANs) that can reach up to 20 Mbps. However, this standard does not support energy efficiency and other QoS requirements [35]. The IEEE 802.15.4 standard is considered as an energy efficient protocol but only capable for low data rate applications and services [35]. The latest IEEE 802.15.6 defines MAC and PHY layers for low-power consumption implant devices, whereas the security requirements of data authentication and encryption have not been well defined [20]. The Zigbee IP is the first open standard for the IPv6 standard, which enriches the WBAN services by adding network and security layers. However, recently, the Zigbee Alliance decided to incorporate standards from IETF into its technical specifications [20].

I2O WBANs can be divided into two categories; (1) every sensor node is provided with a unique entity or; (2) the network is an entity and accessible via a coordinator node that has full information about the network [50]. I2O WBAN sensor nodes are attached on, or within the human body area to process sensed data with wireless transfer to a coordinator connected to the internet [3]. Considering QoS-aware requirements, this paper presents a potential QoS-aware protocol design for I2O WBAN with the following characteristics:

(1) The I2O WBAN consists of at least one full function device (such as a smartphone) and a series of reduced function small-size sensor nodes, which are assigned a unique ID.
(2) Each sensor node senses different physical parameters to reduce the address configuration cost.
(3) A relay strategy is considered, which decreases system configuration delays and minimizes the overall length of communication distance within I2O WBAN.

6. Conclusions

In this paper, we first propose and analyze an efficient and accurate PL model for the 3D human I2O communication system at 2.45 GHz. Due to the limitation of the implanted sensor batteries, we have investigated and compared several established high efficiency modulation schemes. The threshold SNRs of BPSK, QPSK, 16PSK and 16QAM are approximately 11 dB, 13 dB and 15.5 dB and 18 dB respectively, when an acceptable predetermined BER of 10^{-3} is adopted. Results demonstrate that the communication system can achieve satisfactory performance at relatively high data rates of 30 Mbps over distances of up to 1.6 m. Alternatively it can provide an highly reliable communication link for longer distances, at lower data rates (0.25 to 5 Mbps) by adopting the BPSK modulation technique. Based on the proposed I2O WBAN system, an energy efficient incremental routing protocol has been implemented and compared with the existing two-relay strategy. Simulation results demonstrate that our proposed data routing technique could significantly improve the performance of network lifetime, throughput and propagation delay. Open issues and standardization within the I2O WBAN area are summarized and explored as an inspiration towards developments in health IoT applications. Our future work involves radio channel measurements in biological phantoms to validate the PL model and communication system performance. Moreover, this can be extended to animal studies by multi-disciplinary collaboration with clinical professionals and biologists in order to overcome the difficulties of performing measurements in humans.

Author Contributions: This paper is completed by Yangzhe Liao and Chenyao Bai under the guidance of supervisors Mark S. Leeson and Matthew D. Higgins. Yangzhe Liao and Chenyao Bai conceived and designed the research topics and analyzed the data. Yangzhe Liao wrote the paper under the help of Mark S. Leeson.

Conflicts of Interest: The authors declare no conflicts of interest.

References

1. Islam, S.M.R.; Kwak, D.; Kabir, M.H.; Hossain, M.; Kwak, K.S. The internet of things for health care: A comprehensive survey. *IEEE Access* **2015**, *3*, 678–708. [CrossRef]
2. Fan, Y.J.; Yin, Y.H.; Xu, L.D.; Zeng, Y.; Wu, F. IoT-based smart rehabilitation system. *IEEE Trans. Ind. Inform.* **2014**, *10*, 1568–1577.

3. Liao, Y.; Leeson, M.S.; Higgins, M.D. Flexible quality of service model for wireless body area sensor networks. *IET Healthc. Technol. Lett.* **2016**, *3*, 12–15. [CrossRef] [PubMed]

4. Movassaghi, S.; Abolhasan, M.; Lipman, J.; Smith, D.; Jamalipour, A. Wireless body area networks: A survey. *IEEE Commun. Surv. Tutor.* **2014**, *16*, 1658–1686. [CrossRef]

5. Liao, Y.; Leeson, M.S.; Higgins, M.D. A communication link analysis based on biological implant wireless body area networks. *Appl. Comput. Electromagn. Soc. J.* **2016**, *31*, 619–628.

6. Galluccio, L.; Melodia, T.; Palazzo, S.; Santagati, G.E. Challenges and implications of using ultrasonic communications in intra-body area networks. In Proceedings of the 9th Annual Conference on Wireless On-demand Network Systems and Services (WONS), Courmayeur, Italy, 9–11 January 2012.

7. Fang, Q.; Lee, S.; Permana, H.; Ghorbani, K.; Cosic, I. Developing a wireless implantable body sensor network in MICS band. *IEEE Trans. Inform. Technol. Biomed.* **2011**, *15*, 567–576. [CrossRef] [PubMed]

8. IEEE standard for safety levels with respect to human exposure to radio frequency electromagnetic fields, 3 KHz to 300 GHz amendment 1: Specifies ceiling limits for induced and contact current, clarifies distinctions between localized exposure and spatial peak power density. Available online: http://ieeexplore.ieee.org/servlet/opac?punumber=10830 (accessed on 11 July 2016).

9. The International Commission on Non-Ionizing Radiation Protection. Guidelines for limiting exposure to time-varying electric, magnetic, and electromagnetic fields (up to 300 GHz). *Health Phys.* **1998**, *74*, 494–522.

10. Ntouni, G.D.; Lioumpas, A.S.; Nikita, K.S. Reliable and energy-efficient communications for wireless biomedical implant systems. *IEEE J. Biomed. Health Inform.* **2014**, *18*, 1848–1856. [CrossRef] [PubMed]

11. Darwish, A.; Hassanien, A.E. Wearable and implantable wireless sensor network solutions for healthcare monitoring. *Sensors* **2011**, *11*, 5561–5595. [CrossRef] [PubMed]

12. Elias, J. Optimal design of energy-efficient and cost-effective wireless body area networks. *Ad Hoc Netw.* **2014**, *13*, 560–574. [CrossRef]

13. Zhu, S.; Langley, R. Dual-band wearable textile antenna on an EBG substrate. *IEEE Trans. Antennas Propag.* **2009**, *57*, 926–935. [CrossRef]

14. Kurup, D.; Joseph, W.; Vermeeren, G.; Martens, L. Path loss model for in-body communication in homogeneous human muscle tissue. *Electron. Lett.* **2009**, *45*, 453–454. [CrossRef]

15. Kurup, D.; Joseph, W.; Vermeeren, G.; Martens, L. In-body path loss model for homogeneous human tissues. *IEEE Trans. Electromagn. Compat.* **2012**, *54*, 556–564. [CrossRef]

16. Computer Simulation Technology. Available online: https://www.cst.com/ (accessed on 4 April 2016).

17. Roman, K.L.L.; Vermeeren, G.; Thielens, A.; Joseph, W.; Martens, L. Characterization of path loss and absorption for a wireless radio frequency link between an in-body endoscopy capsule and a receiver outside the body. *EURASIP J. Wirel. Commun. Netw.* **2014**, *2014*, 1–10.

18. Hannan, M.A.; Abbas, S.M.; Samad, S.A.; Hussain, A. Modulation techniques for biomedical implanted devices and their challenges. *Sensors* **2012**, *12*, 297–319. [CrossRef] [PubMed]

19. Elhadj, H.B.; Chaari, L.; Kamoun, L. A survey of routing protocols in wireless body area networks for healthcare applications. *Int. J. E Health Med. Commun.* **2012**, *3*, 1–18. [CrossRef]

20. Cao, H.; Leung, V.; Chow, C.; Chan, H. Enabling technologies for wireless body area networks: A survey and outlook. *IEEE Commun. Mag.* **2009**, *47*, 84–93. [CrossRef]

21. Javaid, N.; Ahmad, A.; Khan, Y.; Khan, Z.A.; Alghamdi, T.A. A relay based routing protocol for wireless in-body sensor networks. *Wirel. Personal Commun.* **2015**, *80*, 1063–1078. [CrossRef]

22. Deepak, K.; Babu, A.V. Improving energy efficiency of incremental relay based cooperative communications in wireless body area networks. *Int. J. Commun. Syst.* **2015**, *28*, 91–111. [CrossRef]

23. Kurup, D.; Vermeeren, G.; Tanghe, E.; Joseph, W.; Martens, L. In-to-out body antenna-independent path loss model for multilayered tissues and heterogeneous medium. *Sensors* **2015**, *15*, 408–421. [CrossRef] [PubMed]

24. Dielectric Properties of Body Tissues. Available online: http://niremf.ifac.cnr.it/tissprop/ (accessed on 29 March 2016).

25. Kurup, D.; Joseph, W.; Vermeeren, G.; Martens, L. Specific absorption rate and path loss in specific body location in heterogeneous human model. Microwaves. *IET Antennas Propag.* **2013**, *7*, 35–43. [CrossRef]

26. Liu, X.; Hin, T.K.; Heng, C.H.; Gao, Y.; Toh, W.D.; Cheng, S.J.; Je, M. A 103 pJ/bit multi-channel reconfigurable GMSK/PSK/16-QAM transmitter with band-shaping. In Proceedings of the 2014 IEEE Asian Solid-State Circuits Conference (A-SSCC), Kaohsiung, Taiwan, 10–12 November 2014.

27. Kim, S.R.; Ryu, H.G. Analysis and design of QAPM modulation based on multi-carrier using compressive sensing for low power communication. In Proceedings of the 2012 International Conference on ICT Convergence (ICTC), Jeju, Korea, 15–17 October 2012.
28. Cheffena, M. Performance evaluation of wireless body sensors in the presence of slow and fast fading effects. *IEEE Sens. J.* **2015**, *15*, 5518–5526. [CrossRef]
29. Wang, J.; Fujiwara, T.; Kato, T.; Anzai, D. Wearable ECG based on impulse radio type human body communication. *IEEE Trans. Biomed. Eng.* **2015**. [CrossRef] [PubMed]
30. Liao, Y.; Leeson, M.S.; Higgins, M.D. An in-body communication link based on 400 MHz MICS band wireless body area networks. In Proceedings of the 2015 IEEE 20th International Workshop on Computer Aided Modelling and Design of Communication Links and Networks (CAMAD), Guildford, UK, 7–9 September 2015.
31. Anzai, D.; Aoyama, S.; Yamanaka, M.; Wang, J. Impact of spatial diversity reception on SAR reduction in implant body area networks. *IEICE Trans. Commun.* **2012**, *95*, 3822–3829. [CrossRef]
32. Nagata, Y.; Hiraoka, M.; Akuta, K.; Abe, M.; Takahashi, M.; Jo, S.; Nishimura, Y.; Masunaga, S.; Fukuda, M.; Imura, H. Radiofrequency thermotherapy for malignant liver tumors. *Cancer* **1990**, *65*, 1730–1736. [CrossRef]
33. Kailas, A.; Ingram, M.A. Wireless aspects of telehealth. *Wirel. Personal Commun.* **2009**, *51*, 673–686. [CrossRef]
34. Yousaf, S.; Javaid, N.; Khan, Z.A.; Qasim, U.; Imran, M.; Iftikhar, M. Incremental relay based cooperative communication in wireless body area networks. *Procedia Comput. Sci.* **2015**, *52*, 552–559. [CrossRef]
35. Yigitel, M.A.; Incel, O.D.; Ersoy, C. QoS-aware mac protocols for wireless sensor networks: A survey. *Comput. Netw.* **2011**, *55*, 1982–2004. [CrossRef]
36. Javaid, N.; Ahmad, A.; Nadeem, Q.; Imran, M.; Haider, N. Im-Simple: Improved stable increased-throughput multi-hop link efficient routing protocol for wireless body area networks. *Comput. Hum. Behav.* **2015**, *51*, 1003–1011. [CrossRef]
37. Sandhu, M.; Javaid, N.; Jamil, M.; Khan, Z.; Imran, M.; Ilahi, M.; Khan, M. Modeling mobility and psychological stress based human postural changes in wireless body area networks. *Comput. Hum. Behav.* **2015**, *51*, 1042–1053. [CrossRef]
38. Hayajneh, T.; Almashaqbeh, G.; Ullah, S.; Vasilakos, A.V. A survey of wireless technologies coexistence in WBAN: Analysis and open research issues. *Wirel. Netw.* **2014**, *20*, 2165–2199. [CrossRef]
39. Kathuria, M.; Gambhir, S. Quality of service provisioning transport layer protocol for WBAN system. In Proceedings of the 2014 International Conference on Optimization, Reliability, and Information Technology (ICROIT), Haryana, India, 6–8 February 2014.
40. Otto, C.; Milenkovic, A.; Sanders, C.; Jovanov, E. System architecture of a wireless body area sensor network for ubiquitous health monitoring. *J. Mob. Multimed.* **2006**, *1*, 307–326.
41. Razzaque, M.A.; Javadi, S.S.; Coulibaly, Y.; Hira, M.T. QoS-aware error recovery in wireless body sensor networks using adaptive network coding. *Sensors* **2014**, *15*, 440–464. [CrossRef] [PubMed]
42. Monowar, M.M.; Hassan, M.M.; Bajaber, F.; Al-Hussein, M.; Alamri, A. Mcmac: Towards a MAC protocol with multi-constrained QoS provisioning for diverse traffic in wireless body area networks. *Sensors* **2012**, *12*, 15599–15627. [CrossRef] [PubMed]
43. IEEE/IOS Healthcare IT Standards. Available online: https://standards.ieee.org/findstds/standard/healthcare_it.html (accessed on 30 April 2016).
44. European Union Framework Programme 7 'WBAN Architecture and Open Middleware'. Available online: http://daphne-fp7.eu/sites/default/files/D3.2%20DAPHNE_WBAN_Open_Middleware.pdf (accessed on 30 April 2016).
45. Picazo-Sanchez, P.; Tapiador, J.E.; Peris-Lopez, P.; Suarez-Tangil, G. Secure publish-subscribe protocols for heterogeneous medical wireless body area networks. *Sensors* **2014**, *14*, 22619–22642. [CrossRef] [PubMed]
46. Rushanan, M.; Rubin, A.D.; Kune, D.F.; Swanson, C.M. Sok: Security and privacy in implantable medical devices and body area networks. In Proceedings of the 2014 IEEE Symposium on Security and Privacy (SP), San Jose, CA, USA, 18–21 May 2014.
47. Wu, W.H.; Bui, A.A.; Batalin, M.A.; Au, L.K.; Binney, J.D.; Kaiser, W.J. Medic: Medical embedded device for individualized care. *Artif. Intell. Med.* **2008**, *42*, 137–152. [CrossRef] [PubMed]
48. Animal Experimentation the Facts. Available online: http://www.bbc.co.uk/ethics/animals/using/facts.shtml (accessed on 30 April 2016).

49. IEEE Standards Association 802.15.6-2012-Part 15.6: Wireless Body Area Networks. Available online: https://standards.ieee.org/findstds/standard/802.15.6-2012.html (accessed on 30 April 2016).
50. Reina, D.G.; Toral, S.L.; Barrero, F.; Bessis, N.; Asimakopoulou, E. The role of ad hoc networks in the internet of things: A case scenario for smart environments. In *Internet of Things and Inter-Cooperative Computational Technologies for Collective Intelligence*; Springer: Berlin, Germany, 2013; pp. 89–113.
51. Ghamari, M.; Janko, B.; Sherratt, R.S.; Harwin, W.; Piechockic, R.; Soltanpur, C. A survey on wireless body area networks for e-healthcare systems in residential environments. *Sensors* **2016**, *16*, 831. [CrossRef] [PubMed]

electronics

Article

A Galvanic Coupling Method for Assessing Hydration Rates

Clement Ogugua Asogwa *, Stephen F. Collins, Patrick Mclaughlin and Daniel T.H. Lai

College of Engineering & Science, Victoria University, Melbourne, VIC 8001, Australia;
stephen.collins@vu.edu.au (S.F.C.); patrick.mclaughlin@vu.edu.au (P.M.); daniel.lai@vu.edu.au (D.T.H.L.)
* Correspondence: clement.asogwa@live.vu.edu.au; Tel.: +61-3-9919-5047

Academic Editors: Enzo Pasquale Scilingo and Gaetano Valenza
Received: 30 May 2016; Accepted: 1 July 2016; Published: 13 July 2016

Abstract: Recent advances in biomedical sensors, data acquisition techniques, microelectronics and wireless communication systems opened up the use of wearable technology for ehealth monitoring. We introduce a galvanic coupled intrabody communication for monitoring human body hydration. Studies in hydration provide the information necessary for understanding the desired fluid levels for optimal performance of the body's physiological and metabolic processes during exercise and activities of daily living. Current measurement techniques are mostly suitable for laboratory purposes due to their complexity and technical requirements. Less technical methods such as urine color observation and skin turgor testing are subjective and cannot be integrated into a wearable device. Bioelectrical impedance methods are popular but mostly used for estimating total body water with limited accuracy and sensitive to 800 mL–1000 mL change in body fluid levels. We introduce a non-intrusive and simple method of tracking hydration rates that can detect up to 1.30 dB reduction in attenuation when as little as 100 mL of water is consumed. Our results show that galvanic coupled intrabody signal propagation can provide qualitative hydration and dehydration rates in line with changes in an individual's urine specific gravity and body mass. The real-time changes in galvanic coupled intrabody signal attenuation can be integrated into wearable electronic devices to evaluate body fluid levels on a particular area of interest and can aid diagnosis and treatment of fluid disorders such as lymphoedema.

Keywords: galvanic coupling; signal attenuation; hydration rates; body fluid level

1. Introduction

Assessment of human body composition is fundamental to the understanding of body physiological and metabolic processes. Body fluid contributes up to 60 percent of the mass of the human body. The body's fluid state is affected by both endogenous processes, such as body metabolism, and exogenous factors such as climatic changes, exercise, disease, and diet. Investigations into hydration are required because they help identify or quantify ill-health and understanding poor exercise performance. Consequences of excessive fluid losses or inadequate fluid intake can include hypohydration, urinary infections, reduction in cognitive function, reduction of cellular metabolism and mortality [1]. The body fluid shifts between the intracellular and the extracellular tissues to maintain water balance. This movement follows an osmotic gradient in order to ensure that optimal concentrations of electrolytes and non-electrolytes are maintained in the cells, tissues, plasma, and interstitial fluid. Two adverse conditions can be identified: hyperhydration, in which there is excess water in the body, and hypohydration or dehydration, when there is less than the normal amount of water to meet the body's requirement [1] and accurate and easy estimates of hydration levels with cost effective technology are essential to policy makers in setting public health priorities [2], for doctors and clinicians to classify body fluid and cell mass conditions of healthy persons and patients with certain diseases [3] and for individuals, especially the elderly who are at higher risk of dehydration [3,4]. Further research evidence on the elderly showed that older people who were dehydrated at admission were more likely to die than their counterparts [5].

An electrical signal passing through the human body is strongly affected by the size of the tissue, available fluid and its dielectric properties. Tissues have a high ability to store electrical energy. High frequencies are affected by human body antenna effects and possible radiation. Our frequency range lies between 800 kHz and 1.3 MHz, which lies within the β dispersion region that is related to the cellular structure of biological materials [6] and can penetrate into the extracellular and intracellular tissue spaces. This also falls within the frequency range (5–1000 kHz) used for whole-body fluid analysis with bioimpedance spectroscopy method [7]. The electrical conduction at this frequency is affected by the amount of water solute available in the tissue spaces. Loss of body fluid, for example, during exercise could cause up to 6% increase in muscle impedance [8]. Thus, by coupling a low frequency electrical signal galvanically on the body, the signal passing through the tissue will vary in attenuation to the changes in the water level. Consequently, the attenuation (negative gain) of the signal amplitude will depend on the composition of the tissue in terms of the amount of water present at the time, the tissue muscle-fat ratio and the input signal frequency. External factors such as the type of electrode, the distance between the connecting electrodes and environmental conditions can be experimentally controlled. Similarly, signal changes due to the effects of limbs around joints [9] can be minimised by avoiding measurement on joint areas.

Techniques for assessing hydration include intrusive and non-intrusive methods. The intrusive method requires intravenous access to the body and is usually performed by trained personnel such as technicians, doctors or nurses. This system requires in vivo access and testing of the blood and is regarded as a later indicator of dehydration rather than a warning system that informs a quick preventative measure [10]. Physical signs such as urine color observation, urine specific gravity test, and body weight changes are some examples of non-intrusive methods for assessing human body hydration [11]. This method gives oversimplified results and poorly sensitive to changes in dehydration [12]. Wearable electronics that measure perspiration metabolites [13] can only estimate the physiological state of an individual's body fluid level under sweat and not without sweat secretion. Moreover, these techniques can not be used to target fluid disorder on a specific part of the body. The purpose of this work is to introduce a galvanic coupled signal propagation method for assessing hydration rates that can also be used for assessing fluid changes on a particular area of the body. Our results are comparable to previous methods, but present an easier wearable alternative to hydration. We argue that undertaking a composite testing of our proposed system alongside known urinary markers of dehydration is vital to validating our proposed idea. This paper is organised are follows: Section 2 details the modification of our previous circuit model. Section 3 is our methodology and Section 4 is the results. This is followed by discussion in Section 5 and conclusion in Section 6.

2. Modification on Previous Circuit Model

Our group developed a human body circuit model [14], which was later improved by introducing a dynamic tissue impedance $Z_F(t)$. We defined hydration as the process of gaining tissue fluid and proposed in [15] a circuit model of real time human body hydration with time variable fluid impedance (Figure 1), in which Z_{ES} is the impedance of the contact interface to the body at the transmitter and the receiver nodes, Z_T corresponds to the transverse impedance, while Z_b is the cross impedance. V_i is the transmit voltage, while V_0 is the output voltage at the receiver end with load R_l. Z_L is the longitudinal impedance of the transmission path consisting of the skin, fat, muscle, and bone, body fluid, cortical bone and bone marrow. We defined the variable impedance due to hydration as, [15]:

$$Z_F(t) = Z_{f0} - Z_w(1 - e^{-\frac{t}{\tau}}),\qquad(1)$$

where t is the time for the change in impedance to occur, Z_{f0} is the impedance at time $t = 0$ just before hydration begins, Z_w is the impedance resulting from the water consumed and the ratio $\frac{t}{\tau}$ is a characteristic that predicts the rate of hydration. τ is the time constant that characterises the metabolic process of a particular individual. We found that given an initial fluid volume V_{ib} before hydration and V_w amount of fluid consumed, the body will hydrate to a fluid volume V_b given as:

$$V_b = V_{ib} + V_w e^{\frac{t}{\tau}}; t = 0; V_b = V_{ib}, \tag{2}$$

and would reach the state of water balance at time t_f.

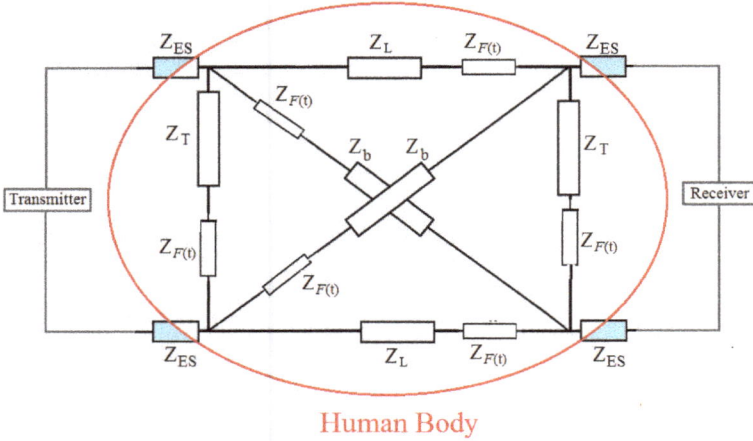

Figure 1. Circuit diagram of galvanic coupled intrabody signal with variable impedance component from variable changes in human body fluid [15], Copyright 2016, IEEE.

Based on this, we propose that the increase in the volume of body fluid, as hydration occurs, increases the volume of tissue fluid which will result in a gain in body weight by an amount equivalent to

$$wt_g = V_w e^{\frac{t}{\tau}}, \tag{3}$$

where wt_g is gain in weight, since short term changes in body weight can be attributed to loss or gain of body water and 1 mL of water has a mass of one gram [16]. By representing the anthropometric parameter contributing to the longitudinal impedance between the transmitter and the receiver electrode pairs by the cross-sectional area of the muscle-fat, θ, as a ratio. We have

$$\theta = \frac{A_m}{A_f}, \tag{4}$$

where A_m and A_f are the cross-sectional areas of muscle and fat, respectively, to the distance between the transmitter and receiver electrode pairs. We set $0 < \theta < 1$—if we assume high θ corresponded to low fat index (low BMI) and low θ corresponded to high fat index (high BMI). BMI by definition is body weight (wt) divided by square of height (h^2), unit is (kg/m^2)

$$BMI = \frac{wt}{h^2}. \tag{5}$$

Assuming no change in height, since all experimental protocols and measurements were completed within 14 h for each participating adult, then,

$$wt \propto BMI, \tag{6}$$

and a change in wt due to hydration or dehydration will also result in a change in BMI,

$$\Delta wt \propto \Delta BMI. \tag{7}$$

Therefore, θ can be defined in terms of the changes in real body weight. We know from [17,18] that short term changes in body mass are associated with changes in human body hydration state given by Equation (3). $wt_g = \Delta wt$. If we assume a BMI mainly due to fat mass, then θ is inversely proportional to BMI. By these definitions:

$$\Delta wt \propto \Delta \frac{1}{\theta}, \tag{8}$$

or

$$\Delta wt\theta = k. \tag{9}$$

Similarly,

$$\Delta wt \propto \Delta G, \tag{10}$$

where ΔG is the change in gain (negative attenuation) of the electrical signal as a result of the change in the body hydration state, measured in dB/minute. Δwt is related to θ by a proportionality constant k. If Δwt and θ are biological constants, then the constant of proportionality k which affects the biological behaviour of the body under hydration is also biological and, from Equations (9) and (10), k is a metabolic process equivalent to τ, which, by Mifflin-St. Jeor equation [19], is related to resting metabolic rate (RMR). A dynamic change in the impedance caused by a change in the human body hydration state would result in a change of the impedance of Z_T, Z_L, and Z_b as following:

$$\acute{Z}_T = Z_T + Z_F(t), \tag{11}$$

$$\acute{Z}_L = Z_L + Z_F(t), \tag{12}$$

$$\acute{Z}_b = Z_b + Z_F(t). \tag{13}$$

Similarly, $G(f; t; wt; \tau)$.

Thus, the signal attenuation $G(f, t, wt, \tau)$ of a galvanically coupled circuit passing through the human body can be expressed in terms of frequency f, time t , change in body weight wt, and a time-dependent constant τ related to RMR [19]. This expression is similar to our previous expression of the gain G of a galvanic coupled circuit (Figure 1), with the transfer function as shown in [15]. We argued that if Equation (5) defined BMI in terms of excess body fat, then individuals with high BMI will, on average, have a lower hydration rates, will keep water longer in the body, and hence take longer to urinate, and the reverse is true if defined in terms of excess muscle mass. We shall use this relationship to empirically determine the attenuation per unit volume of water consumed in Section 3.

3. Methodology

3.1. Equipment

The measurement set up is as shown in Figure 2. We used a hand held refractometer, URICON-NE, Cat. No. 2722 with measurement uncertainty of 0.001 from ATAGO Co., Ltd.,Itabashi-ku, Tokyo, Japan to measure the urine specific gravity of the urine samples provided by the participants. Urine specific gravity (SPG) measures the ratio of the density of urine relative to the density of pure water. A specific gravity greater than one means the fluid is denser than water [20]. Urine specific gravity measurements usually range from 1.002 to 1.030. Minimal dehydration ranges from 1.010 to 1.020 with increasing severity of dehydration from 1.020 and upwards. A specific gravity of 1.030 and upwards is regarded as highly severe and values below 1.010 are classified as hyperhydration [20–22]. Other factors such as glucose level and drug use that may affect specific gravity readings were countered by ensuring that participants' diets prior to experiment were controlled and exclusive of supplements, were not taking medical drugs, and are healthy. Moreover, the experiment started after an overnight 12-h fast. We also set the specific gravity measurements as a second test for determining individual hydration level and used it in conjunction with body mass changes for our investigation. We used Hanna digital thermometer model number HI-98509 (Manufactured by Hanna Instruments, Woonsocket, RI, USA,) to measure the temperature of the urine samples. We measured the urine specific gravity of the urine samples at the required temperature of 20.0 °C . A mini Pro Vector Network Analyser (VNA), frequency range 100 kHz to 200 MHz, manufactured by Mini Radio Solutions, Poland, baluns (Coaxial RF transformers, FTB-1-1+, turns ratio of one, manufactured by Mini-Circuits (Brooklyn, NY, USA), and frequency range 0.2–500 MHz), and round pre-gelled self-adhesive Ag/AgCl snap single electrodes (1 cm diameter, manufactured by Noraxon (Noraxon Inc.,Scottsdale, AZ, USA) were used with 20 cm as the separation distance between the transmit and receive electrodes [23] to measure the signal attenuation. The Noraxon self-adhesive Silver/Silver-Chloride electrodes (Ag/AgCl) are preferred because they are designed for both research and clinical use, contain hypoallergenic gel, can be used for two hours of measurement, and reduce the effects of motion artifacts and refection compared to polarizable electrodes. The baluns are used to electrically isolate the two ports of the VNA to ensure the return current does not make a loop through the common ground of the two ports. An off-the-shelf electronic WeightWatchers weight tracking & body composition monitor model number WW125A (made in China) with measurement uncertainty ±50 g was used to measure changes in body mass of the subjects. Subjects wore light clothing and were barefoot. Subject height was measured to the nearest 0.5 cm measured against the wall bare footed and heels together with buttocks, shoulders, and head touching the vertical wall surface and clear horizontal marking sighted. Participants were given time to read and understand the experiment procedures and their consents were obtained in line with the approved procedures of the Victoria University Human Research Ethics Committee.

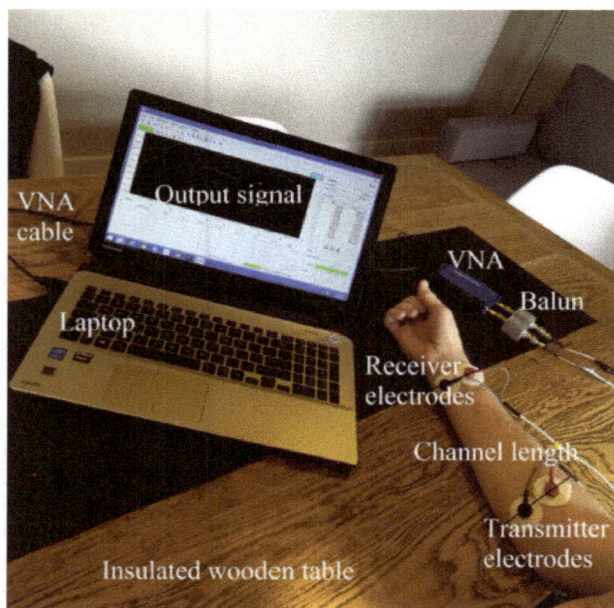

Figure 2. Intrabody signal propagation connected galvanically to measure signal attenuation after 600 mL fluid intake and after urinating on 12 male and eight female healthy adults.

3.2. Experiment I: Hydration Testing

Twenty subjects consisting of 12 males and eight females participated in this experiment. We defined two experimental protocols in this test. First, the hydration measurement protocol, which is preceded by fluid abstinence after supper (latest 10.00 pm) until 10 a.m. to induce dehydration. We measured the level of dehydration on each subject by testing the specific gravity of urine sample 1 collected prior to the start of the experiment. In addition, 600 mL of water was given to each subject to drink and the rate of hydration measured 5 min after intake—this is because water appears in plasma and blood cells within 5 min after consumption [24]. All of the participants sat on a plastic chair and were told not to move as much as humanly possible. The measured arm rested on a wooden table (Figure 2) was insulated to ensure no current leakage and to ensure movement artefacts were minimized. In the second protocol, we determined the rate of dehydration after the subjects had urinated following the consumption of 600 mL of water. We also recorded the elapsed time to produce urine by each subject. We established the pre and post hydration states of each subject by testing the individual urine specific gravity of both urine samples 1 and 2 with the hand held refractometer. We observed the urine colour changes and took measurements of the body mass differences with the electronic floor scale. The body mass of each participant was measured as W_0 before drinking, W_1 immediately after drinking and W_2 after urinating. We used a volumetric cylinder to measure the volume of urine samples produced after the 600 mL of water intake. Both the hydration and dehydration measurements were measured by taking five measurements of signal attenuation at 5 min intervals, and the average was used. The change in post drink weight and post urination weight was observed and recorded against the refractometer readings and the changes in signal attenuation. Subjects were not permitted to do rigorous exercises during the intervals throughout the period of the experiment, and all measurements were carried out at 10.00 am and average room temperature of 25 ± 0.1 °C was maintained throughout. Interference and background noise was minimised by switching off electronic devices and wireless systems around the vicinity. We also isolated communication cables away from power packs and the laptop operated at battery mode. All measurements followed the approved procedures of the Victoria University Human Research Ethics Committee, approval number HRE 14-122.

3.3. Experiment II: Sensitivity Test By Empirical Measurement

Three subjects consented to participate further in this experiment. The experiment was performed on three random days and completed in three weeks. We set the control for the sensitivity test as the average value of the signal attenuation measured for a given period of time before fluid intake. We define our sensitivity as the smallest amount of water consumed that causes a galvanic coupled intrabody signal to amplify beyond the control level after fluid intake. Both the control and the sensitivity test were performed the same day and under the same condition. To measure this, we extended the pre-drink, post drink and the measurements after urinating to 30 min. This is because we observed in experiment I that, while many subjects indicated hydration within twenty minutes, we want to observe, if there was any evidence of hydration occurring after 20 min. The 30 min pre-drink measurement was to serve as a baseline or control test. The sensitivity test followed the process narrated in experiment I, with a variation in the amount of water consumed by the subjects ranging from 100 mL, 250 mL to 300 mL on each day of the experiment. The result is reported in Section 4 for both hydration and dehydration stages.

3.4. Experiment III: Sensitivity Test By Simulation

Experiment II showed the minimum amount of water to be detected as 100 mL. We test this theoretically using our circuit model Figure 1 with the same parameters for the anthropometric measurements of the arm [25] in which a 50 mm arm radius, has the thickness of body fluid layer as 23 mm estimated from [26,27]. After consuming 100 mL of water, the maximum gain will occur when all the water consumed is retained within the 20 cm channel length. This will increase the fluid layer thickness to 26 mm, so that for a 100 mL fluid in intake, $wt_g = 0.1$ kg. Using the transfer function of G derived in [15]. We have the sensitivity after consumption of 100 mL of water maximum signal gain occurring when the 100 mL are absorbed within the 20 cm channel length.

4. Results

Table 1 presents the data from the twenty subjects. Firstly, the weighing scale measurements differed slightly from expected results after 600 mL of fluid intake, i.e., $W_1 - W_0 \neq 600$ g $\pm$ (uncertainty in measurement) in some cases. However, the change in weight after urinating, $W_2 - W_1$, corresponded to the volume of urine produced for most of the subjects. Similarly, urine specific gravity (SPG) measurement decreased from SPG1, measurement after fasting, to SPG2, measurement after fluid intake, as expected. This means that the fluid intake produced rehydration and lowering of the urine density. In a healthy person, the kidney regulates water balance by conserving water or getting rid of excess water relative to the requirement for a healthy water balance [28]. When the amount of water consumed is large enough to reduce the concentration of blood plasma, a urine more dilute than blood plasma is produced; on the other hand, when the available water is too small to dilute the blood plasma concentration, a more concentrated urine than the blood plasma is produced [29]. Higher specific gravity values indicates higher dehydration. These instances are reflected in our result. Therefore, we shall match the changes in the urine specific gravity of the subjects with the differences in weight between W_2 and W_0, and the measured attenuation after an intrabody signal is transmitted galvanically, as explained in the experiment procedure, for our analysis. The average observation on the elapsed time between fluid intake and urination increases with increase in body mass index.

Table 1. Effect of hydration on body weight and urine specific gravity (SPG) on 20 subjects.

Subject	BMI (kg/m^2)	W_0 (kg)	W_1 (kg)	W_2 (kg)	Volume of Urine (mL)	Elapsed Time (Minutes)	SPG1 before Drink	SPG2 after Drink
A	29.3	87.60	88.15	88.10	75	118	1.021	1.018
B ***	20.8	56.70	57.30	57.10	190	61	1.030	1.007
C	24.2	72.55	73.10	72.60	250	56	1.025	1.010
D **	31.4	83.15	83.55	83.15	340	76	1.016	1.010
E	33.1	93.40	93.90	93.50	305	111	1.017	1.008
F *	28.5	75.65	76.15	75.30	360	95	1.020	1.005
G **	23.5	62.80	63.20	62.80	330	60	1.016	1.010
H	31.4	101.00	101.60	101.40	150	70	1.020	1.015
I	26.4	81.65	82.10	81.85	220	99	1.023	1.014
J	36.5	104.25	104.75	104.35	325	125	1.021	1.016
K	22.9	76.00	76.55	76.30	100	70	1.019	1.016
L	23.7	73.50	74.20	73.80	300	51	1.020	1.007
M ***	25.6	95.30	95.70	95.50	200	60	1.031	1.007
N	25.9	78.60	79.10	78.80	175	86	1.024	1.010
O *	42.5	122.8	123.30	122.25	400	155	1.011	1.010
P	24.4	64.00	64.50	63.90	250	77	1.021	1.005
Q	21.9	60.30	60.75	60.50	125	78	1.017	1.014
R	23.7	66.20	66.80	66.40	350	93	1.017	1.004
S	24.0	67.70	68.30	68.00	250	62	1.023	1.006
T	26.1	92.35	92.60	92.50	175	47	1.020	1.008

* $W_2 < W_0$ Hyper hydration; ** $W_2 = W_0$ Optimal hydration; *** $W_2 > W_0$ Dehydrated.

Subject specific cases were observed based on the assumption that the water consumed was retained or excreted based on the initial body hydration level and the need to maintain homeostasis water balance [28]. In Case 1 Figure 3, $W_2 < W_0$ = Hyper hydration (example, subjects F and O). Subjects weighed less than the base line weight after urinating and initial specific gravity is low. Subjects produced the largest volume of urine after the specified drink. This suggests excess water in the subjects before the $600mL$ intake, thus $W_2 < W_0$. Absorption was observed after subjects had urinated. In Case 2 Figure 4, $W_2 = W_0$ = Optimal hydration (example, subjects D and G). Here, subjects' baseline weight was the same as the weight after urinating, and the subjects produced a large amount of urine. The fluid intake after the state of water balance was reached did not cause immediate hydration; therefore, tissue absorption and de-absorption was not continuous and water intake did not cause significant change in signal attenuation. Fluid abstinence before 10.00 a.m. did not make these groups of subjects dehydrate.

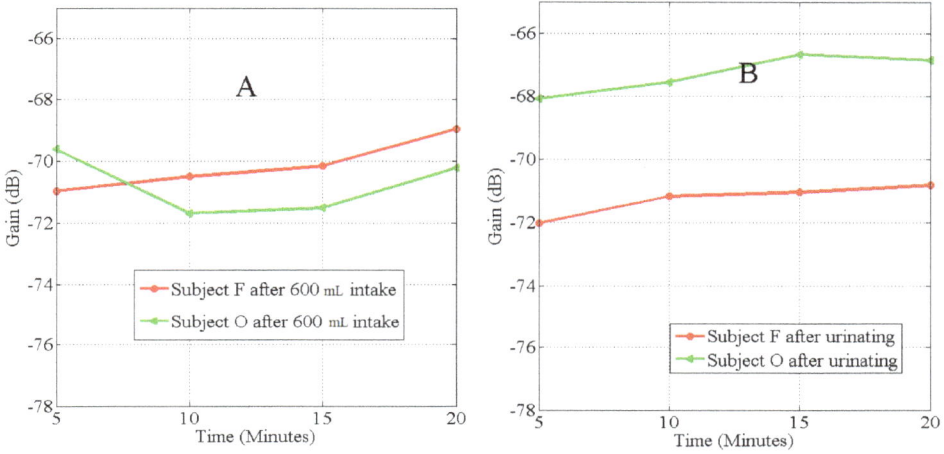

Figure 3. Graph of the rate of hydration (**A**), and dehydration (**B**), after 600 mL fluid intake on subjects F and O observed at 1.2 MHz, $W_2 < W_0$. Subjects were hyper hydrated by the protocol.

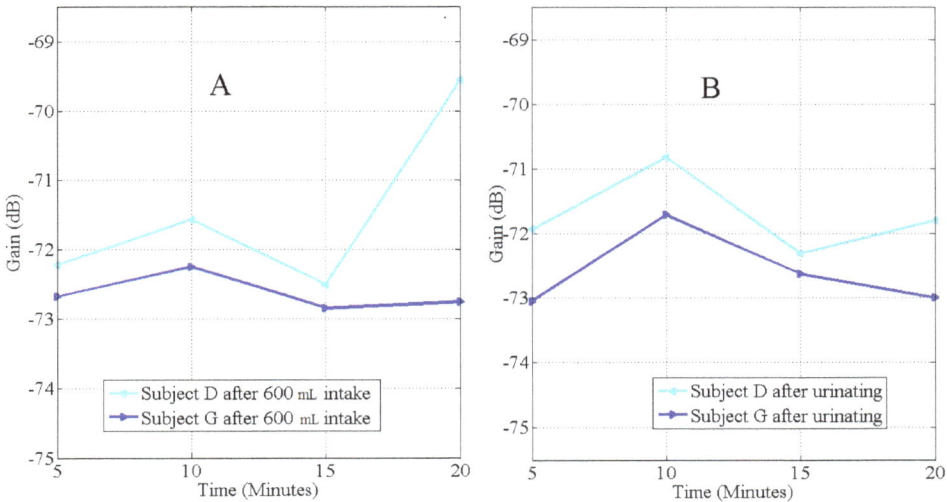

Figure 4. Graph of the rate of hydration (**A**), and dehydration (**B**), after 600 mL of fluid intake on subjects D and G at 900 kHz, $W_2 = W_0$. Subjects were normally hydrated by the protocol.

In Case 3 Figure 5, $W_2 > W_0$ = Severe dehydration (example, subjects B and M). Subjects had urine specific gravity that reflected extreme water loss, and the time taken to urinate was high compared to individual BMI. After urinating, dehydration occurred and was observed at different times.

The rest of the subjects were grouped as Case 4 Figure 6, $W_2 > W_0$, mild dehydration based on the urine specific gravity reading SPG1 measured before fluid intake. Among this group, the maximum rate of hydration was 0.44 dB/min occurring in subject N, SPG1 = 1.024, while the minimum rate occurred at 0.02 dB/min on subject Q, SPG1 = 1.017. After urination, the maximum rate of dehydration occurred at 0.40 dB/min with subject L, SPG1 = 1.020, while the minimum rate was 0.11 dB/min occurring in subject C, SPG1 = 1.025.

Figure 7 is the graph of the subject's BMI against time taken for individual metabolic process to complete and process urine. The figure shows that BMI is related with the time it takes to process urine and is also related with the rate of hydration and dehydration in accordance with our previous findings [15]. Figure 8 depicts the empirical results of the sensitivity of a galvanic coupled signal to detect hydration due to body fluid intake, while Figure 9 is the simulation result for a 100 mL maximum fluid absorption with our circuit model.

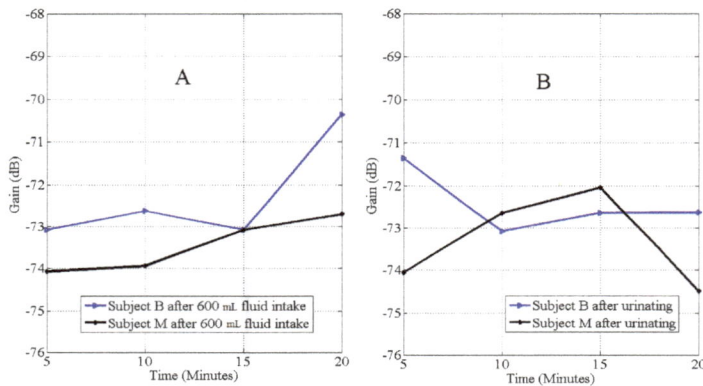

Figure 5. Graph of the rate of hydration (**A**), and dehydration (**B**), after 600 mL of fluid intake on subjects B and M observed at 900 kHz, $W_2 < W_0$. The protocol produced severe dehydration on subjects B and M. After urinating, subject B dehydrated and stopped after 10 min, while subject M started dehydration after 15 min. Both showed longer periods of re-absorption.

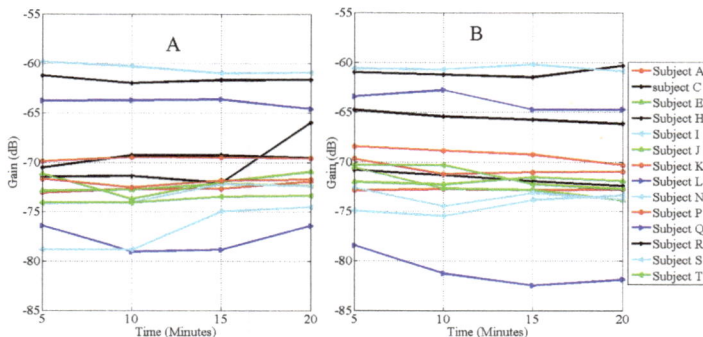

Figure 6. Graph of the rate of hydration (**A**), and dehydration (**B**), after 600 mL of fluid intake on subjects A,C,E,H,I,J,K,L,N,P,Q,R,S and T observed at 900 kHz. $W_2 < W_0$. The protocol produced mild dehydration on subjects.

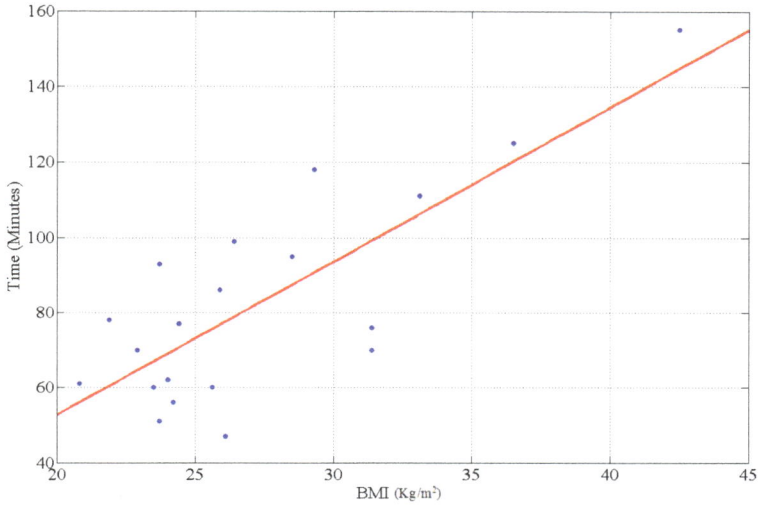

Figure 7. Graph of the relation between subject specific body mass index (BMI) and the time it took to urinate after consuming 600 mL of water.

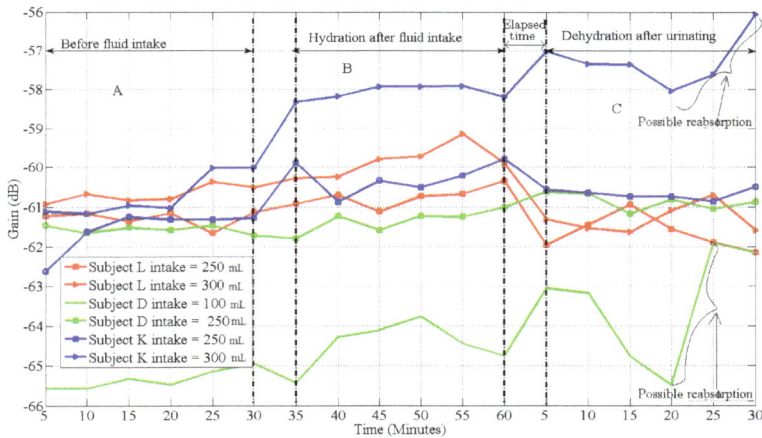

Figure 8. This graph represents the sensitivity test for an intrabody signal measured on three subjects for three random days at 900 kHz, before drinking (**A**), after drinking 100 mL, 250 mL and 300 mL amounts of water (**B**), and after urinating (**C**). From 5 min–30 min is the average attenuation measured before drink. After drinking, the attenuation was measured from 35–60 min. The gap between measurement after drink and urination is a variable time that elapsed before each subject urinates. After urinating, the attenuation was again measured from 5 min to 30 min. Measurements were taken at 900 kHz and subject specific parameters are: Subject D height = 166 cm and BMI = 29.83 on day 1, Subject K height = 182 cm and BMI 23.81 on day 2, and subject L height = 176 cm and BMI 23.34 on day 3.

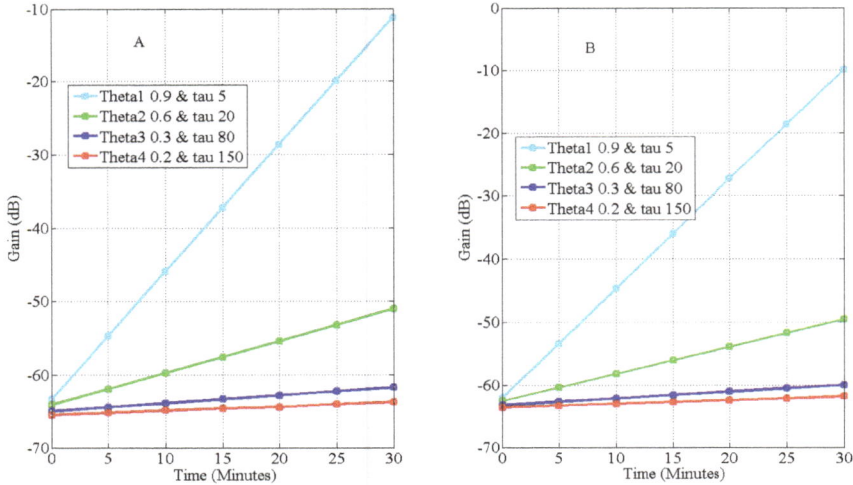

Figure 9. Simulated sensitivity of a galvanic coupled intrabody at 900 kHz at different combinations of τ and θ. (**A**) is before 100 mL fluid intake and thickness = 23 mm, while (**B**) is after 100 mL intake, assuming maximum concentration of this amount within 20 cm inter-electrode distance, fluid thickness = 26 mm. Predicted increase in attenuation is 1.5 dB for τ = 20 and θ = 0.6, and 1.70 dB for τ = 80 and θ = 0.3 and 1.91 dB for τ = 150 and θ = 0.2

Subjects D, K, and L, height 166 cm, 182 cm and 176 cm, respectively, participated in the control and sensitivity tests.

Tables 2–4 are the control and sensitivity results. The tables show that for the three random days tested, the variation in the subjects' body mass was below 2 kg, and the change in body weight corresponded to the quantity of water consumed by the subjects on each day of the experiment. The weight also decreased after the subjects had urinated and the urine specific gravity dropped after water was consumed.

Table 2. Control and sensitivity test, Day I.

	Fluid Intake = 100 mL					
Subject	BMI (kg/m^2)	W_0 (kg)	W_1 (kg)	W_2 (kg)	SPG1 before Drink	SPG2 after Drink
D	29.83	82.20	82.30	82.20	1.014	1.014
K	23.58	78.10	78.20	78.20	1.022	1.021
L	23.44	72.60	72.70	72.60	1.023	1.021

Table 3. Control and sensitivity test, Day II.

	Fluid Intake = 250 mL					
Subject	BMI (kg/m^2)	W_0 (kg)	W_1 (kg)	W_2 (kg)	SPG1 before Drink	SPG2 after Drink
D	29.65	81.70	81.90	81.80	1.016	1.012
K	23.81	77.95	78.20	78.10	1.025	1.018
L	23.27	72.10	72.40	72.20	1.021	1.010

Table 4. Control and sensitivity test, Day III.

Subject	BMI (kg/m^2)	W_0 (kg)	W_1 (kg)	W_2 (kg)	SPG1 before Drink	SPG2 after Drink
			Fluid Intake = 300 mL			
D	29.90	82.40	82.70	82.55	1.014	1.014
K	23.81	78.00	78.20	78.10	1.023	1.021
L	23.34	72.30	72.60	72.35	1.018	1.007

The control and sensitivity tests, Figure 8, show the variation between repeated measurements on three subjects on three random days for the first 30 min and the hydration as the subjects consume different amounts of water. The standard deviation between average data points on the reported data for the three subjects who consented to participate and were used for control or baseline values ranged from 0.08 dB–0.18 dB for subject D, 0.47 dB–0.58 dB for subject K and 0.21 dB–0.22 dB for subject L. The deviation from the baseline on the subjects after 100 mL intake are: subject D increased to 1.31 dB, while subjects K and L increased by 2.75 dB and 0.77 dB, respectively, after 250 mL. As the amount of water consumed increased, hydration increased evidenced by the increase in signal gain. For example, subject D increased by 4.4 dB when the amount consumed increased from 100 mL to 250 mL. After consuming 300 mL, the change in signal gain from 250 mL to 300 mL on subjects K and L are 1.57 dB and 0.64 dB, respectively. The measured attenuations after drink and after urination fall within baseline values, which were measured after fluid abstinence and before intake. The variation in the degree of measured attenuation between the three subjects is a result of individual specific body composition, initial hydration state and specific body metabolism.

Our simulation results show that, for maximum absorption of 100 mL within 20 cm channel length, the gain for different combinations of τ and θ are: $\tau = 20$ and $\theta = 0.6$, gain is 1.5 dB, and when $\tau = 80$ and $\theta = 0.3$, gain is 1.71 dB, and for $\tau = 150$ and $\theta = 0.2$, gain is 1.91 dB. The theoretical dependences of the sensitivity are the individual metabolic function τ, muscle-fat ratio, an equivalent of body mass index represented as θ and initial fluid level. Empirically, it is susceptible that a contributor to the sensitivity is height, which was not considered in the simulation. Others are subject specific endogenous processes or metabolism and initial body fluid states.

5. Discussion

5.1. Hydration Measurement Techniques

5.1.1. Urine Colour Indices

Urine color is attributed to the level of concentration of soluble waste substances in the urine. Higher concentration of soluble wastes may indicate a level of dehydration because the human body, in a healthy state, constantly tries to maintain homoeostatic water balance. Thus, with the loss of body water, urine colour changes in proportion to the level of dehydration, and darkens as dehydration increases or as the concentration of soluble waste increases in the urine [29]. Armstrong et al. [21] found that urine colour can, in some cases, indicate a person's hydration state because the changes in urine colour coincides with other techniques for measuring hydration such as plasma osmolality and urine specific gravity. However, there is no standard urine colour index to match a given magnitude of an individual hydration or dehydration state, especially when little changes occur that are not easily noticeable by observation. A 2015 report by Fortes et al. [12] showed that physical signs such as urine color and saliva are not sufficient for detecting changes in dehydration level. Our experiment uses the attenuation of a propagating electrical signal amplitude as a result of fluid absorbed through hydration or lost through dehydration to assess changes in body fluid level. Our sensitivity test showed that up to 1.30 dB gain can be detected in the arm when 100 mL of water is consumed, depending on the initial fluid requirement of an individual.

5.1.2. Change in Body Weight

The human body mass contains 60%–70% water [27] with dynamic systems of gaining or losing water. Short term changes in body weight are attributed to loss or gain of body water because 1 mL of water has a mass of one gram [16]. Changes in body mass are often used to quantify water gain or loss in a clinical measurement. Consequently, we used the differences in body weight to deduce the amount of water retained in the body and to relate it with the changes in the intrabody signal passing through the body. We note that, as the body hydrates, only a proportion of the amount consumed comes to the arm where we measured. Our sensitivity test investigated the different amount of water required by each subject before it could be detected in the arm. Results show that the signal attenuation decreased as the body weight increased by the fluid consumed and increased as the body weight decreased due to loss of water (Figure 8).

5.1.3. Refractometry

With the loss of body water, urine specific gravity increases due to increase in the concentration of urinary waste products [30]. The concentration is determined by the amount of urinary waste per unit volume of urine. Urine specific gravity measures the ratio of the density of urine relative to the density of pure water. A specific gravity greater than 1 means the fluid is denser than water [22]. Urine specific gravity measurements usually range from 1.002 to 1.030. Our results show that subjects with high SPG values usually had high hydration rates with the galvanic coupling method.

5.2. Intrabody Signal Propagation Method

Our proposed intrabody signal propagation method uses time varying changes in galvanic coupled signal amplitude to predict hydration rates. The decrease in attenuation (hydration) or increase (dehydration) during each observation matched the changes in the body weight recorded for all of the 20 subjects as well as the changes in the urine specific gravity and the visual observations on the urine colour.

We observed four distinct cases of human tissue hydration. We defined W_2 as Weight after urination and W_0 as weight before drink and grouped the cases as following:

Case 1: $W_2 < W_0$ = Hyper hydration;
Case 2: $W_2 = W_0$ = Optimal hydration;
Case 3: $W_2 > W_0$ = Severe dehydrated;
Case 4: $W_2 > W_0$ = Mild dehydration.

The graphs show that hydration started and ceased at different times per subject: 30% recorded hydration levels 10 min after drinking, while 25% recorded hydration levels at 5 min after drinking. Subjects with high SPG1 usually had higher rates of hydration, while subjects with very low SPG1 had lower rates of hydration. Thus, different individuals have different hydration rates and the specific gravity reading indicated an individual's initial dehydration state. Similarly, the amount of water consumed was too much for some subjects and just appropriate for others. These observations explain why the state of water balance is reached at different times and is dependent upon the initial body fluid level in each subject [15]. Figure 7 show that individuals with high BMI have longer time-dependent metabolic processes, and lower rates of hydration than people with low BMI, in line with our previous findings. The measured hydration rates lie within the range bounded by the mean value of the attenuation on an individual's baseline measurement. From Figure 8, the signal amplitude changed in proportion with the quantity of water absorbed by the tissues. Similarly, fluid consumption and losses caused changes in the body mass measurement in line with Equations (2) and (3). The variations in the signal amplitude were caused by the absorption of water by the tissues, which caused an increase due to hydration or reduction in body weight due to loss of water by evaporation, metabolism and urination. The signal amplitude increased more prominently as the volume of fluid consumed increased. For instance, within the first 5 min, when the amount of water consumed by subject

D increased from 100 mL to 250 mL, the signal attenuation decreased from −65.5 dB to −61.9 dB. The attenuation decreased as the amount of water consumed increased. This was reversed in the dehydration cycle measured after urination.

A review of bioimpedance and plasma osmolality methods indicates that intrabody methods, as shown in our experiment, are sensitive to detecting mild body fluid changes. Plasma osmolality measures the amount of osmoles (Osm) of solute per kilogram of solvent (osmol/kg) and is regarded as one of the best methods of estimating the level of dehydration in an individual [31]. The osmolality of blood increases with dehydration and decreases with increasing hydration. A problem with this method, however, is that it does not track body fluid changes quickly when fluid loss is below 3% of body mass [11,21]. This implies that, for a 70 kg adult, plasma osmolality could only detect changes in body fluid level until the water loss causes up to 2.1 kg decrease in body weight. Our technique is sensitive to 100 mL of fluid intake or 0.1 kg change in body weight due to hydration. Similarly, bioimpedance method tracks total body water from estimates of extracellular and intracellular fluid volumes [32]. However, although the bioimpedance method has been validated for estimating total body water [33], in 2007, Armstrong argued that bioelectrical impedance methods may not be accurate when the amount of water loss is less than 800–1000 mL [31]. Our sensitivity test showed that intrabody method could detect changes in body fluid level for as little as 100 mL of fluid intake or water loss as observed in subject D. Our sensitivity test suggests that the sensitivity increases with decrease in height. In addition, the magnitude of the signal attenuation from theoretical simulation changed from −65.98 dB to −63.27 dB for $\tau = 80$ and $\theta = 0.3$, similar to empirical measurements Figures 8 and 9. Our simulation results, in Figure 9, predicted a lowering of signal attenuation by 1.50 dB to as much as 1.91 dB, depending on individual anthropometric ratio and body metabolism, when 100 mL is optimally absorbed in the arm. Our empirical measurements detected 1.31 dB on subject D, BMI 29.83 kg/m^2 and height 166 cm. Thus, the empirical measurement is evidenced by simulation results. However, a challenge to the empirical measurements is that the majority of the participants were uncomfortable trying to remain still during the 20 min period in the first experiment, which was required to minimise movement artefacts and external effects. Thus, only three subjects consented to participate in the control and sensitivity test when it was extended to 30 min of measurements. In addition, we argue that the initial hydration state of the subjects with SPG readings should not be the gold standard [31] for assessing human body hydration state.

6. Conclusions

In this paper, we show that galvanic coupled intrabody signal propagation can be used to measure the rate of human body hydration with sensitivity as low as 100 mL of fluid intake or loss, and the system is non-invasive and hygienic. Our studies with 12 male and eight female volunteers show that the rate of hydration depended more on individual metabolic requirements, initial hydration level, and body mass index. Hydration rates are not constant but are affected by the immediate body physiological state and metabolic equilibrium, which, in turn, determines how long it takes to change from one fluid state to another. To use this technique, a baseline amplitude of the signal variations of an individual is required. This capacity makes it potentially applicable for monitoring changes in body fluid level, targeted body fluid disorder and the response of tissues when monitored for targeted fluid level variations. Using galvanic coupled intrabody communication means that the technology can be integrated into existing wearable devices in a cost effective way. In the future, we will further investigate the effects due to anthropometric parameters on the sensitivity.

Acknowledgments: This project is part of a PhD research project by Clement Ogugua Asogwa with no external funding beside the scholarship allowance received by the first author as a PhD candidate.

Author Contributions: Clement Ogugua Asogwa and Daniel TH Lai conceived and designed the experiments which was performed by the student (first author) and wrote the major draft; Patrick MacLaughlin, Stephen Collins and Daniel TH Lai assisted with the ethics application and review of the paper.

Conflicts of Interest: The authors declare no conflict of interest.

References

1. Armstrong, L.E.; Pumerantz, A.C.; Fiala, K.A.; Roti, M.W.; Kavouras, S.A.; Casa, D.J.; Maresh, C.M. Human hydration indices: Acute and longitudinal reference values. *Int. J. Sport Nutr.* **2010**, *20*, 145–153.
2. Gandy, J. Water intake: Validity of population assessment and recommendations. *Eur. J. Nutri.* **2015**, *54*, 11–16.
3. Ibrahim, F.; Thio, T.H.G.; Faisal, T.; Neuman, M. The application of biomedical engineering techniques to the diagnosis and management of tropical diseases: A review. *Sensors* **2015**, *15*, 6947–6995.
4. Agostoni, C.V.; Bresson, J.L.; Fairweather-Tait, S. Scientific opinion on dietary reference values for water. *EFSA J.* **2010**, *8*, 1–47.
5. El-Sharkawy, A.M.; Watson, P.; Neal, K.R.; Ljungqvist, O.; Maughan, R.J.; Sahota, O.; Lobo, D.N. Hydration and outcome in older patients admitted to hospital (The HOOP prospective cohort study). *Age Ageing* **2015**, *44*, 943–947.
6. Schwan, H.P. Electrical properties of tissue and cell suspensions. *Adv. Biol. Med. Phys.* **1957**, *5*, 147–209.
7. Wabel, P.; Chamney, P.; Moissl, U.; Jirka, T. Importance of whole-body bioimpedance spectroscopy for the management of fluid balance. *Blood Purif.* **2009**, *27*, 75–80.
8. Rothlingshofer, L.; Ulbrich, M.; Hahne, S.; Leonhardt, S. Monitoring change of body fluid during physical exercise using bioimpedance spectroscopy and finite element simulations. *J. Electr. Bioimped.* **2011**, *2*, 79–85.
9. MirHojjat, S.; Kibret, B.; Lai, D.T.H.; Faulkner, M. A survey on intrabody communications for body area network applications. *IEEE Trans. Biomed. Eng.* **2013**, *60*, 2067–2079.
10. Mentes, J.C.; Wakefield, B.; Culp, K. Use of a Urine Color Chart to Monitor Hydration Status in Nursing Home Residents. *Biol. Res. Nurs.* **2006**, *7*, 173–203.
11. Francesconi, R.P.; Hubbard, R.W.; Szlyk, P.C.; Schnakenberg, D.; Carlson, D.; Leva, N.; Sils, I.; Hubbard, L.; Pease, V.; Young, J. Urinary and hematologic indexes of hypohydration *J. Appl. Physiol.* **1987**, *62*, 1271–1276.
12. Fortes, M.B.; Owen, J.A.; Raymond-Barker, P.; Bishop, C.; Elghenzai, S.; Oliver, S.J.; Walsh, N.P. Is this elderly patient dehydrated? Diagnostic accuracy of hydration assessment using physical signs, urine, and saliva markers. *J. Am. Med. Dir. Assoc.* **2015**, *16*, 221–228.
13. Gao, W.; Emaminejad, S.; Nyein, H.Y.Y.; Challa, S.; Chen, K.; Peck, A.; Fahad, H.M.; Ota, H.; Shiraki, H.; Kiriya, D.; et al. Fully integrated wearable sensor arrays for multiplexed in situ perspiration analysis. *Nature* **2016**, *529*, 509–514.
14. Kibret, B.; Seyedi, M.; Lai, D.T.H.; Faulkner, M. Investigation of Galvanic Coupled Intrabody Communication using Human Body Circuit Model. *IEEE J. Biomed. Health Inform.* **2014**, *62*, 1196–1206.
15. Asogwa, C.O.; Teshome, A.K.; Lai, D.T.H.; Collins, S.F. A Circuit Model of Real Time Human Body Hydration. *IEEE Trans. Biomed. Eng.* **2016**, *63*, 1239–1247
16. Lentner, C. Geigy scientific tables: Units of measurement, body fluids, composition of the body, nutrition. *Basle Ciba-Geigy* **1981**, *1*.
17. Kavouras, S. Assessing hydration status. *Curr. Opin. Clin. Nutr. Metab. Care* **2002**, *5*, 519–524.
18. Shirreffs, S.M. Markers of hydration status. *Eur. J. Clin. Nutr.* **2003**, *57*, S6–S9.
19. Mifflin, M.D.; St Jeor, S.T.; Hill, L.A.; Scott, B.J.; Daugherty, S.A.; Koh, Y.O. A new predictive equation for resting energy expenditure in healthy individuals. *Am. J. Clin. Nutr.* **1990**, *51*, 241–247.
20. Armstrong, L.E. Hydration assessment techniques. *Nutr. Res.* **2005**, *63* (Suppl. 1), S40–S54.
21. Armstrong, L.E.; Soto, J.A.H.; Hacker, F.T., Jr.; Douglas, J.C.; Kavouras, S.A.; Maresh, C.M. Urinary indices during dehydration, exercise, and rehydration. *Int. J. Sport Nutr.* **1998**, *8*, 345–355.
22. Armstrong, L.E.; Maresh, C.M.; Castellani, J.W.; Bergeron, M.F.; Kenefick, R.W.; LaGasse, K.E.; Riebe, D. Urinary indices of hydration status. *Int. J. Sport Nutr.* **1994**, *4*, 265–279.
23. Asogwa, C.O.; Seyedi, M.; Lai, D.T. A preliminary investigation of human body composition using galvanically coupled signals. In Proceedings of the 9th International Conference on Body Area Networks, ICST (Institute for Computer Sciences, Social-Informatics and Telecommunications Engineering), London, UK, 29 September 2014; pp. 346–351.

24. Peronnet, F.; Mignault, D.; Du Souich, P.; Vergne, S.; Le Bellego, L.; Jimenez, L.; Rabasa-Lhoret, R. Pharmacokinetic analysis of absorption, distribution and disappearance of ingested water labelled with D2O in humans. *Eur. J. Appl. Physiol.* **2012**, *112*, 2213–2222.

25. Wegmueller, M.S. Intra-body communication for biomedical seensor networks. Ph.D. Dissertation, ETH Zurich, Zurich, Switzerland, 2007.

26. Skelton, H. The storage of water by various tissues of the body. *Arch. Intern. Med.* **1927**, *40*, 140–152.

27. Jequier, E.; Constant, F. Water as an essential nutirent: the physiological basis of hydration. *Eur. J. Clin. Nutr.* **2009**, *64*, 115–123.

28. Bankir, L.; Bouby, N.; Trinh-Trang-Tan, M.M. The role of the kidney in the maintenance of water balance. *Baillière Clin. Endoc.* **1989**, *3*, 249–311.

29. Sands, J.M.; Layton, H.E. The physiology of urinary concentration: An update. *Semin. Nephrol.* **2009**, *29*, 178–195.

30. Oppliger, R.A.; Magnes, S.A.; Popowskim, L.A.; Gisolfi, C.V. Accuracy of urine specific gravity and osmolality as indicators of hydration status. *Int. J. Sport Nutr. Exerc. Metab.* **2005**, *15*, 236–251.

31. Armstrong, L.E. Assessing hydration status: the elusive gold standard. *J. Am. Coll. Nutr.* **2007**, *26*, 575S–584S.

32. Lukaski, H.C. Applications of Bioelectrical Impedance Analysis: A Critical Review. In *In Vivo Body Composition Studies*; Springer US: NY, USA, 1990; pp. 365–374.

33. Van Loan, M.D.; Withers, P.; Matthie, J.; Mayclin, P.L. Use of bioimpedance spectroscopy to determine extracellular fluid, intracellular fluid, total body water, and fat-free mass. In *Human Body Composition*; Springer US: NY, USA, 1993; pp. 67–70.

electronics

MDPI

Review

A Comparative Review of Footwear-Based Wearable Systems

Nagaraj Hegde, Matthew Bries and Edward Sazonov *

Department of Electrical and Computer Engineering, The University of Alabama; Tuscaloosa, AL 35487, USA;
nhegde@crimson.ua.edu (N.H.); mdbries@crimson.ua.edu (M.B.)
* Correspondence: esazonov@eng.ua.edu; Tel.: +1-205-348-1981

Academic Editor: Enzo Pasquale Scilingo and Gaetano Valenza
Received: 14 July 2016; Accepted: 1 August 2016; Published: 10 August 2016

Abstract: Footwear is an integral part of daily life. Embedding sensors and electronics in footwear for various different applications started more than two decades ago. This review article summarizes the developments in the field of footwear-based wearable sensors and systems. The electronics, sensing technologies, data transmission, and data processing methodologies of such wearable systems are all principally dependent on the target application. Hence, the article describes key application scenarios utilizing footwear-based systems with critical discussion on their merits. The reviewed application scenarios include gait monitoring, plantar pressure measurement, posture and activity classification, body weight and energy expenditure estimation, biofeedback, navigation, and fall risk applications. In addition, energy harvesting from the footwear is also considered for review. The article also attempts to shed light on some of the most recent developments in the field along with the future work required to advance the field.

Keywords: accelerometry; energy expenditure; energy harvesting; footwear; gait; plantar pressure; wearable sensors

---

1. Introduction

Footwear is an irreplaceable part of human life across the globe. While the initial necessity was purely to protect the feet [1], they have also become a symbol of style and personality [2]. Footwear acts as the interface between the ground and the wearer's foot. Lots of information can be gleaned from observing this interaction. Attempts to capture this information by integrating sensing elements and electronics in the footwear began in the 1990s, both for academic research purposes and in commercial products [3]. In recent times, development of low power, wireless, unobtrusive and socially acceptable wearable computing systems has become an increasingly important research topic. This trend is aided by the exponential growth in the electronics industry, which is driving rapid advancements in microfabrication processes, wireless communication, and sensor systems.

The applications for footwear-based systems range from simple step counting solutions to more advanced systems intended for use in rehabilitation programs for disabled subjects. Footwear-based systems available on the market or in research laboratories today vary in their sensor modalities and data acquisition methodologies in order to meet different application requirements. Typically, these systems consist of pressure sensors for plantar pressure measurement, inertial sensors (accelerometer and/or gyroscope) for movement detection and a wired or wireless connection for data acquisition. The signal processing of this collected data varies depending on the application, can range from lightweight signal processing methodologies (for example, binary decision trees) running on a handheld device to complex signal processing/machine learning models (for example, Support Vector Machines) running on a PC.

Several vital biomechanical parameters can be estimated using sensors placed in the footwear. For example, by placing pressure-sensitive elements in the footwear, foot plantar pressure can be measured. By utilizing pressure-sensitive elements along with inertial sensors, several gait parameters can be calculated. Additionally, by placing actuators in the footwear and measuring gait patterns, one can generate biofeedback to assist patients suffering from stroke. The same set of pressure sensors and inertial sensors can also be used in tracking posture and activity recognition and energy expenditure estimation. These and other important applications have driven footwear wearable technology to its present day state and continue to drive the technology even further.

In this work we review advancements in footwear-based wearable systems based on their target application scenarios. Applications described in the work include those that focus on gait monitoring, plantar pressure measurement, posture and activity classification, body weight and energy expenditure estimation, biofeedback, navigation, and fall risk applications. For each application, example systems are taken from published research and consumer products. Keywords such as 'gait monitoring systems', 'plantar pressure measurement', and others were searched in databases such as IEEE Xplore, PubMed, and Google Scholar. Literature that described portable or wearable systems that have sensors embedded in the shoes, insoles, sandals, or socks were included and the stationary systems were excluded from the review.

The article discusses the existing systems with respect to their hardware, sensor modalities, modes of data acquisition and data processing methodologies. Merits/demerits of each system are also pointed out. The work also attempts to shine a light on some of the most recent advancements happening in the field, as well as on the future direction for footwear-based wearable systems.

2. Application Scenarios for Footwear-Based Wearable Systems

2.1. Gait Analysis

A person's walk is characterized by their gait, which involves a repetitious sequence of limb motions to move the body forward while simultaneously maintaining stability [4]. Having a normal gait allows someone to remain agile so that they may easily change directions, walk up or down stairs, and avoid obstacles. Patients with neuromuscular disorders are likely to have abnormal gaits and suffer in their ability to perform locomotive activities. Objective measurement and analysis of gait patterns can help in the rehabilitation of such disabled individuals.

Figure 1a shows an illustration of an instrumented insole developed for gait monitoring by Crea et al. [5]. There are two kinds of parameters that are computed in gait monitoring applications: temporal and spatial. Some of the examples of the temporal gait parameters are cadence, stance time, step time, single support time, and double support time; while step length and stride length are examples for spatial gait parameters. Gait monitoring is one field of wearable computing where there are a considerably high number of footwear-based systems deployed. Many such systems are compared in Table 1. There are force plates available for gait analysis [6], and there are also systems that make use of the Kinect [7,8]; but footwear-based solutions are much better suited for uncontrolled free living conditions outside the laboratory environment. Footwear is also an ideal location to measure the gait parameters as these applications measure the parameters involved in the movement of foot.

By utilizing pressure-sensitive elements, such as force sensitive resistors (FSR), for gait monitoring, temporal parameters such as cadence, step time, stance time and others can be computed. This is done utilizing the heel strike and toe off time events (Figure 1b). The gait monitoring applications extract gait event information from the changes in pressure sensor readings and not the absolute pressure. Hence, the pressure sensing does not need high spatial resolution, so only a few pressure elements are used in such applications. High pressure measurement precision is also not needed, and for that reason, sometimes these are called foot switches.

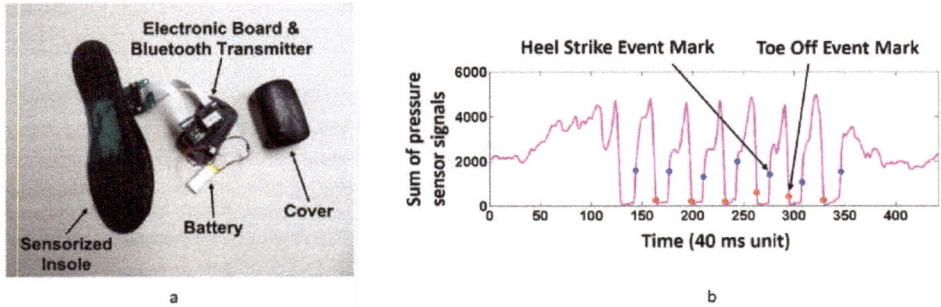

Figure 1. (**a**) Instrumented footwear for gait monitoring presented by and Crea et al. [5]; (**b**) marking of heel strike and toe off time instances from the pressure sensor signals.

For computation of temporal gait parameters, a higher sampling frequency means better time resolution, resulting in higher accuracy in the computed parameters. However, there is a tradeoff between battery life, sampling frequency, and accuracy (applicable to all of the systems discussed in this work). A higher sensor sampling frequency will result in higher power consumption by the system, resulting in a shortened battery life; while a lower sampling frequency may decrease the accuracy of the system. The work in [5] has used 18.75 Hz, however it has not yet been fully validated. In [10], it was shown that 25 Hz is sufficient for gait parameter extraction for walking at the speed of 2 km/h or less. However, our ongoing work suggests that for computation of parameters such as double support time, especially during walking at speeds more than 4 km/h or running, 25 Hz is insufficient. Sampling frequencies of greater than 50 Hz, as reported in [9], will provide a better accuracy for all the gait parameter computations.

Inertial sensors such as an accelerometer in conjunction with gyroscope can be used in the computation of distance or elevation. Hence, these inertial sensors are utilized in the computation of spatial gait parameters such as step length or stride length, as shown in [9].

Table 1. Comparison of footwear-based gait monitoring systems.

	Bamberg et al. [9]	Sazonov et al. [10]	Chen et al. [11]	Mariani et al. [12]	Leunkeu et al. [13]	Rampp et al. [14]
Sensing element	FSR, polyvinylidene fluoride strip, accelerometer, gyroscope	FSR, accelerometer	FSR, accelerometer, gyroscope	Accelerometer and gyroscope	Parotec plantar pressure insoles	Accelerometer and gyroscope
Sampling frequency	75 Hz	25 Hz	50 Hz	200 Hz	150 Hz	102.4 Hz
Data transmission method	RF to PC	Bluetooth to Smartphone	RF to PC	SD card logged	SD card logged	Data logged using Shimmer 2R® [15]
Data analysis method	PC post processing to compute gait parameters	PC post processing to compute gait parameters	PC post processing to predict abnormal gait	PC post processing to compute gait parameters	PC post processing	PC post processing
Real time gait monitoring	NA	NA	NA	NA	NA	NA
Clinical/Validation Study	Computed gait parameters of 10 healthy subjects and five Parkinson Disease (PD) patients	Computed gait parameters of 16 healthy subjects and seven post stroke patients	NA	Computed gait parameters of 10 PD patients and 10 healthy subjects	Computed gait parameters of 15 Cerebral Palsy (CP) and 10 normal children	Stride parameter calculation 116 geriatric patients
Gait analysis performed	Computed maximum pitch, minimum pitch, stride length, stride time, % stance time	Computed cadence, step time, cycle time, swing %, stance %, single support %, double support %	Classified 5 different gait types: Normal, toe in, toe out, over supination and heel walking	Computed turning angle, stride velocity, stride length, swing width, path length	Computed step duration, double support time, ground contact time, velocity, step frequency and stride length	Computed stride length, stride time, swing time, and stance time
Gold Standard Comparison	Validated against MGH BMLs Selspot II	Validated against GaitRite	NA	Validated against optical system by Vicon Motion Systems Ltd	NA	Validated against GaitRite
Accuracy	Highest mean percentage change of 15.6% in maximum pitch and least mean percentage change of 6.5% in stride length	Highest relative error of 18.7% for step time and least relative error of 2.7% for cycle time	Highest accuracy of 97% for detecting over supination and least accuracy of 82.3% for toe in	Stride velocity and stride length accuracy $\pm$ precision of 2.8 ± 2.4 cm/s and 1.3 ± 3.0 cm	NA	Correlation of 0.93 and 0.95 between GaitRite and this system in stride length and stride time. Absolute error of stride length was 6.26 cm in normal walking

Table 1. *Cont.*

	Kong et al. [16]	Liu et al. [17]	González et al. [18]	Crea et al. [5]	Wu et al. [19]	Ferrari et al. [20]
Sensing element	Custom air pressure sensor	Triaxial force sensors for measuring GRF and COP	FSR and accelerometer	Optoelectronic sensing	Fiber based pressure sensors	Accelerometer and gyroscope
Sampling frequency	200 Hz	100 Hz	50 Hz	18.75 Hz	30 Hz	200 Hz
Data transmission method	NI Compactrio® data logger	Storing data in MCU's SRAM and offline uploading to PC	Bluetooth to Smartphone	Bluetooth to PC	SD card logging	Bluetooth to smartphone
Data analysis method	PC Post processing	PC post processing	PC Post processing	PC Post processing	PC post processing	PC post processing
Real time gait monitoring	NA	NA	Android smartphone	LabVIEW user interface in PC	NA	Android smartphone
Clinical/Validation Study	NA	Validated on seven healthy subjects.	Validated on six healthy subjects.	Validated on 2 healthy subjects	NA	Validated on 12 healthy subjects and 16 PD subjects
Gait analysis performed	Fuzzy logic based gait phase abnormality detection	Average coefficient of variation for three-directional GRF to evaluate extrinsic gait variability	A fuzzy rule-based inference algorithm to detect each of the gait phases.	Computed stance and swing duration of both feet; duration of the double-support phases; and step cadence of both feet.	Methodology for local randomized selective sensing based on pressure signal maps.	Computation of step length, stride time, stride length and stride velocity.
Gold Standard Comparison	NA	Validated against Kyowa force plate and optical motion analysis system by NAC Image Tech	NA	Validated against AMTI force plate	NA	Validated against GaitRite
Accuracy	Proposed sensing unit showed a repeatability of 97%. Abnormal gait monitoring results weren't quantified	RMS error of 7.2% ± 0.8% and 9.0% ± 1% for transverse component of ground reaction force and 1.5% ± 0.9% for vertical component	92% cross validation accuracy for the probabilistic classifier	Pearson correlation 0.89 ± 0.03 with the force plate	Normalized mean square error of the proposed sensing methodology is within 10%, compared against actual signal	Over the 1314 strides, the total root mean square difference on step length estimation between this system and gold standard was 2.9%

Only a couple of the works discussed the expected battery life for their systems. In [5] it was reported that a battery life of about 20 h is achievable, which can enable almost two days of wear in real life situations. In [19], 10 h of usage time on single charge is reported, which can enable one day of wear on full charge. However, in [19], having this particular lithium polymer battery under the foot may be potentially hazardous.

From connectivity perspective, Bluetooth is quite commonly used in many of the footwear-based systems as done in [5,10,18] and others. A lower power consumption version of Bluetooth, named Bluetooth Low Energy (BLE), is coming to prominence in the recent years and there are a few footwear-based systems that have utilized BLE, [21,22], which will be discussed in the later sections. The study in [21] has reported more than two orders of magnitude in power savings when utilizing BLE compared to traditional Bluetooth.

With regards to data processing, many gait monitoring applications in their current form rely on collecting the data from human subject experiments and a PC performing post processing, as in most of the cases as shown in Table 1. Real-time gait monitoring and visual feedback can help in clinical applications and González et al. [18], Ferrari et al. [20], and Crea et al. [5] have taken this into account.

There are a few works that have used gait information obtained from footwear-based systems in pattern recognition. Huang et al. [23] identified a human subject based on the wearer's gait against eight other human subjects. Jochen Klucken et al. [24] were able to successfully distinguish PD patients from healthy subjects with an accuracy of 81%. Jens Barth et al. [25] have presented a methodology to search for patterns matching a pre-defined stride template from footwear sensor data, to automatically segment single strides from continuous movement sequences.

All of the footwear-based solutions listed are research prototypes and some of them [9,11] are primarily suited for laboratory studies. On the other hand, Sensoria® has developed commercial instrumented socks for gait monitoring [26]. The associated smartphone application running on iOS or Android is intended to help runners and provides real-time feedback on foot landing patterns, cadence and other important gait characteristics. Technical characteristics of the Sensoria product were not available at the time of this review.

Systems in [5,11,19,23] would need further human subject studies as they have not yet been fully tested. Additionally, it is important to note that none of the above-described systems have undergone a longitudinal free-living study, which is essential to understand the gait behavior of wearers in community living. Development and full-fledged validation of a socially acceptable, user friendly, and reliable footwear-based gait monitor well suited for longitudinal studies is still an open challenge in the field that needs to be addressed.

2.2. Plantar Pressure Measurement

Plantar pressure is the pressure distribution between the foot and the support surface during everyday locomotion activities. Foot plantar pressure measurement applications focus on measuring of the pressure distribution between the foot and the support surface. Figure 2a shows an illustration of a plantar pressure map during one stance phase (heel strike to heel off) of a healthy individual [27].

The foot and ankle provide the support and flexibility for weight bearing and weight shifting activities such as standing and walking. During such functional activities, plantar pressure measurement provides an indication of foot and ankle functions. Plantar pressure measurement has been recognized as an important area in the assessment of patients with diabetes [28]. The information derived from plantar pressure measurement can also assist in identification and treatments of the impairments associated with various musculoskeletal and neurological disorders [29]. Hence, plantar pressure measurement is important in the area of biomedical research for gait and posture analysis [11,30,31], sport biomechanics [32,33], footwear and shoe insert design [34], and improving balance in the elderly [35], among other applications.

For all the above applications, there are solutions that utilize non-wearable systems, such as force plates and force mapping systems [36], but footwear is an ideal location for such measurements. Footwear-based platforms also offer much higher portability and can potentially enable monitoring outside of the laboratory, in uncontrolled, free living applications. Almost all the footwear-based applications reviewed in this work have some form of plantar pressure sensing elements built in to them; however, in this section we place an emphasis on the footwear systems that deal explicitly with plantar pressure measurement. These systems are compared in Table 2. Figure 2b shows the F-scan® system by Tekscan, Inc. (Boston, MA, USA) [37].

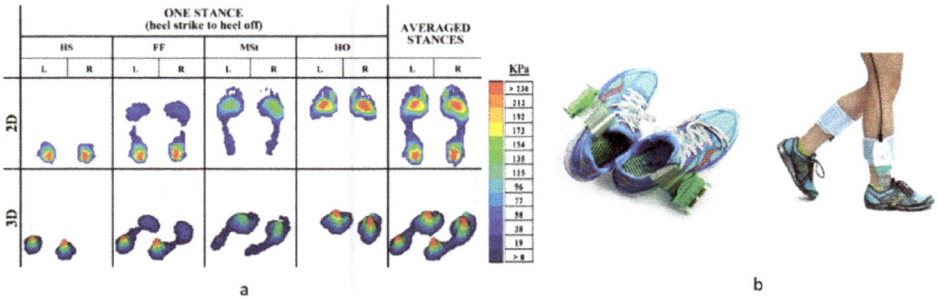

Figure 2. (**a**) Plantar pressure map during one stance [27]. HS—heel strike, FF—foot flat, MSt—mid stance, HO—heel off; and (**b**) F-Scan® System, courtesy of Tekscan, Inc. (Boston, MA, USA).

As seen in the Table 2, the sensing nodes for plantar pressure measurement application are much denser compared to those used in gait monitoring or activity monitoring, with the F-scan® system using as many as 960 sensing elements. This is because plantar pressure measurement applications demand the estimation of the absolute pressure that is exerted at different locations of the foot; while in activity and gait monitoring applications [7,30], it is more important to capture the relative changes in the pressure levels than the actual pressure values as previously discussed. F-scan® [29] can be used for the pressure ranges of 345 to 825 kPa, while Pedar® from Novel, gmbh (Munich, Germany) [31] can be used in the range of 15–600 kPa or 30–1200 kPa.

Since plantar pressure measurement applications are concerned more with the absolute pressure measurement and not the events, time resolution in the data from such systems do not play as vital a role as in gait monitoring. Sampling frequencies of as low as 13 Hz up to 750 Hz have been used in systems as shown in Table 2.

Table 2. Comparison of plantar pressure measurement systems.

	Adin Ming et al. [38]	Lin Shu et al. [39]	Saito et al. [40]	TekScan F-Scan® [37]	Novel Pedar® [41]	Orpyx LogR® [42]
Sensing element	Piezo resistive material, total 75 nodes	Resistive fabric sensor array, six sensor array	Pressure sensitive conductive rubber	Fabricated resistive insole, 960 sensing elements	Capacitive sensing element, 256 nodes	The system has custom-built force sensor array of 8 sensors
Device usage	Academic research prototype	Academic research prototype	Academic research prototype	Commercial	Commercial	Commercial
Sampling frequency	13 Hz	100 Hz	50 Hz	Up to 750 Hz *	78 Hz	100 Hz
Sensor data transmission method	Bluetooth	Bluetooth	Wired to PC	PC tethering, data logging, or Bluetooth	Bluetooth/SD card logging	Bluetooth
Visualization method	Real time visualization of plantar pressure distribution in PC	Real-time visualization of mean pressure, peak pressure, center of pressure (COP), and speed of COP, in PC and smartphone	Visualization of plantar pressure distribution in PC after data logging	Real-time visualization of plantar pressure distribution in PC	Real time visualization of plantar pressure distribution in PC	Real time visualization of plantar pressure distribution in iPhone
Validation	Validated against the standard force plate and the measured plantar forces showed R^2 value of 0.981	Relative mean difference of 5% in plantar pressure against standard force plate	Validated against F-scan. difference in computed plantar pressure varied from −4% to 18%	Multiple validation studies as reported in [43–47] have validated the system	Multiple validation studies [46–48] have validated the system	A validation study in [49] reported r^2 of 0.86 in plantar pressure measurement against (undisclosed) gold standard

* Up to 100 Hz when wireless data collection is used.

A potential concern with the pressure sensing elements in the footwear is their drift over time, which will become important when the systems are used in real life settings for long periods of time. A periodic recalibration would be needed to obtain repeatability as done for the case of F-scan® in [45,50]. Studies in [46,47] have suggested that capacitive sensing based Novel Pedar® has higher repeatability and accuracy when compared to the resistive sensing based Tekscan F-scan® systems. However, out of all the systems reviewed, F-scan offers the highest spatial resolution with 960 sensing elements in the insole. The F-scan® system has a reported 2 h of battery life, which is rather low. The LogR® by Orpyx, Inc. (Calgary, AB, Canada) [42] has reported 8–12 h of battery life, long enough to last through a regular day of wear.

All of the systems listed in Table 2 have connectors on the insoles which need to be connected to the microcontroller unit (MCU) outside of the footwear. This can limit usability in free living conditions by causing the footwear look out of the ordinary and feel uncomfortable. The system presented in [40] has a wired PC interface, which can limit use even in a controlled laboratory setup.

These above solutions measure the vertical force component of the ground reaction force (GRF), and Liu et al. [51] have presented a wearable force plate system for the continuous measurement of tri-axial ground reaction force in biomechanical applications. This system not only measures the vertical component of the GRF during ambulatory phase, but also measures transverse components of the GRF (anterior-posterior and medial-lateral). However, comfort levels wearing such systems made with tri-axial force sensors under heel may be questionable, as the sensor itself is 5 mm tall and it would be rather uncomfortable below the heel. Under the arch of the foot may be a better option for placing these sensors.

2.3. Posture and Activity Recognition, and Energy Expenditure Estimation (EE)

The ever increasing problem of obesity has brought immense importance to study in the field of posture and activity recognition and energy expenditure estimation. Weight gain is caused by a sustained positive energy balance, where daily energy intake is greater than daily energy expenditure. This is typically caused by living a sedentary lifestyle [52,53]. In [14] it was reported that obese individuals spend more time seated and less time ambulating than lean individuals. More than one third of U.S. adults are obese [54] and quantifying posture and activity allocation to help keep track of energy expenditure utilizing wearable sensors is quickly becoming a part of weight management programs. The applications extend beyond weight management programs, as posture and activity monitoring is an important aspect in the rehabilitation programs for post-stroke individuals [55]. Posture and activity classification is a large part of the consumer electronics industry's fitness segment. Fitness applications running on smartphones, smart watches, and fitness trackers are becoming common parts of modern daily life [56–60]. Almost all of these solutions are based on accelerometry and there are many published works on using accelerometer for posture and activity recognition [61–68]. There are also comparisons of commercially available accelerometry based activity and energy expenditure estimation (EE) monitors in [69–71]. In this section however, we focus only on footwear-based solutions used for posture and activity recognition, body weight, and EE estimation.

There are several footwear-based systems for posture and activity recognition purposes, and an example of these are the SmartShoe system developed by Sazonov et al. [72] which have been validated extensively for posture and activity monitoring. They have been used with healthy subject groups [30,72,73], people with stroke [55,74] and children with cerebral palsy [75]. The most recent incarnation of the SmartShoe systems, named SmartStep by Hegde et al., has shown the capability to be accurate in posture and activity classification [21,76,77]. Chen et al. have designed a foot-wearable interface for locomotion mode recognition based on contact force distribution [78]. Kawsar et al. have developed a novel activity detection system using plantar pressure sensors and a smartphone [79]. Table 3 contains the comparison of these systems. Figure 3a shows a picture of SmartStep insole [80] and the associated Android application for daily activity monitoring.

Table 3. Comparison of footwear-based posture and activity recognition systems.

	Sazonov et al. [55,72–74]	Hegde et al. [21,76,77]	Chen et al. [78]	Kawsar et al. [79]
Sensing element	Five Interlink force sensitive sensors FSR402 in the insole, accelerometer in the shoe clip on	Three Interlink force sensitive sensors FSR402, accelerometer, gyroscope all in the insole	Four FlexiForce A401 force sensor from Tekscan	Eight Fabric pressure sensors reported in [39]. Accelerometer and Gyroscope in the smartphone
Sampling frequency	25 Hz	25 Hz–75 Hz	100 Hz	37 Hz
Data transmission method	Bluetooth to smartphone	BLE to smartphone	RF module to PC	Bluetooth to smartphone
Data processing method	PC post processing for activity classification using neural network [81], decision trees [74], and support vector machines (SVM) [81]	PC post processing for activity classification utilizing multinomial logistic discrimination (MLD) [80]	PC post processing for activity classification using decision trees [78] and linear discriminant analysis [78]	Four different decision trees to classify activity from four sets of sensors (left shoe, right shoe, accelerometer, and gyroscope. Majority voting to decide the activity [79]
Activities classified	Sitting, standing, walking, upstairs, downstairs, cycling	Initial validation study classified sitting, standing, walking and cycling	Sitting, standing, level walking, obstacle clearance, upstairs, downstairs	Sitting, standing, walking, running
Real time activity feedback	Windows smartphone	Android smartphone	PC post processing and no real time feedback	Android smartphone
Clinical/Validation Study	Validated on stroke subjects [55,82] and healthy subjects [73,81]	Initial validation study on five healthy subjects [80]	Validated on five healthy subjects and one subject with amputee	NA
Accuracy	~99%	96%	98.4%	~99%

209

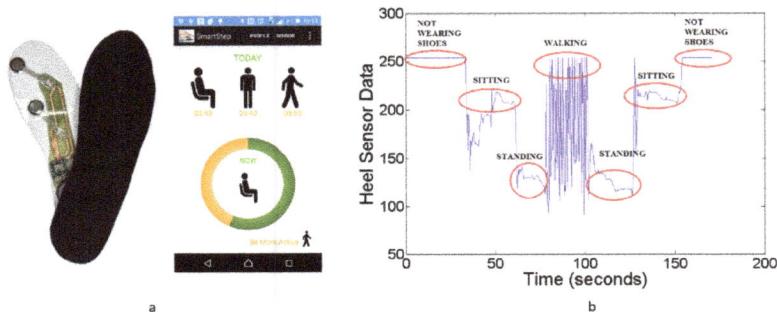

Figure 3. (**a**) SmartStep system by Hegde et al. [80]; and (**b**) heel pressure sensor signals for daily living activities [76].

All of the systems listed make use of various pressure sensors in order to determine the activity the user is undergoing. Pressure sensors can help to handle the ambiguity between weight bearing and non-weight bearing activities such as sitting and standing postures that cannot easily be determined using only accelerometry. An example of a pressure sensor signal located at heel for different daily living activities [76] is shown in Figure 3b. All of the systems utilize motion sensors such as accelerometers and the systems in [77,79] have a gyroscope as well.

Similar to the gait monitoring systems, the pressure sensitive elements in the footwear for activity monitoring are used for marking the events. However, since changes in activity do not happen too often, a lower sampling frequency can also be used in activity classification to save battery power. In [83] it was shown that for accelerometer based daily activity classification, a 15 Hz sampling rate can provide a 85% classification accuracy. When it comes to footwear-based systems, as shown by works of Sazonov et al. even 1 Hz sampling rate can be used in monitoring of the daily activities with 93% accuracy [72], and as shown in [55,72–74], a 25 Hz sampling frequency can provide a 99% accurate activity classification. Other described systems also have a very high accuracy (>96% across any system presented) in discriminating between the daily living activities of sitting, standing, walking, running, and cycling.

One of the limitations of footwear-based systems in activity monitoring is that they cannot be used in classifying upper body activities. To monitor upper body activities, additional sensors need to be worn on locations of the upper body. An example of such a system is the one presented by Ryan et al., who have used a wrist worn accelerometer along with the footwear system to classify daily living activities of ascending stairs, descending stairs, doing the dishes, vacuuming, and folding laundry, along with many athletic activities [84].

A long battery life is very important for the usability of these systems. Reports in [21,72,85] have discussed the battery life of the corresponding systems and the SmartStep system in [21], stands out among these, with more than four days of operations on single charge in certain modes. The other two systems had battery lives of approximately 5 h on a single charge.

None of the above systems have undergone longitudinal free living studies. There are many factors that need to be effectively addressed to enable such studies in community living environments. Social acceptability, user friendly operation, comfort for wear and unobtrusiveness are some of the important challenges that need attention. SmartStep [21,77] has tried to address many of these challenges in footwear-based systems, with its socially acceptable design, low power usage, and completely unobtrusive form factor.

In terms of data processing, as reported in [81], support vector machines (SVM), being computationally expensive, are not suitable for implementing in portable electronic devices, such as smartphones, for real-time activity classification purposes. The study in [81] also reports that activity prediction models based on multinomial logistic discrimination (MLD) is computationally less expensive in terms of required memory space and execution time, and performs equally well in terms of accuracy, as compared to SVM. Binary decision trees [74,78,79], being a light-weight classifier, can possibly enable the implementation of predictive models on the sensor itself. This can potentially reduce the power consumption at the sensor node that would occur during the wireless connection events for raw data transmission.

With respect to energy expenditure (EE) estimation, there are relatively few footwear-based systems that are targeted for such applications. The SmartShoe platforms have been extensively validated in EE (in controlled laboratory environments) [81,86]. SmartShoes were compared against other accelerometry-based wearable devices and have proven to be equally or more accurate [87]. There are also several commercial footwear-based systems that can be used in EE, such as Lechal systems [88] and Lenovo SmartShoe [89], track their user's EE. The Lenovo ones have yet to enter the market.

In general, most of the present day solutions predict EE in terms of a steady state (sitting/standing/walking etc.). However, daily life is a mixture of both steady states and the continuous transitions between them. It is important to be able to quantify these in-between states in order to better estimate energy expenditure. Hence, we expect to see more and more research from a data processing perspective, to try to more accurately estimate EE in daily life. This may lead to the inclusion of several different sensors alongside activity predictors (heart rate, breathing rate and others) and novel data processing techniques, similar or better to the ones presented in [90] in order to better estimate EE.

Excessive body weight is the factor which defines obesity and body weight is also one of the most significant factors in calculating energy expenditure. Self-reporting of body weight can be highly erroneous [91]; hence, objective and autonomous measurements of body weight can help in accurate EE estimation and obesity treatment programs. Footwear is an ideal location for automatic body weight estimation systems since all of the body's weight is placed upon the feet when standing. A few footwear-based systems have been used for body weight estimation as done in [92,93]. The study in [92] reported root-mean squared error of 10.52 kg in estimating body weight of 9 study subjects, while the one in [93] reported an average overestimation error of 16.7 kg in estimating body weight of 10 study subjects. These works validate the approach, but there is a room for improving the accuracy of the footwear-based systems in estimating body weight.

In all, footwear-based systems have yet to become matured in the field of daily energy expenditure estimation. The problem with measuring daily energy expenditure utilizing footwear systems is that people, on average, may wear footwear for 12 h of their whole day. This leaves a good 50% of the activities outside the purview of such sensor systems. Though one may argue that the majority of the remaining 12 h may be spent sleeping (6–8 h), to measure accurate daily living energy expenditure, one may have to use other sensor systems in conjunction with footwear-based ones. These may include smartphones or smart watches, which are generally used even when someone is not wearing footwear at home. Another potential concern with footwear-based wearable systems for use in daily living is that people usually tend to use multiple pairs of footwear. Insole based systems would be much more practical, as users can insert the insole into any of their shoes that they want to wear. However, the actions of taking out the insole from the shoe and inserting into another may not be a very comfortable act and future research work in this area needs to address this challenge.

2.4. Biofeedback

The sensing technologies embedded in the footwear in conjunction with real-time feedback mechanisms can be deployed in rehabilitation programs of many health conditions. For example, utilizing the real-time gait information retrieved from the footwear, post stroke individuals can be given feedback to improve the asymmetry in their walking. An example of real-time monitoring of stroke patient's gait and generation of active feedback to improve the asymmetry is depicted in Figure 4a. The feedback can be delivered in visual, auditory, or tactile manners. Several footwear-based systems have been presented in biofeedback applications and are presented in Table 4.

Table 4. Comparison of footwear-based biofeedback systems.

	Orpyx [42,49]	Khoo et al. [94]	Donovan et al. [97]	Hegde et al. [95]	Bamberg et al. [96]
Sensing element	Array of eight custom pressure sensors	Six FSR sensors	Pedar-x plantar pressure system and EMG	Two FSR sensors	Ten FSR sensors
Sampling frequency	Not reported	Not reported	100 Hz	20 Hz	114 Hz
Data transmission method	Wireless to smartwatch	Wired connection to MCU mounted on the lower back	Wired data logger	Wired connection to MCU mounted on the waist	Wireless to a portable PC
Feedback methodology	Smartwatch alerts when dangerous pressure levels are detected, so the user can modify behavior and avoid foot damage	Tactile and auditory feedback to correct the gait asymmetry	Auditory biofeedback to reduce the plantar pressure in the area of lateral forefoot	Tactile and auditory feedback to correct the gait asymmetry	Auditory feedback when the symmetry ratio is less than one
Computation method	Computation of plantar pressure levels for diabetic patients	Real-time computation of stance time difference; swing difference and generation of active feedback	Real-time computation of the lateral column plantar pressure and generation of active biofeedback	Real-time computation of stance time difference and generation of active feedback	Real-time Matlab program running on the PC that computes stance time and gait symmetry ratio
Clinical/Validation Study and results	NA	Validation on four healthy subjects and preliminary validation on post stroke patients. Gait parameters validated against gold standard (results not quantified)	Validated on nine subjects with chronic ankle instability. Pronounced reductions in peak pressure and pressure time integral of the lateral midfoot and lateral forefoot with the biofeedback	Validated on single stroke subject. Subject showed increased symmetry (step time differential improved by 48% standard deviation for the same increased by 88%)	Validated on three stroke patients. One subject reduced trunk sway by 85.5%, and the other subject reducing trunk sway by 16.0% and increasing symmetry ratio toward unity by 26.5%

a

b

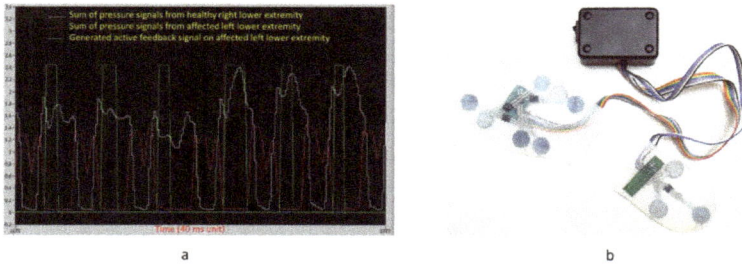

Figure 4. (**a**) Pressure sensor signals for asymmetric walking of a post-stroke subject and generation of active feedback [95]; and (**b**) biofeedback system presented by Bamberg et al. for stroke patients [96].

Orpyx® have insole-based wearable systems that have been clinically validated [49] that give biofeedback to diabetic patients on a smartwatch, based on their plantar pressure profile [42]. This system makes use of a neurological rewiring phenomena in the brain, termed neuroplasticity. Khoo et al. [94] and, independently, Hegde et al. [95], and Bamberg et al. [96] (Figure 4b) have developed biofeedback devices for post-stroke patients to improve their gait asymmetry. Donovan et al. have worked on a shoe-based biofeedback device to assist people with chronic ankle instability [97].

Compared to gait and activity monitoring systems, biofeedback systems need to process the sensor data in real-time and also generate active feedback in real-time. Hence, these systems do not make use of elaborate sets of sensing elements. All of them use only pressure sensing elements as seen in Table 4, with Donovan et al. [97] being the only exception which uses EMG sensing along with pressure sensing. The sampling frequency can be as low as 20 Hz as shown by Hegde et al. [95], to save battery power. The computation methodologies need to be lightweight and cannot be running on a PC for providing real-time feedback.

All of the presented systems make use of auditory or tactile feedback mechanisms in order to communicate with the user. Accuracies of the systems in [94–96] seem comparable. Additionally, all of these were externally wired systems (which could potentially get in the way of their purpose). A full-fledged insole system would be much more attractive for real world usage. In [97], during experiments, the device threshold was adjusted using a small screw driver to turn the trimpot's dial. A more autonomous system than this would be attractive and make the device easier to use. Furthermore, all of these systems need further human subject validation.

Studies in [95–97] have showed significant improvements to the patients' health conditions. These results indicate that footwear-based systems could indeed be of great use in rehabilitation and should be further pursued.

2.5. Fall Risk Assessment and Fall Detection Applications

One of the primary causes for the disorders referenced in the sections above is aging. By 2040, the number of elderly people in U.S. is projected to be 21.7% of the total population [98], and the risk of falling in older adults is an important social problem to be addressed [99]. Systems that monitor the individual's gait over a long period of time and predict risk of falling are termed as fall risk assessment systems [100]. On the other hand, a fall detection system is a real time assistive device which has a main objective of alerting when a fall event occurs. We review such footwear-based systems in Table 5.

Doukas et al. have developed an advanced fall detection system based upon movement and sound data [101]. Wiisel is an advanced insole-based system that is used for fall risk applications [22]. Otis et al. [102] have come up with an efficient home-based falling risk assessment test using a smartphone and instrumented insole. Sim et al. [103] have worked on a fall detection algorithm for the elderly using acceleration sensors on the shoes. Majumder et al. have implemented a real-time smartshoe and smartphone-based fall risk prediction and prevention system [104]. Figure 5a shows the Wiisel system that is used for fall risk applications.

Table 5. Comparison of footwear-based systems in fall risk applications.

	Doukas et al. [101]	Wiisel [22]	Otis et al. [102]	Sim et al. [103]	Majumder et al. [104]
Sensing element	Accelerometers and microphone	Textile based smart insole, 14 pressure sensors, accelerometer, gyroscope	FSR, accelerometer and bending sensor	Accelerometer	Piezo resistive sensors and inertial sensors from smartphone
Sampling frequency	NA	30 Hz	1000 Hz	225 Hz	25 Hz
Data transmission method	ZigBee to PC	BLE to smartphone	Bluetooth to Android	Bluetooth to PC	Wi-Fi to iPhone
Data analysis method	Short time Fourier transform and spectrogram analysis of the data to detect fall incidents. The classification of the sound and movement data is performed using Support Vector Machines	Data is transmitted from insoles to smartphone to back end server. Stand-alone program at the server analyses the gait and predicts fall risk	Proposed an automatic version of One-Leg Standing (OLS) score, based on COP measurements, for risk of falling assessment	Resultant acceleration signal is averaged and a threshold is used to predict the risk of falling	Tilt-invariant calculations on accelerometer and gyroscope data and usage of decision trees to classify high risk
Clinical/Validation Study	Three human subject validation while they performed walk, walk and fall, walk and run. 100% accuracy for fall detection and 96.72% for walk and run detection	Validated on 54 elderly participants [105], results NA	Twenty-three subject human subject study [106] including seven elderly and four PD subjects. Results suggest that the risk of falling depends on the type of ground	Six subject test. 81.5% sensitivity	Fifteen subject study. Subject dependent individual model has high accuracy but group model has accuracy of only 72%

Figure 5. (**a**) Wiisel footwear system for fall risk application; and (**b**) Lechal system® for by Ducere Technologies Pvt. Ltd. (Secunderabad, Andhra Pradesh, India).

Accelerometer is the common sensing element across all the systems used in this field as seen in Table 5. Otis et al. [102] and Sim et al. [103] have used relatively high sampling frequencies, with Otis et al. using 1000 Hz, though the reports did not substantiate the use.

Making the footwear systems comfortable for daily wear is a challenge and Wiisel insoles [22] have addressed this. These insoles are industrially built and with their built in wireless charging, this solution be used to benefit the elderly. However, people need to remove the insoles from their shoes in order to charge them. This is due to a limitation of the current version of the wireless charging standard Qi, which allows a maximum of 1.5 cm distance between the wireless power transmitter and receiver. This might be increased to 4 or 5 cm in the future and will remove the need for taking out the insoles for charging purposes. Systems other than Wiisel in this section are all laboratory prototypes.

If the systems have some kind of real time feedback, such as to call for help, that would be beneficial. Only [104] has any kind of real-time feedback implemented and adding this will be the next logical step for footwear systems used in fall risk applications.

Of the above mentioned systems, the one proposed by Doukas et al. [101] was the most accurate with 100% accuracy for fall detection. In that system, usage of microphone data along with accelerometer was novel. On the other hand, Sim et al. [103] system's 81.5% sensitivity is considerably low given the severity of misclassification.

2.6. Navigation and Pedestrian Tracking Systems

Another interesting area where footwear-based systems are being utilized is in the field of navigation assistive technologies and pedestrian tracking. These systems are aimed at providing assistance for the vision impaired, guiding emergency first responders, and work in augmented reality applications. In general, systems targeted for tracking utilize inertial sensors mounted on the footwear alongside GPS or radar; while navigation systems use actuators along with the sensors to guide the user in real-time. The key challenge for devices in this field is to be able to provide accurate location information without the need of a pre-installed infrastructure. In Table 6 several of these systems are reviewed.

Table 6. Comparison of footwear-based navigation systems.

	Schirmer et al. [107]	Bebek et al. [108]	Castaneda et al. [109]	Foxlin [110]	Lechal System® [88]
Sensors	Accelerometer, gyroscope, magnetometer, compass, smartphone GPS	Accelerometer, gyroscope, magnetometer heel of the shoe. capacitive pressure sensor at the heel	Accelerometer and gyroscope	Accelerometer, gyroscope, magnetometer	Insole pressure sensors not detailed. The electronics is mentioned to be having motion sensors
Actuators	Two vibration motors	No actuator	No actuator	No actuator	Haptic or vibratory feedback
Sensor sampling frequency	NA	NA	NA	300 Hz	NA
Data transmission method	BLE to iOS	Wired to laptop	Wired to laptop	RF to laptop	BLE to iOS or Android
Data analysis method	Phone computes walking path and turns, and communicates with the shoes to trigger the actuators	IMU bias compensation, and computing of position after zero velocity update (using the pressure sensor signals) and slope correction	Fuzzy logic procedure for better foot stance phase detection and an indirect Kalman filter for drift correction based on the zero-updating measurement	Zero velocity update of accelerometer and gyroscope, magnetometer calibration followed by Geomagnetic modeling and heading drift Correction	NA
Real time analysis	Different vibration patterns for different paths (front, back, left, right)	Future Work	Real time analysis in laptop	Future Work	Can provide turn-by-turn navigation feedback
Validation Study	Twenty-one subject study showed that 99.7% of the time users correctly identified the path and turns as fed back by the shoes	Six walking experiments of half hour each, average relative error 0.35% in the final position tracked by the system	Three walking experiments, average relative error of 0.55% in the final position tracked by the system	Single user, 118 m walking indoor with 0.06% error and 741 m outdoor experiment with 0.3% error	NA

All of the described systems in Table 6 make use of accelerometer and gyroscope for computing the navigation path. These applications demand high temporal resolution in sensor sampling (as demonstrated by Foxlin [110] with 300 Hz), because of complex computation (such as Kalman filtering [110]) and integration that needs to be performed to accurately determine the path.

Schirmer et al. [107] and Lechal System [88] provide feedback for real-time navigation. In order to use the system in [107], the user will need initial training to be able to understand the actuator patterns for front and back movement guidance, as they only use two actuators.

The Lechal® system by Ducere Technologies Pvt. Ltd. (Secunderabad, Andhra Pradesh, India) [88], being a commercial system, is industrially built (Figure 5b). Though initially intended as an assistance for a blind subject population, it can also be used by all for navigating purposes. Footwear feedback levels in this system are user configurable and it also provides options for configuration using voice commands.

The wired connection to a laptop limits the use of systems in [109,110] for practical applications. Reported error in navigating a predefined path is very much comparable in all these systems and is less than 1% in the final position tracked by these systems.

2.7. Other Enabling Technologies

Energy harvesting from footwear is an area which has been of interest for a long time. As electronics are becoming smaller and smaller, the recent trend in wearable technology is to move towards smaller batteries or battery-less systems and to tap the energy needed from the human body or its motions. As early as 1995, Starner [41] reported that 67 watts of power are available in the heel movement of a 68 kg person who is walking at a pace of two steps per second. Even a fraction of that energy, if harvested, can easily power today's low power electronics. Many attempts are underway to tap the energy from body heat and the force produced during locomotive activities.

In Table 7 we compare several footwear-based systems that make use of energy harvesting methods. Shenck et al. have devised a methodology for energy scavenging with shoe-mounted piezoelectrics [111]. Orecchini et al. have come up with an inkjet-printed RFID system for scavenging walking energy [112]. Zhao et al. have developed a shoe-embedded piezoelectric energy harvester for wearable sensors [113]. Kymissis et al. have tested three different energy harvesting elements in their work [114]. Meier et al. have presented a piezoelectric energy-harvesting shoe system for podiatric sensing [115].

Table 7. Comparison of footwear-based energy harvesting systems.

	Shenck et al. [111]	Orecchini et al. [112]	Zhao et al. [113]	Kymissis et al. [114]	Meier et al. [115]
Energy harvesting element	Piezoelectric lead zirconate titanate (PZT)	Piezoelectric pushbutton	Polyvinylidene difluoride (PVDF)	PZT, PVDF, and rotary magnetic generator	Vibrational transducer and piezoelectric transducer
Placement of the energy harvesting element	Insole	Underneath the shoe heel	Insole	Insole for PZT and PVDF, under the shoe for rotary magnetic generator	Shoe heel
Validation application scenario	Shoe-powered RF tag system	Self-powered RFID shoe	NA	A self-powered RF Tag System	Self-powered gait data capture system
Salient features	One of the first practical systems demonstrating the feasibility of the approach	Emphasis was put on designing RF antenna, in the shape of a logo to make the system socially acceptable	Flexible and thin insole platform	Compared the efficiency and practicality of 3 different energy harvesting elements	Were able to run a data acquisition at 5 Hz from harvested energy
Reported harvested energy	8.4 mW in a 500-kohm load at 0.9 Hz walking pace	833 μJ (test conditions not reported)	1 mW during a walk at a frequency of 1 Hz	From PZT 1.8 mW, from PVDF 1.1 mW, from rotary magnetic generator 0.23 W	10–20 μJ of energy per step

The energy can be harvested by vibration, compression, or bending produced in the footwear while the wearer performs motion activities such as walking. Piezoelectric lead zirconate titanate (PZT) and polyvinylidene difluoride (PVDF) are quite commonly used energy harvesting elements. PZT is a ceramic material, while PVDF is a plastic material.

From the study in [114] it is concluded that, even though mechanical systems such as the rotary magnetic element generate two orders of magnitude more energy than other systems, they are difficult to integrate into footwear. PZT and PVDF are more compact elements and can be much more easily integrated. The combination of the harvesting element and the placement of the element determine the resultant harvested energy.

Among the described works, Nathan S. Shenck [111] seems to be the one with the highest-reported power generated, but the system appears to be a laboratory prototype. The Rich Meier et al. solution [115] involves alteration of the shoes, which would limit its usage in generic footwear systems.

Though the amount of power generated by these systems is quite low, many of these systems were able to provide enough energy to drive low power RFID systems. For the systems demanding higher power, the resulting power from energy harvesting can be utilized in supplementing the power provided by the battery, to extend its runtime.

3. Recent Trends and Future

3.1. Socially Acceptable and User Friendly Solutions

It is important that footwear-based systems are as discrete and user friendly as possible. Most of the systems that are discussed in this work require a wireless/wired MCU placed outside the footwear. For free living daily life studies this may be a concern, and we are seeing work being done to move towards full-fledged insole-based systems. Insoles are now being equipped with all of the sensing elements, battery, recharging circuitry, and wireless interfaces [22,77,88,91]. Systems in [49,88] have wireless charging capabilities which make them more user friendly.

The SmartStep [77] system has an over-the-air firmware upgrade feature, which can be used to easily configure the system for use in different application scenarios. The concept of real-time data collection and offline transmission of SmartStep, presented in [77], can be attractive for the elderly population as they do not need to carry smartphones with them to use the system. In this scenario, the sensor data is logged in the system's flash memory during wear and later is transferred to the base station when the insoles are being charged. These trends will hopefully continue and can help researchers, as well as users, to better utilize such systems.

3.2. Footwear as Internet of Things (IoT) Devices and Big Data

IoT is a field that is redefining the way people interact with their environment in their daily lives. IoT can enable every object we interact with (for example: key chains, coffee mug, cloths, appliances, and many more) to be a sensor and a minicomputer connected to the internet [116]. We foresee that footwear is going to become a part of the IoT revolution soon and can help people become more connected and help them better manage their lives. The application scenarios such as gait and activity monitoring, fall risk/fall detection, and others, can take advantage of such infrastructure. An example use case can be that the individual with asymmetric walking (caused by a neurological disorder), wears IoT-enabled footwear, which sends the gait parameters to the physician in a distant place in real-time. The infrastructure can also allow the physician to give feedback to the individual based on the progress.

The next problem which arises after the systems are ready for daily usage and are a part of the IoT, is how to handle the enormous amount of data coming in from these systems. Novel data processing techniques, which do not only deal with the data from one set of sensors/systems, but from multiple sensors in a smart environment, will be gaining more and more traction from researchers. This greater expanse of data will help to make better informed decisions. We also foresee that footwear-based systems are going to play an important role in the remote monitoring of disabled and elderly people in the future.

3.3. Advanced Study Approaches for Footwear-Based Systems

From a research perspective, many of the footwear-based systems have not undergone longitudinal free-living studies. Footwear are subjected to enormous amounts of wear and tear, and the electronics built into them need to be able to withstand this for a long period of time. This may be one of the reasons why Nike and some other footwear manufacturers have stopped producing their SmartShoe product lines [117]. We foresee that, in the coming years, there will be more and more longitudinal studies in free-living conditions conducted, in each of the different application scenarios.

3.4. Affective Computing

Affective computing is a field of technology, in which systems can determine the users' mood and emotions, based upon his or her behavior sampled through different physiological factors, and adjust a smart environment to suit their mood. There are wearable systems that monitor users joy, stress, frustration, and other moods/emotions utilizing heart rate monitoring sensors, electroencephalogram (EEG) sensors, electro dermal activity (EDA) sensors, and others [118].

From footwear-based systems perspective, there was some initial work being done with this by Lenovo with their Smartshoes [89], that display a person's mood on a small screen embedded directly on the footwear, though technical details were not available at the time of this review. Additionally, there is a Kickstarter campaign on its way to make shoes that can change their entire appearance with a smartphone application [119]. Using affective computing, it would be possible to one day change the shoes color or display to reflect the user's mood.

4. Conclusions

In this work we reviewed footwear-based wearable systems based on their target application. Existing footwear-based solutions from academic research as well as commercial ones in the areas of gait monitoring, plantar pressure measurement, posture and activity classification, body weight and energy expenditure estimation, biofeedback, fall risk applications, navigation, along with footwear-based energy harvesting solutions were detailed. The article also discussed sensor technology, data acquisition, signal processing techniques of different footwear-based systems along with critical discussion on their merits and demerits. Additionally, we attempted to shine a light on recent trends and future technological pathways for footwear-based solutions.

Author Contributions: NH and ES designed the research; NH conducted the literature review and segmentation based on their target application. NH, ES and MB contributed in writing the review article and proof reading. All the authors read and approved the final manuscript.

Conflicts of Interest: The authors declare no conflict of interest.

References

1. History of Footwear. Available online: http://www.footwearhistory.com/ (accessed on 2 December 2015).
2. Shoes Reveal Personality Traits. Available online: http://www.scientificamerican.com/article/shoes-reveal-personality-traits/ (accessed on 2 December 2015).
3. DigiBarn Weird Stuff: Puma RS Computer Tennis Shoes (Pedometer, 1980s). Available online: http://www.digibarn.com/collections/weirdstuff/computer-tennis-shoes/ (accessed on 8 January 2016).
4. Perry, J. *Gait Analysis: Normal and Pathological Function*; SLACK: Thorofare, NJ, USA, 1992.
5. Crea, S.; Donati, M.; De Rossi, S.M.M.; Oddo, C.M.; Vitiello, N. A Wireless Flexible Sensorized Insole for Gait Analysis. *Sensors* **2014**, *14*, 1073–1093. [CrossRef] [PubMed]
6. GAITRite Systems—Portable Gait Analysis. Available online: http://www.gaitrite.com/ (accessed on 2 December 2015).

7. Full Body Gait Analysis with Kinect—Microsoft Research. Available online: https://www.microsoft.com/en-us/research/publication/full-body-gait-analysis-with-kinect/ (accessed on 7 July 2016).
8. Xu, X.; McGorry, R.W.; Chou, L.; Lin, J.; Chang, C. Accuracy of the Microsoft Kinect™ for measuring gait parameters during treadmill walking. *Gait Posture* **2015**, *42*, 145–151. [CrossRef] [PubMed]
9. Bamberg, S.J.M.; Benbasat, A.Y.; Scarborough, D.M.; Krebs, D.E.; Paradiso, J.A. Gait analysis using a shoe-integrated wireless sensor system. *IEEE Trans. Inf. Technol. Biomed.* **2008**, *12*, 413–423. [CrossRef] [PubMed]
10. Lopez-Meyer, P.; Fulk, G.D.; Sazonov, E.S. Automatic detection of temporal gait parameters in poststroke individuals. *IEEE Trans. Inf. Technol. Biomed.* **2011**, *15*, 594–601. [CrossRef] [PubMed]
11. Chen, M.; Huang, B.; Xu, Y. Intelligent shoes for abnormal gait detection. In Proceedings of the IEEE International Conference on Robotics and Automation, 2008 (ICRA 2008), Pasadena, CA, USA, 19–23 May 2008; pp. 2019–2024.
12. Mariani, B.; Jiménez, M.C.; Vingerhoets, F.J.G.; Aminian, K. On-shoe wearable sensors for gait and turning assessment of patients with Parkinson's disease. *IEEE Trans. Biomed. Eng.* **2013**, *60*, 155–158. [CrossRef] [PubMed]
13. Nsenga Leunkeu, A.; Lelard, T.; Shephard, R.J.; Doutrellot, P.-L.; Ahmaidi, S. Gait cycle and plantar pressure distribution in children with cerebral palsy: Clinically useful outcome measures for a management and rehabilitation. *NeuroRehabilitation* **2014**, *35*, 657–663. [PubMed]
14. Rampp, A.; Barth, J.; Schülein, S.; Gaßmann, K.-G.; Klucken, J.; Eskofier, B.M. Inertial sensor-based stride parameter calculation from gait sequences in geriatric patients. *IEEE Trans. Biomed. Eng.* **2015**, *62*, 1089–1097. [CrossRef] [PubMed]
15. Shop | Shimmer—Wearable Sensor Tech. Available online: http://www.shimmersensing.com/shop/ (accessed on 26 July 2016).
16. Kong, K.; Tomizuka, M. A Gait Monitoring System Based on Air Pressure Sensors Embedded in a Shoe. *IEEE/ASME Trans. Mechatron.* **2009**, *14*, 358–370. [CrossRef]
17. Liu, T.; Inoue, Y.; Shibata, K. A wearable ground reaction force sensor system and its application to the measurement of extrinsic gait variability. *Sensors* **2010**, *10*, 10240–10255. [CrossRef] [PubMed]
18. González, I.; Fontecha, J.; Hervás, R.; Bravo, J. An Ambulatory System for Gait Monitoring Based on Wireless Sensorized Insoles. *Sensors* **2015**, *15*, 16589–16613. [CrossRef] [PubMed]
19. Wu, Y.; Xu, W.; Liu, J.J.; Huang, M.-C.; Luan, S.; Lee, Y. An Energy-Efficient Adaptive Sensing Framework for Gait Monitoring Using Smart Insole. *IEEE Sens. J.* **2015**, *15*, 2335–2343. [CrossRef]
20. Ferrari, A.; Ginis, P.; Hardegger, M.; Casamassima, F.; Rocchi, L.; Chiari, L. A Mobile Kalman-Filter Based Solution for the Real-Time Estimation of Spatio-Temporal Gait Parameters. *IEEE Trans. Neural Syst. Rehabil. Eng.* **2016**, *24*, 764–773. [CrossRef] [PubMed]
21. Hegde, N.; Sazonov, E. SmartStep: A Fully Integrated, Low-Power Insole Monitor. *Electronics* **2014**, *3*, 381–397. [CrossRef]
22. WIISEL. Available online: http://www.wiisel.eu/ (accessed on 15 July 2015).
23. Huang, B.; Chen, M.; Huang, P.; Xu, Y. Gait Modeling for Human Identification. In Proceedings of the 2007 IEEE International Conference on Robotics and Automation, Roma, Italy, 10–14 April 2007; pp. 4833–4838.
24. Klucken, J.; Barth, J.; Kugler, P.; Schlachetzki, J.; Henze, T.; Marxreiter, F.; Kohl, Z.; Steidl, R.; Hornegger, J.; Eskofier, B.; et al. Unbiased and Mobile Gait Analysis Detects Motor Impairment in Parkinson's Disease. *PLoS ONE* **2013**, *8*, e56956. [CrossRef] [PubMed]
25. Barth, J.; Oberndorfer, C.; Pasluosta, C.; Schülein, S.; Gassner, H.; Reinfelder, S.; Kugler, P.; Schuldhaus, D.; Winkler, J.; Klucken, J.; et al. Stride Segmentation during Free Walk Movements Using Multi-Dimensional Subsequence Dynamic Time Warping on Inertial Sensor Data. *Sensors* **2015**, *15*, 6419–6440. [CrossRef] [PubMed]
26. Sensoria Fitness. Available online: http://www.sensoriafitness.com/ (accessed on 3 December 2015).
27. Wafai, L.; Zayegh, A.; Woulfe, J.; Aziz, S.M.; Begg, R. Identification of Foot Pathologies Based on Plantar Pressure Asymmetry. *Sensors* **2015**, *15*, 20392–20408. [CrossRef] [PubMed]
28. Mueller, M.J.; Hastings, M.; Commean, P.K.; Smith, K.E.; Pilgram, T.K.; Robertson, D.; Johnson, J. Forefoot structural predictors of plantar pressures during walking in people with diabetes and peripheral neuropathy. *J. Biomech.* **2003**, *36*, 1009–1017. [CrossRef]

29. Plantar Pressure Assessment | Physical Therapy Journal. Available online: http://ptjournal.apta.org/content/80/4/399 (accessed on 7 July 2016).

30. Sazonov, E.S.; Bumpus, T.; Zeigler, S.; Marocco, S. Classification of plantar pressure and heel acceleration patterns using neural networks. In Proceedings of the 2005 IEEE International Joint Conference on Neural Networks, 2005 (IJCNN 2005), Montreal, QC, Canada, 31 July–4 August 2005; Volume 5, pp. 3007–3010.

31. Morris, S.J.; Paradiso, J.A. Shoe-integrated sensor system for wireless gait analysis and real-time feedback. In Proceedings of the 24th Annual Conference and the Annual Fall Meeting of the Biomedical Engineering Society EMBS/BMES Conference (Second Joint), Houston, TX, USA, 23–26 October 2002; Volume 3, pp. 2468–2469.

32. Queen, R.M.; Haynes, B.B.; Hardaker, W.M.; Garrett, W.E. Forefoot loading during 3 athletic tasks. *Am. J. Sports Med.* **2007**, *35*, 630–636. [CrossRef] [PubMed]

33. Gioftsidou, A.; Malliou, P.; Pafis, G.; Beneka, A.; Godolias, G.; Maganaris, C.N. The effects of soccer training and timing of balance training on balance ability. *Eur. J. Appl. Physiol.* **2006**, *96*, 659–664. [CrossRef] [PubMed]

34. Mueller, M.J. Application of plantar pressure assessment in footwear and insert design. *J. Orthop. Sports Phys. Ther.* **1999**, *29*, 747–755. [CrossRef] [PubMed]

35. Begg, R.; Palaniswami, M. *Computational Intelligence for Movement Sciences: Neural Networks and Other Emerging Techniques*; Idea Group Inc. (IGI): Calgary, AB, Canada, 2006.

36. Zebris Medical GmbH Gait Analysis. Available online: http://www.zebris.de/english/medizin/medizin-ganganalyse.php#fdm (accessed on 1 December 2015).

37. F-Scan System. Available online: https://www.tekscan.com/products-solutions/systems/f-scan-system (accessed on 1 December 2015).

38. Tan, A.M.; Fuss, F.K.; Weizman, Y.; Woudstra, Y.; Troynikov, O. Design of Low Cost Smart Insole for Real Time Measurement of Plantar Pressure. *Procedia Technol.* **2015**, *20*, 117–122. [CrossRef]

39. Shu, L.; Hua, T.; Wang, Y.; Qiao Li, Q.; Feng, D.D.; Tao, X. In-shoe plantar pressure measurement and analysis system based on fabric pressure sensing array. *IEEE Trans. Inf. Technol. Biomed.* **2010**, *14*, 767–775. [PubMed]

40. Saito, M.; Nakajima, K.; Takano, C.; Ohta, Y.; Sugimoto, C.; Ezoe, R.; Sasaki, K.; Hosaka, H.; Ifukube, T.; Ino, S.; et al. An in-shoe device to measure plantar pressure during daily human activity. *Med. Eng. Phys.* **2011**, *33*, 638–645. [CrossRef] [PubMed]

41. Pedar. Available online: http://www.novel.de/novelcontent/pedar (accessed on 1 December 2015).

42. SurroSense Rx. Available online: http://orpyx.com/pages/surrosense-rx (accessed on 4 December 2015).

43. Coda, A.; Santos, D. Repeatability and Reproducibility of the F-Scan System in Healthy Children. *J. Orthop. Rheumatol. Sports Med.* **2015**, *1*, 104. [CrossRef]

44. Luo, Z.P.; Berglund, L.J.; An, K.N. Validation of F-Scan pressure sensor system: A technical note. *J. Rehabil. Res. Dev.* **1998**, *35*, 186–191. [PubMed]

45. Hamzah, H.; Osman, N.A.A.; Hasnan, N. Comparing Manufacturer's Point Calibration and Modified Calibration Setup for F-Scan Insole Sensor System: A Preliminary Assessment. In Proceedings of the 4th Kuala Lumpur International Conference on Biomedical Engineering 2008, Kuala Lumpur, Malaysia, 25–28 June 2008; Osman, N.A.A., Ibrahim, F., Abas, W.A.B.W., Rahman, H.S.A., Ting, H.-N., Eds.; Springer Berlin Heidelberg: Berlin, Germany, 2008; pp. 424–427.

46. Quesada, P.M.; Rash, G.S.; Jarboe, N. Assessment of pedar and F-Scan revisited. *Clin. Biomech. (Bristol Avon)* **1997**, *12*, S15. [CrossRef]

47. Giacomozzi, C. Appropriateness of plantar pressure measurement devices: A comparative technical assessment. *Gait Posture* **2010**, *32*, 141–144. [CrossRef] [PubMed]

48. Putti, A.B.; Arnold, G.P.; Cochrane, L.A.; Abboud, R.J. Normal pressure values and repeatability of the Emed ST4 system. *Gait Posture* **2008**, *27*, 501–505. [CrossRef] [PubMed]

49. Ferber, R.; Webber, T.; Everett, B.; Groenland, M. Validation of plantar pressure measurements for a novel in-shoe plantar sensory replacement unit. *J. Diabetes Sci. Technol.* **2013**, *7*, 1167–1175. [CrossRef] [PubMed]

50. Cazzola, D.; Trewartha, G.; Preatoni, E. Time-based calibrations of pressure sensors improve the estimation of force signals containing impulsive events. *Proc. Inst. Mech. Eng. Part. P J. Sports Eng. Technol.* **2014**, *228*, 147–151. [CrossRef]

51. Liu, T.; Inoue, Y.; Shibata, K. A wearable force plate system for the continuous measurement of triaxial ground reaction force in biomechanical applications. *Meas. Sci. Technol.* **2010**, *21*, 85804–85812. [CrossRef]

52. Hill, J.O.; Wyatt, H.R.; Reed, G.W.; Peters, J.C. Obesity and the environment: Where do we go from here? *Science* **2003**, *299*, 853–855. [CrossRef] [PubMed]

53. Levine, J.A.; Lanningham-Foster, L.M.; McCrady, S.K.; Krizan, A.C.; Olson, L.R.; Kane, P.H.; Jensen, M.D.; Clark, M.M. Interindividual variation in posture allocation: Possible role in human obesity. *Science* **2005**, *307*, 584–586. [CrossRef] [PubMed]

54. Adult Obesity Facts. Available online: http://www.cdc.gov/obesity/data/adult.html (accessed on 2 December 2015).

55. Fulk, G.D.; Sazonov, E. Using sensors to measure activity in people with stroke. *Top. Stroke Rehabil.* **2011**, *18*, 746–757. [CrossRef] [PubMed]

56. S Health -Fitness Diet Tracker—Android Apps on Google Play. Available online: https://play.google.com/store/apps/details?id=com.sec.android.app.shealth&hl=en (accessed on 2 December 2015).

57. iOS 9—Health. Available online: http://www.apple.com/ios/health/ (accessed on 2 December 2015).

58. Apple Watch. Available online: http://www.apple.com/watch/ (accessed on 2 December 2015).

59. Fitbit Official Site for Activity Trackers & More. Available online: https://www.fitbit.com/ (accessed on 2 December 2015).

60. LG G Watch R. Available online: http://www.lg.com/us/smart-watches/lg-W110-lg-watch-r (accessed on 5 August 2016).

61. Uslu, G.; Baydere, S. RAM: Real Time Activity Monitoring with feature extractive training. *Expert Syst. Appl.* **2015**, *42*, 8052–8063. [CrossRef]

62. Riou, M.-È.; Rioux, F.; Lamothe, G.; Doucet, É. Validation and Reliability of a Classification Method to Measure the Time Spent Performing Different Activities. *PLoS ONE* **2015**, *10*, e0128299. [CrossRef] [PubMed]

63. González, S.; Sedano, J.; Villar, J.R.; Corchado, E.; Herrero, Á.; Baruque, B. Features and models for human activity recognition. *Neurocomputing* **2015**, *167*, 52–60. [CrossRef]

64. Gerd Krassnig, D.T. User-friendly system for recognition of activities with an accelerometer. In Proceedings of the 2010 4th International Conference on Pervasive Computing Technologies for Healthcare, Munich, Germany, 22–25 March 2010; pp. 1–8.

65. Mathie, M.J.; Coster, A.C.F.; Lovell, N.H.; Celler, B.G. Detection of daily physical activities using a triaxial accelerometer. *Med. Biol. Eng. Comput.* **2003**, *41*, 296–301. [CrossRef] [PubMed]

66. Mathie, M.J.; Celler, B.G.; Lovell, N.H.; Coster, A.C.F. Classification of basic daily movements using a triaxial accelerometer. *Med. Biol. Eng. Comput.* **2004**, *42*, 679–687. [CrossRef] [PubMed]

67. Karantonis, D.M.; Narayanan, M.R.; Mathie, M.; Lovell, N.H.; Celler, B.G. Implementation of a Real-Time Human Movement Classifier Using a Triaxial Accelerometer for Ambulatory Monitoring. *IEEE Trans. Inf. Technol. Biomed.* **2006**, *10*, 156–167. [CrossRef] [PubMed]

68. Nishkam Ravi, N.D. Activity Recognition from Accelerometer Data. *AAAI* **2005**, *3*, 1541–1546.

69. Ferguson, T.; Rowlands, A.V.; Olds, T.; Maher, C. The validity of consumer-level, activity monitors in healthy adults worn in free-living conditions: A cross-sectional study. *Int. J. Behav. Nutr. Phys. Act.* **2015**, *12*. [CrossRef] [PubMed]

70. Lee, J.-M.; Kim, Y.; Welk, G.J. Validity of consumer-based physical activity monitors. *Med. Sci. Sports Exerc.* **2014**, *46*, 1840–1848. [CrossRef] [PubMed]

71. Spierer, D.K.; Hagins, M.; Rundle, A.; Pappas, E. A comparison of energy expenditure estimates from the Actiheart and Actical physical activity monitors during low intensity activities, walking, and jogging. *Eur. J. Appl. Physiol.* **2011**, *111*, 659–667. [CrossRef] [PubMed]

72. Sazonov, E.S.; Fulk, G.; Hill, J.; Schutz, Y.; Browning, R. Monitoring of posture allocations and activities by a shoe-based wearable sensor. *IEEE Trans. Biomed. Eng.* **2011**, *58*, 983–990. [CrossRef] [PubMed]

73. Tang, W.; Sazonov, E.S. Highly Accurate Recognition of Human Postures and Activities through Classification with Rejection. *IEEE J. Biomed. Health Inform.* **2014**, *18*, 309–315. [CrossRef] [PubMed]

74. Zhang, T.; Fulk, G.D.; Tang, W.; Sazonov, E.S. Using decision trees to measure activities in people with stroke. In Proceedings of the 2013 35th Annual International Conference of the IEEE Engineering in Medicine and Biology Society (EMBC), Osaka, Japan, 3–7 July 2013; pp. 6337–6340.

75. Zhang, T.; Lu, J.; Uswatte, G.; Taub, E.; Sazonov, E.S. Measuring gait symmetry in children with cerebral palsy using the SmartShoe. In Proceedings of the 2014 IEEE Healthcare Innovation Conference (HIC), Seattle, WA, USA, 8–10 October 2014; pp. 48–51.

76. Sazonov, E.S.; Hegde, N.; Tang, W. Development of SmartStep: An insole-based physical activity monitor. In Proceedings of the 2013 35th Annual International Conference of the IEEE Engineering in Medicine and Biology Society (EMBC), Osaka, Japan, 3–7 July 2013; pp. 7209–7212.

77. Hegde, N.; Sazonov, E. SmartStep 2.0—A completely wireless, versatile insole monitoring system. In Proceedings of the 2015 IEEE International Conference on Bioinformatics and Biomedicine (BIBM), Washington, DC, USA, 9–12 November 2015; pp. 746–749.

78. Chen, B.; Wang, X.; Huang, Y.; Wei, K.; Wang, Q. A foot-wearable interface for locomotion mode recognition based on discrete contact force distribution. *Mechatronics* **2015**, *32*, 12–21. [CrossRef]

79. Kawsar, F.; Hasan, M.K.; Love, R.; Ahamed, S.I. A Novel Activity Detection System Using Plantar Pressure Sensors and Smartphone. In Proceedings of the 2015 IEEE 39th Annual Computer Software and Applications Conference (COMPSAC), Taichung, Taiwan, 1–5 July 2015; Volume 1, pp. 44–49.

80. Hegde, N.; Sazonov, E.S.; Melanson, E. Development of a Real Time Activity Monitoring Android Application Utilizing SmartStep. In Proceedings of the 2016 38th Annual International Conference of the IEEE Engineering in Medicine and Biology Society (EMBC), Orlando, FL, USA, 16–20 August 2016.

81. Sazonov, E.; Hegde, N.; Browning, R.C.; Melanson, E.L.; Sazonova, N.A. Posture and activity recognition and energy expenditure estimation in a wearable platform. *IEEE J. Biomed. Health Inform.* **2015**, *19*, 1339–1346. [CrossRef] [PubMed]

82. Fulk, G.D.; Edgar, S.R.; Bierwirth, R.; Hart, P.; Lopez-Meyer, P.; Sazonov, E. Identifying activity levels and steps of people with stroke using a novel shoe-based sensor. *J. Neurol. Phys. Ther. JNPT* **2012**, *36*, 100–107. [CrossRef] [PubMed]

83. Maurer, U.; Smailagic, A.; Siewiorek, D.P.; Deisher, M. Activity recognition and monitoring using multiple sensors on different body positions. In Proceedings of the International Workshop on Wearable and Implantable Body Sensor Networks (BSN'06), Cambridge, MA, USA, 3–5 April 2006; p. 116.

84. Edgar, S.R.; Fulk, G.D.; Sazonov, E.S. Recognition of household and athletic activities using SmartShoe. In Proceedings of the 2012 Annual International Conference of the IEEE Engineering in Medicine and Biology Society, San Diego, CA, USA, 28 August–1 September 2012; pp. 6382–6385.

85. Benocci, M.; Rocchi, L.; Farella, E.; Chiari, L.; Benini, L. A wireless system for gait and posture analysis based on pressure insoles and Inertial Measurement Units. In Proceedings of the 2009 3rd International Conference on Pervasive Computing Technologies for Healthcare, London, UK, 1–3 April 2009; pp. 1–6.

86. Sazonova, N.; Browning, R.C.; Sazonov, E. Accurate prediction of energy expenditure using a shoe-based activity monitor. *Med. Sci. Sports Exerc.* **2011**, *43*, 1312–1321. [CrossRef] [PubMed]

87. Dannecker, K.L.; Sazonova, N.A.; Melanson, E.L.; Sazonov, E.S.; Browning, R.C. A comparison of energy expenditure estimation of several physical activity monitors. *Med. Sci. Sports Exerc.* **2013**, *45*, 2105–2112. [CrossRef] [PubMed]

88. Lechal. Available online: http://lechal.com/insoles.html (accessed on 3 December 2015).

89. Smart Shoe. Available online: https://us.vibram.com/article-smart-shoe.html (accessed on 3 December 2015).

90. Gjoreski, H.; Kaluža, B.; Gams, M.; Milić, R.; Luštrek, M. Context-based ensemble method for human energy expenditure estimation. *Appl. Soft Comput.* **2015**, *37*, 960–970. [CrossRef]

91. Villanueva, E.V. The validity of self-reported weight in US adults: A population based cross-sectional study. *BMC Public Health* **2001**, *1*, 11. [CrossRef] [PubMed]

92. Sazonova, N.A.; Browning, R.; Sazonov, E.S. Prediction of Bodyweight and Energy Expenditure Using Point Pressure and Foot Acceleration Measurements. *Open Biomed. Eng. J.* **2011**, *5*, 110–115. [CrossRef] [PubMed]

93. Hellstrom, P.; Folke, M.; Ekström, M. Wearable Weight Estimation System. *Procedia Comput. Sci.* **2015**, *64*, 146–152. [CrossRef]

94. Khoo, I.-H.; Marayong, P.; Krishnan, V.; Balagtas, M.N.; Rojas, O. Design of a biofeedback device for gait rehabilitation in post-stroke patients. In Proceedings of the 2015 IEEE 58th International Midwest Symposium on Circuits and Systems (MWSCAS), Fort Collins, CO, USA, 2–5 August 2015; pp. 1–4.

95. Hegde, N.; Fulk, G.D.; Sazonov, E.S. Development of the RT-GAIT, a Real-Time feedback device to improve Gait of individuals with stroke. In Proceedings of the 2015 37th Annual International Conference of the IEEE Engineering in Medicine and Biology Society (EMBC), Milan, Italy, 25–29 August 2015; pp. 5724–5727.

96. Yang, L.; Dyer, P.S.; Carson, R.J.; Webster, J.B.; Bo Foreman, K.; Bamberg, S.J.M. Utilization of a lower extremity ambulatory feedback system to reduce gait asymmetry in transtibial amputation gait. *Gait Posture* **2012**, *36*, 631–634. [CrossRef] [PubMed]

97. Donovan, L.; Feger, M.A.; Hart, J.M.; Saliba, S.; Park, J.; Hertel, J. Effects of an auditory biofeedback device on plantar pressure in patients with chronic ankle instability. *Gait Posture* **2016**, *44*, 29–36. [CrossRef] [PubMed]

98. Aging Statistics. Available online: http://www.aoa.acl.gov/Aging_Statistics/index.aspx (accessed on 7 May 2015).

99. NIHSeniorHealth: Falls and Older Adults—Causes and Risk Factors. Available online: http://nihseniorhealth.gov/falls/causesandriskfactors/01.html (accessed on 2 December 2015).

100. Schwenk, M.; Hauer, K.; Zieschang, T.; Englert, S.; Mohler, J.; Najafi, B. Sensor-derived physical activity parameters can predict future falls in people with dementia. *Gerontology* **2014**, *60*, 483–492. [CrossRef] [PubMed]

101. Doukas, C.; Maglogiannis, I. Advanced patient or elder fall detection based on movement and sound data. In Proceedings of the 200 Second International Conference on Pervasive Computing Technologies for Healthcare, Tampere, Finland, 30 January–1 February 2008; pp. 103–107.

102. Ayena, J.C.; Chapwouo, T.L.D.; Otis, M.J.-D.; Menelas, B.A.J. An efficient home-based risk of falling assessment test based on Smartphone and instrumented insole. In Proceedings of the 2015 IEEE International Symposium on Medical Measurements and Applications (MeMeA), Turin, Italy, 7–9 May 2015; pp. 416–421.

103. Sim, S.Y.; Jeon, H.S.; Chung, G.S.; Kim, S.K.; Kwon, S.J.; Lee, W.K.; Park, K.S. Fall detection algorithm for the elderly using acceleration sensors on the shoes. In Proceedings of the 2011 Annual International Conference of the IEEE Engineering in Medicine and Biology Society (EMBC), Boston, MA, USA, 30 August–3 September 2011; pp. 4935–4938.

104. Majumder, A.J.A.; Zerin, I.; Uddin, M.; Ahamed, S.I.; Smith, R.O. SmartPrediction: A Real-time Smartphone-based Fall Risk Prediction and Prevention System. In Proceedings of the 2013 Research in Adaptive and Convergent Systems RACS '13; ACM: New York, NY, USA, 2013; pp. 434–439.

105. Documents | WIISEL. Available online: http://www.wiisel.eu/?q=content/documents (accessed on 20 June 2016).

106. Otis, M.J.-D.; Menelas, B.J. Toward an augmented shoe for preventing falls related to physical conditions of the soil. In Proceedings of the 2012 IEEE International Conference on Systems, Man, and Cybernetics (SMC), Seoul, Korea, 14–17 October 2012; pp. 3281–3285.

107. Schirmer, M.; Hartmann, J.; Bertel, S.; Echtler, F. Shoe Me the Way: A Shoe-Based Tactile Interface for Eyes-Free Urban Navigation. In Proceedings of the 17th International Conference on Human-Computer Interaction with Mobile Devices and Services MobileHCI '15; ACM: New York, NY, USA, 2015; pp. 327–336.

108. Bebek, O.; Suster, M.A.; Rajgopal, S.; Fu, M.J.; Huang, X.; Çavuşoğlu, M.C.; Young, D.J.; Mehregany, M.; van den Bogert, A.J.; Mastrangelo, C.H. Personal Navigation via High-Resolution Gait-Corrected Inertial Measurement Units. *IEEE Trans. Instrum. Meas.* **2010**, *59*, 3018–3027. [CrossRef]

109. Castaneda, N.; Lamy-Perbal, S. An improved shoe-mounted inertial navigation system. In Proceedings of the 2010 International Conference on Indoor Positioning and Indoor Navigation (IPIN), Zurich, Switzerland, 15–17 September 2010; pp. 1–6.

110. Foxlin, E. Pedestrian tracking with shoe-mounted inertial sensors. *IEEE Comput. Graph. Appl.* **2005**, *25*, 38–46. [CrossRef] [PubMed]

111. Shenck, N.S.; Paradiso, J.A. Energy scavenging with shoe-mounted piezoelectrics. *IEEE Micro* **2001**, *21*, 30–42. [CrossRef]

112. Orecchini, G.; Tentzeris, M.M.; Yang, L.; Roselli, L. Smart Shoe: An autonomous inkjet-printed RFID system scavenging walking energy. In Proceedings of the 2011 IEEE International Symposium on Antennas and Propagation (APSURSI), Spokane, WA, USA, 3–8 July 2011; pp. 1417–1420.

113. Zhao, J.; You, Z. A shoe-embedded piezoelectric energy harvester for wearable sensors. *Sensors* **2014**, *14*, 12497–12510. [CrossRef] [PubMed]

114. Kymissis, J.; Kendall, C.; Paradiso, J.; Gershenfeld, N. Parasitic power harvesting in shoes. In Proceedings of the Second International Symposium on Wearable Computers, 1998 Digest of Papers, Pittsburgh, PA, USA, 19–20 October 1998; pp. 132–139.

115. Meier, R.; Kelly, N.; Almog, O.; Chiang, P. A piezoelectric energy-harvesting shoe system for podiatric sensing. In Proceedings of the 2014 36th Annual International Conference of the IEEE Engineering in Medicine and Biology Society (EMBC), Chicago, IL, USA, 26–30 August 2014; pp. 622–625.
116. Internet of Things. Available online: https://en.wikipedia.org/wiki/Internet_of_things (accessed on 3 August 2016).
117. The Rise and Fall of the Smart Shoe—and Why They Could Be on the Way Back. Available online: http://www.wareable.com/running/smart-shoes-875 (accessed on 5 December 2015).
118. MIT Media Lab: Affective Computing Group. Available online: http://affect.media.mit.edu/projects.php (accessed on 26 July 2016).
119. ShiftWear: Customize Your Kicks. Available online: https://www.indiegogo.com/projects/shiftwear-customize-your-kicks/#/story (accessed on 5 December 2015).

electronics

MDPI

Review

A Fabric-Based Approach for Wearable Haptics

Matteo Bianchi [1,2]

[1] Department of Advanced Robotics, Istituto Italiano di Tecnologia, via Morego 30, Genova, 16163 Italy; matteobianchi23@gmail.com; Tel.: +39-050-2217050

[2] Research Center "E. Piaggio", University of Pisa, Largo Lucio Lazzarino 1, Pisa 56122, Italy

Academic Editors: Enzo Pasquale Scilingo and Gaetano Valenza
Received: 24 May 2016; Accepted: 20 July 2016; Published: 26 July 2016

Abstract: In recent years, wearable haptic systems (WHS) have gained increasing attention as a novel and exciting paradigm for human–robot interaction (HRI). These systems can be worn by users, carried around, and integrated in their everyday lives, thus enabling a more natural manner to deliver tactile cues. At the same time, the design of these types of devices presents new issues: the challenge is the correct identification of design guidelines, with the two-fold goal of minimizing system encumbrance and increasing the effectiveness and naturalness of stimulus delivery. Fabrics can represent a viable solution to tackle these issues. They are specifically thought "to be worn", and could be the key ingredient to develop wearable haptic interfaces conceived for a more natural HRI. In this paper, the author will review some examples of fabric-based WHS that can be applied to different body locations, and elicit different haptic perceptions for different application fields. Perspective and future developments of this approach will be discussed.

Keywords: wearable haptic systems; fabrics; cutaneous feedback; softness rendering; force feedback; affective touch

---

1. Introduction

The sense of touch is one of the most fundamental sensory channels for humans. It represents the primary way of interacting with and exploring the external environment, and is one of the most effective channels for social communication [1]. Not surprisingly, a great deal of effort has been devoted to the development of artificial systems, or haptic devices, which enable users to feel external or virtual objects as if they were directly encountered, even if they are remote or not directly accessible by touch, by delivering different types of tactile information and can be used in many application fields (for a review, the interested reader can refer to reference [2]). In recent years, to increase device usability, a novel idea has gained increasing attention: i.e., to move from physically-grounded haptic interfaces, where haptic stimulation is provided with respect to (w.r.t.) operator's ground, towards wearable systems, which can be worn by users while the body-grounded base is moved, as close as possible, to the point of stimulus application [2,3]. This new generation of wearable haptic systems (WHS) [3] can convey tactile cues in a more natural fashion, while being easily worn by users, carried around, and integrated in everyday life. This shift in system design has opened up exciting avenues in many application fields, such as virtual reality and assistive robotics [4–6]. One of the most convincing motivations for this change relies on the possibility to integrate WHS with the human body with minimal constraints [7], thus enabling a more natural investigation of human behavior and human–robot interaction (HRI). For the latter, tele-robotics represents an ideal application field. In this case, wearable devices can be easily worn by human operator and convey to her/him information from the tele-manipulated environment, thus significantly advancing the naturalness of performance. At the same time, the substitution of kinaesthetic force feedback with another form of feedback, such as the one provided by wearable cutaneous devices, represents a useful approach to overcome stability issues in teleoperation. This method is called sensory substitution: Since no kinaesthetic force is fed back to the operator, the haptic loop is intrinsically stable and, thus, no bilateral controller is needed [8]. An additional method to improve the performance of passive teleoperation systems with force reflection integrates kinaesthetic haptic interfaces with wearable cutaneous haptic feedback. This approach proposes to scale down kinaesthetic feedback to satisfy passivity, at the expense of transparency: In this case, to recover performance, cutaneous force information is conveyed through wearable devices [9]. Human–robot co-working is also another promising scenario for WHS, where the cooperative manipulation of robots, together with a human partner, is envisioned. In this case, mutual knowledge of the current reality, achievable through wearable sensing systems and haptic feedback devices, represents a key component to ensure an effective and natural cooperation [10].

1.1. Wearable Haptic Systems: Technologies and Main Characteristics

Looking at the state-of-the-art, it is possible to observe a number of strategies for stimulus delivery through wearable systems, specifically developed to generate vibrations [4], apply forces [11], stimulate skin using pin-arrays [12] or using electrocutaneous feedback [13], and considering different body locations for stimulus application, such as arm [7,14], foot [15], finger [5,16], among the others; the interested reader may refer to references [16,17] (for a review of these topics see Figure 1).

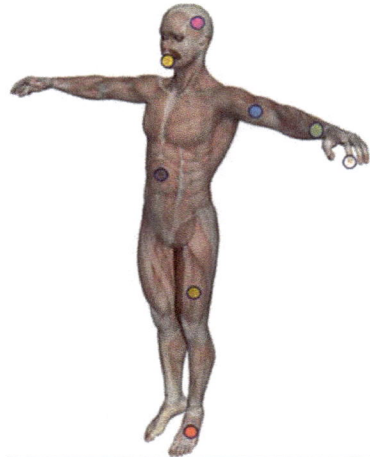

Location	Devices
Finger	[5,16,18]
Wrist	[19,21]
Arm and Forearm	[14,20,22,23]
Tongue and Mouth	[24–26]
Head	[27]
Torso, Trunk, Shoulders	[28–30]
Leg	[31,32]
Foot	[15,33]

Figure 1. Wearable haptic systems and body locations. Finger [5,16,18]; wrist [19,21]; arm and forearm [14,20,22,23]; tongue and mouth [24–26]; head [27]; torso, trunk and shoulders [28–30]; leg [31,32]; foot [15,33].

From a technological point of view, different actuation strategies are usually employed to deliver haptic stimuli in wearable devices, the main ones are summarized in tab:electronics-05-00044-t001.

Table 1. Wearable haptic systems (WHS) main actuation types.

Stimulus/Actuation Type	Devices
Pin-arrays	[12,34]
Electrocutaneous	[13,22,35]
Vibration	[4,7,30]
Deformation/Forces	[5,16,36]
Pneumatic	[31,32]

Each of these strategies comes with pros and cons, which must be considered in light of the applications that WHS are designed for, such as rehabilitation, assistive robotics, guidance, among others—the interested reader can refer to reference [17] for further details. From a physiological point of view, different stimulation modes target the different mechanoreceptors of human skin, of which the main characteristics and response typologies are summarized in tab:electronics-05-00044-t002 (note that all contribute to guarantee stable precision grasp and manipulation [37,38])—for an exhaustive description of mechanoreceptor characteristics, the reader can refer to references [37–40].

Table 2. Human skin mechanoreceptors (adapted from [39,41]).

Mechanoreceptors	Primary Functions
Slowly Adapting type I (SAI)	Very-low-frequency vibration detection; coarse texture perception; pattern/form detection
Fast-adapting type I (FA I)	Low-frequency vibration detection
Fast-adapting type II (FA II)	High-frequency vibration detection; fine texture perception
Slowly Adapting type II (SAII)	Direction of object motion and force due to skin Stretch; finger position

Furthermore, low-threshold mechanosensitive C fibers that can be found in the hairy skin represent the neurobiological substrate for the affective properties of touch [42].

The main objective of this subsection is to provide the reader with a general, but useful, overview on WHS, which can provide the essential tools needed to understand the topics and issues introduced in the following sections.

1.2. Wearable Haptic Systems: Open Issues and Fabric-Based Approaches

The diffusion of WHS, of which the main characteristics are described in the previous subsection, has created new and challenging issues: (1) how to design systems with minimal encumbrance, and (2) how do design systems that allow a natural haptic interaction? To tackle these issues, the author proposes to leverage the intrinsic wearability of fabrics with the objective of engineering artificial devices, which can be easily worn and, at the same time, enable a highly natural HRI. These characteristics have been already widely used in the literature to develop sensors for monitoring human behavior, which can successfully handle the demands of on-body sensing and are also light-weight, such as knitted fabrics (e.g., [43–46]). For this reason, extensive research work has been performed for the definition of mechanical textile properties (which can affect haptic sensation), as well as subjective evaluation of haptic properties of fabrics and textiles (see references [47–50]). In reference [51] the authors integrated electroactive polymeric materials into wearable garments to endow them with sensing and actuation properties, with applications in post stroke rehabilitation and assessment. In reference [52], a textile-based glove with electroactive polymers, acting as force/position sensors and haptic feedback actuators, was presented. In reference [53], the authors proposed an air-inflatable vest that can be remotely triggered to create a sensation that resembles a hug. In reference [50], the authors presented BubbleWrap, a matrix of electromagnetic actuators enclosed in fabric. This system consisted of individually controllable cells, which can expand and contract, thus providing both active and passive haptic feedback. In reference [54], the authors described HAPI (Haptic Augmented Posture Interface) Bands, a set of user-worn bands instrumented with eccentric mass motors to provide vibrotactile feedback for the guidance of static poses. In reference [55], the authors introduced TableHop, a tabletop display that provided controlled self-actuated deformation and vibro-tactile feedback to an elastic fabric surface while retaining the capability of high-resolution visual projection. The surface was made of a highly-stretchable pure spandex fabric that was electrostatically actuated using electrodes mounted on its top or underside. In reference [56], the authors reported on a stretchable glove endowed with vibration motors and bend sensors, which was used to provide sensory feedback during rehabilitation training in a virtual reality environment.

While these examples show the effectiveness of textiles and fabrics as sensing tools or haptic devices, all of them require the integration of additional properties into the fabric structures (e.g., conductive, magnetic properties, or pneumatic, vibrotactile actuation). To the best of the author's knowledge, there is no evidence of wearable haptic systems that completely leverage pure mechanical changes of stress–strain fabric characteristics to deliver tactile cues.

This work aims at reviewing three examples of devices that have been developed by the author, which exploit the mechanical properties of a bi-elastic fabric for conveying tactile cues, at three different body locations, and targeting different application fields. The fabric is called Superbiflex HN, by Mectex S.P.A (Erba, Como, Italy), which exhibits a good range of elasticity and resistance to traction. Body locations, i.e., finger, forearm, and arm, can be used to define a taxonomy for devices, while a broad distinction of haptic systems for discriminative and affective touch can be also applied. Following these classifications, I will start with discriminative haptic devices, first describing W-FYD (Wearable Fabric Yielding Device), a tactile display for softness rendering and multi-cue delivery, which is worn on a user's finger [18] and can be profitably employed for virtual reality, neuroscientific studies, and tele-operation. Then, I will describe CUFF [14], a Clenching Upper-limb Force Feedback wearable device for distributed mechano-tactile stimulation of normal and tangential skin forces, which can be applied on a user's arm and used for prosthetics, tele-operation, and guidance for the blind. Finally, considering affective haptics, I will present a device that can simulate a human caress in terms of force and velocity [57], which can be worn on a subject's forearm, and was proven to effectively elicit an emotional response in users. Without any claim of exhaustiveness, these three systems offer an interesting perspective on the usage of the mechanical properties of simple elastic fabrics for wearable haptics, showing promise in different application domains. A discussion of these applications, and possible future directions, regarding the employment of fabrics as enabling ingredients for a successful development and usage of WHS are also reported.

2. W-FYD

Softness represents one of the most fundamental haptic properties [58], which plays a crucial role for task accomplishment in many contexts, from handling fruit to complex medical procedures. However, looking at WHS in the literature, none of the proposed solutions are able to provide controllable softness information to the user and enable both active and passive touch experiences. To bridge this gap, in reference [18], the author proposed W-FYD (Figure 2), which represents the wearable version of the grounded softness display described in references [59–62]. W-FYD controls the stretching state of a fabric to reproduce different stiffness levels and, for the first time, it can convey softness information, tangential cues, and enable both passive and active haptic exploration in users. The mechanical structure is inspired by the grounded version of the device and is also similar to the one reported in reference [36]. More specifically, two DC motors can vary the stiffness of the fabric and, if independently controlled, provide tangential force (Figure 3). In the active mode, the device is attached to the back of the finger; hence, the only movement the user can perform is the flexion of the distal phalanx, which provokes the indentation of the fabric. To enable the passive mode, an additional degree of freedom is implemented trough a servomotor and a lifting mechanism, which puts the fabric into contact with the user's finger pad (Figure 4). Finally, thanks to the presence of the two independently controlled DC motors, W-FYD is endowed with an additional translational degree of freedom, which can induce the sensation of sliding/slipping on the user's fingertip. In this case, the user wears the system and the DC motors are synchronously moved, so that the fabric slides, right and left, against the user's finger (which is still). The control of the stretching states of the fabric, and, hence, of the stiffness stimulus to be delivered, relies on the characterization of the system. In other terms, the stiffness workspace of the device was characterized at different DC motor positions (each corresponding to a given force-indentation curve, i.e., stiffness characteristic or fabric stretching state)

(Figure 3b). When the device is used on the user's finger, DC motor positions were controlled to reproduce a given softness level, based on the value of fabric indentation (h_a, measured through an infrared sensor, or via the commanded servo-motor position, h_p, in the active and passive modes, respectively (see Figure 4)). More specifically, knowing the relation between DC motor position (θ, in our case $\theta_1 = \theta_2 = \theta$ (see Figure 3a)) and force-indentation curves obtained during the characterization phase, we can command motor positions as $\theta = \frac{1}{m}\left(\gamma\delta^{b-1} - q\right)$, where γ is the stiffness level to be reproduced, according to a generic power function, i.e., $\gamma = \frac{F}{\delta^b}$, F is the force, δ is the indentation, while m and q represent the coefficients that characterize device stiffness workspace for different stretching states (i.e., DC motor positions): $\sigma(\theta) = m\theta + q$ (see Figure 3b).

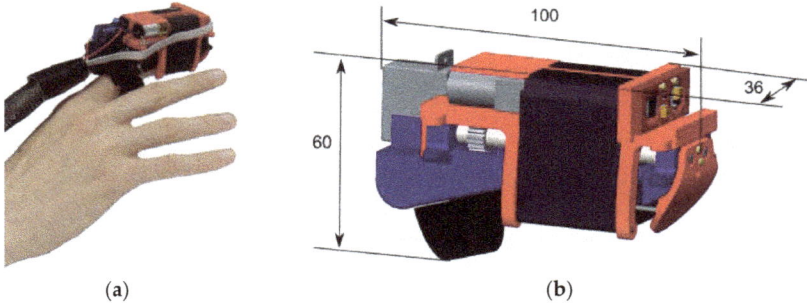

Figure 2. Wearable Fabric Yielding Device (W-FYD) on a user's finger (**a**); W-FYD CAD design and dimensions (in mm) (**b**). Reproduced from [18], Copyright 2016, IEEE.

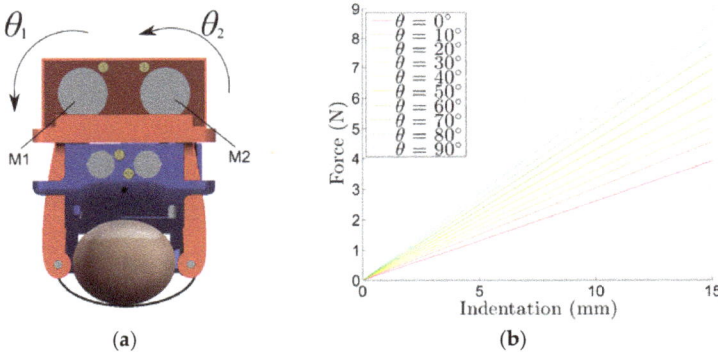

Figure 3. Representation of a finger interacting with the W-FYD (**a**); characterization curves for different motor positions (θ_1 and θ_2) (**b**). Reproduced from [18], Copyright 2016, IEEE.

W-FYD, of which the total mass is 100 g (and dimensions of which are reported in Figure 2), was proven to successfully enable a correct softness discrimination in users, in absolute and relative cognition tasks, as well as an effective identification of the direction of slipping [18].

These results are promising and encourage us to investigate applications of the device in neuroscientific studies, e.g., to study the role of softness information for slip/grasp control or softness perception of non-uniform materials (through multi-digit device implementation), virtual reality, and tele-operation. In the latter case, W-FYD acts as a display mounted on the user's finger for remote-robotic palpation of soft materials, e.g., in robot-assisted medical applications or surgeon training, where the correct rendering of tissue stiffness, and other haptic cues, is extremely important for the success of diagnostic or surgical procedures. In this regard, it is worth to mention that W-FYD was integrated into an augmented physical simulator that allows the real time tracking of artery reproductions and the user's finger, and provides pulse feedback during palpation [63]. Preliminary experiments showed a general consensus among surgeons regarding the realism of the arterial pulse feedback and the usefulness of tactile augmented reality in open-surgery simulators.

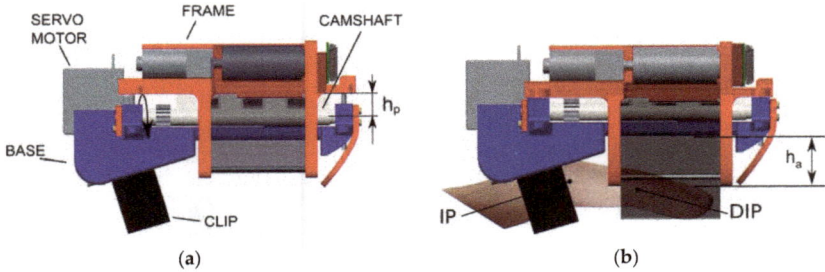

(a)   (b)

Figure 4. Passive mode: The fabric frame is put in contact with the user's finger by the camshaft lifting mechanism and servo-motor, inducing a variation of hp (commanded servo motor position) of the frame (**a**); active mode: The user can indent the fabric by flexing the interphalangeal proximal joint of the index finger (IP), while fabric indentation ha can be measured through a contactless infrared sensor (**b**). Reproduced from [18], Copyright 2016, IEEE.

In all these examples, wearability represents the key factor for successful task accomplishment, or to increase the immersiveness and naturalness of haptic interaction.

3. CUFF

The Clenching Upper-limb Force Feedback (CUFF) is comprised of two DC motors attached to a fabric cuff worn around the arm (Figure 5). The motors can spin in opposite directions to tighten or loosen the band on the arm, thus conveying normal force and pressure cues; they can also spin in the same direction in order to slide the band around the arm, thus inducing skin stretch cues that can easily be associated with directional and navigation information (see Figure 5).

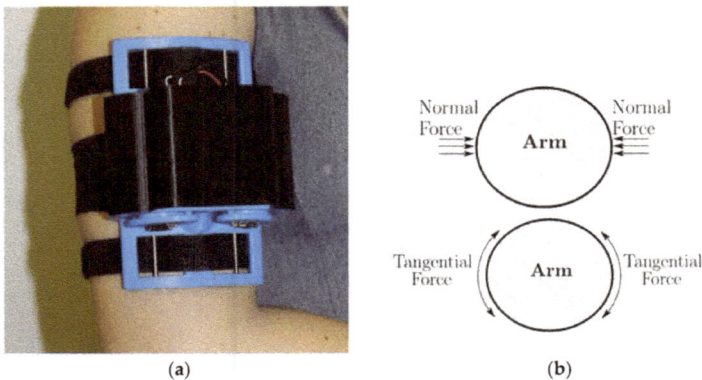

(a)   (b)

Figure 5. Clenching Upper-limb Force Feedback (CUFF) (**a**) and working modes (**b**). The total weight is 494 g and its overall dimensions are 14.5 × 9.7 × 11.6 cm. Reproduced from [14], Copyright 2013, IEEE.

In the first mode, CUFF was used in association with Pisa/IIT SoftHand (SH), an anthropomorphic robotic hand, actuated through a single motor, but capable of adapting for grasping different objects [64]. Because the SH has no built-in force sensors, an estimation of applied force can be obtained based on the current the motor draws. Briefly, the current that the SH absorbs while moving without obstacles is lower than that required to move following contact with an object. The difference between these two currents (residual current) can, thus, be exploited to drive the DC motors of CUFF and to provide force feedback. This is because residual current is proportional to grasp force (see Figure 6). This approach is motivated by the fact that there is a net difference in the absorbed current of the SH motor, in free motion (maximum value of $\approx$ 800 mA) or when the robotic hand grasps an external object (maximum value of $\approx$ 1200 mA). The reconstructed current represents the current absorbed by the motor in free hand motion, which will be subtracted from the current sensed by the μ-controller of the hand, which controls the opening/closing levels of the hand and acts on the current that drives the hand motor.

To achieve this goal, a Look Up Table was implemented as a function of the variables θ, $\dot{\theta}$, $\ddot{\theta}$ (where θ is the angular motor position of SH) and of which the output is the reconstructed free-hand motion current, I_{lut}. For the derivation of the terms of this function, please see reference [14]. The error of the reconstructed current for free-hand motions is under 5%. The *residual current term rI* was then used to compute the reference angular motor position of CUFF motors as: $\theta_{m,ref} = \beta_{CUFF} rI$. In this case, the two angular positions of the CUFF motors are $\theta_{m1} = \theta_{m2} = \theta_m$.

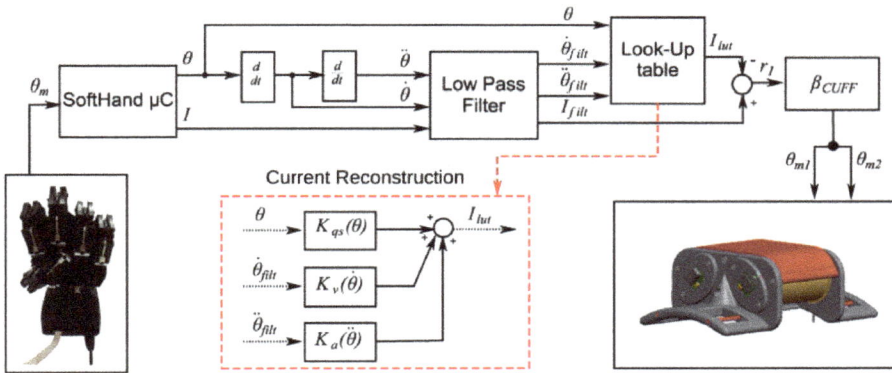

Figure 6. CUFF reproduces the estimated resultant force applied by SoftHand through belt stretching over a user's arm. The suffix filt on signals indicates the measured current, velocity, and acceleration of SH after low pass filtering. Reproduced from [14], Copyright 2013, IEEE.

The CUFF device was studied in applications to evaluate the role of force feedback with the usage of the prosthetic version of SH [65]. Thanks to its high wearability, the CUFF system can be used in conjunction with such a prosthetic device to provide haptic feedback in prosthetic applications or in tele-operation tasks where the modulation of grasping force is crucial. In reference [66], an early version of the device was connected to SH, operated by participants through a human–robot interface and surface electromyography signals. Leveraging the current-based disturbance observer, the interaction forces in contact with the grasped object were estimated and then converted and applied to the upper arm of the user via a custom-made pressure cuff. Such a tactile device was used together with vibrotactile feedback, based on surface irregularities and acceleration signals, which conveyed information on surface properties, contact, and detection of object slippage. Experiments in the evaluation of grasp robustness and intuitiveness of hand control suggested that incorporating these haptic feedback strategies facilitated the execution of safe and stable grasps. In reference [14], CUFF was tested in conjunction with the Pisa/IIT SoftHand, which was controlled using a handle to grasp objects with different stiffness properties. As reported in reference [67], the relationship between the indenting force and the overall rigid displacement (or indentation) between the two bodies can be regarded as an approximation of the kinaesthetic information involved in softness perception. Therefore, in reference [14], participant had information on how much she/he indented the specimens, at least indirectly, through the control of the handle, while CUFF conveyed information on the grip force exerted by the hand. Results of ranking experiments, where softness specimens were sorted in terms of perceived softness, were highly accurate. Furthermore, the tangential skin stretch cue that the CUFF is able to deliver can represent potential directional information, to be investigated in applications where wearability is a mandatory requirement, e.g., the haptic guidance of blind people.

4. Caress-Like Haptic Stimulation

The devices presented in the previous sections are all devoted to elicit discriminative perception of haptic properties in users. In recent years, in parallel with the study of perceptual aspects of touch, the interest in investigating the role of haptic cues in communicating or evoking emotions in humans has also increased. This is not surprising, since touch represents one of the most ancestral human senses [1] and is a profound communication channel for humans, is highly emotionally charged [68], and has immediate affective consequences [69]. Of note, Autonomic Nervous System (ANS) dynamics are strongly affected by emotional changes [70].

The affective touch display in reference [57] exploits the elasticity of fabric to reproduce the haptic stimuli that are commonly conveyed through a human caress. More specifically, the user places a forearm on the forearm support under the fabric layer, of which the extremities are connected to two motors through two rollers. By controlling motor positions and rotation velocity, it is possible to vary the velocity and the strength of the artificial caress on the user's arm. The system is also endowed with a load cell that measures the normal force exerted by the fabric on the forearm. After a calibration phase, where the offset due to forearm weight is removed, the exerted force (i.e., the strength of the caress) can be varied by acting on the two motor positions, which determine how much the fabric is wrapped around the forearm, and, hence, the force exerted on it (maximum force 20 N). Once the desired level of force is achieved, and both motors are in the reference position, the velocity of the caress can be modulated by regulating the velocity of the motors, exploiting a built-in motor position controller and feeding the motors with a sinusoidal input reference trajectory. By setting the frequency and amplitude of the input, the velocity and the amplitude of motor rotation are controlled, respectively (see reference [57] for further details). The maximum angular displacement of the motors from the reference positions is set to 90 deg., while an entire control cycle lasts 1 ms. An overview of the system is shown in Figure 7. Psycho-physiological assessment tests were performed to characterize the capability of the device in eliciting tactually emotional states in humans, using different combinations of velocity and caress strength. The emotional state was expressed in terms of valence and arousal, which represent the two fundamental neurophysiological systems—one related to pleasure/displeasure (valence) and the other to alertness (arousal)—from which all affective states arise [71]. Moreover, the activation of the ANS was also demonstrated through analysis of the electro-dermal activity. Successive studies confirmed that such a caress-like stimuli can significantly affect the dynamics of other ANS-related physiological measurements, such as heart rate variability (HRV) and electro-encephalographic signals [72–74], thus, further demonstrating the effectiveness of this type of haptic stimulation in eliciting emotional responses in users. Of note, this system represents an interesting proof-of-concept, although its wearability needs to be improved to enable users to carry the device around in everyday life. In reference [75], evidence on how the caress-like haptic device can influence physiological measures related to the autonomous nervous system (ANS), which is intimately connected to evoked emotions in humans, were further discussed and related to self-assessment scores of arousal and valence through a pictorial technique known as SAM (Self-Assessment Manikin) [76]. Specifically, a discriminant role of electrodermal response and heart rate variability can be associated to two different caressing velocities, which can also be linked to two different levels of pleasantness. The paper also reported on the implications of these outcomes for HRI, creating a fascinating and novel perspective for the design of haptic interfaces. In this envisioned scenario, ANS-related measurements could be used to assess user's comfort and emotional state in interaction with a given haptic device, and to devise design and control guidelines for the novel generation of haptic systems, which can be commanded to elicit a given emotional state in users. During the administration of haptic stimuli, physiological signals related to ANS dynamics (e.g., the HRV series, respiration dynamics, electrodermal response, etc.) can be recorded and analyzed to infer information on user's emotional status and other parameters, e.g., stress, fatigue, etc.). An interesting point is that these results are consistent across subjects, and, hence, they can be generalized and effectively employed for a large class of human–machine systems, with potential impact on healthcare and social and rehabilitation-assistive robotics. Regarding the latter case, one possible challenging application of the caress-like haptic stimulation might be in clinical rehabilitation scenarios, where it is crucial to have wearable devices to convey repeatable and controllable haptic stimuli. For example, the caress-like haptic system might be employed for patients with Disorders of Consciousness (DOC), where standardized tactile stimuli could evoke a response in the autonomic nervous system. For this reason, it would be important to further increase the wearability of the systems, in terms of portability and usability, for long-term monitoring [57].

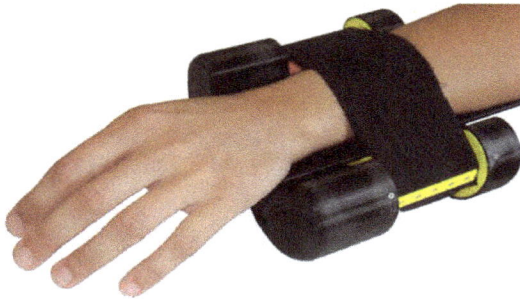

Figure 7. An overview of the haptic system worn by a subject. Reproduced from [57], Copyright 2014, IEEE.

5. Discussion and Conclusions

In this paper, the author has reviewed three main applications of wearable fabric-based devices (tab:electronics-05-00044-t003) that can deliver haptic cues relying only on suitable mechanical modifications of fabric stress–strain characteristics. The reported technologies were validated in experiments with human subjects. W-FYD, for example, was tested in active and passive modes, in absolute and relative cognition tasks, for softness ranking. For the absolute cognition task, participants were asked to associate the stimulus artificially reproduced through W-FYD with its physical counterpart (silicone specimen). Results showed an average accuracy of 88.51% for relative cognition tasks and 84.48% in absolute cognition tasks. For the sliding mode, participants were asked to discriminate the direction of skin stretch on their fingers. The average accuracy was 99%. An average score of 6.67 ± 0.65 using a bipolar Likert-type seven-point scale for the assessment of device capability in inducing slippage sensation was observed. CUFF effectiveness was demonstrated in grasping experiments (in terms of grasp success rate) and softness recognition tasks (through confusion matrices) in conjunction with the SH, as previously described [66]. Finally, a caress-like haptic device was exhaustively investigated in terms of its capabilities in eliciting emotion-related stimulation in users.

These devices can have a significant impact in different fields of HRI. A natural application of wearable systems, given the high level of portability, is virtual and augmented reality, as witnessed for example by results presented in reference [63] for W-FYD. Other important examples can be assistive robotics, e.g., guidance for blind people and force feedback in prosthetics for CUFF, or clinical rehabilitative scenarios, where the possibility to have haptic stimuli conveyed in a controllable and repeatable fashion can be exploited to stimulate ANS—this is the case of caress-like haptic device [77], which can be also used for social interaction through affective cue delivery. Finally, the integration of wearable systems in tele-operation and HRI could represent the key to advance the state-of-the-art and to enable a more natural information exchange from natural to artificial and back again.

Table 3. Main characteristics of reviewed devices.

Name	Dimensions (mm)	Weight (g)	Stimuli	Body Location	Touch	Force Range	Measurements Provided	Stiffness Range	Control Cycle
W-FYD	$100 \times 60 \times 36$	100	Active/passive softness; sliding	Finger	Discriminative	Up to 10 N	Force; motor position; indentation	Up to 0.8 N/mm	≤1ms
CUFF	$145 \times 97 \times 116$	494	Normal-tangential force	Arm	Discriminative	Up to 21 N	Force; motor position	-	≤1ms
Caress	$150 \times 150 \times 80$	560	Velocity; normal force (combination)	Forearm	Affective	Up to 20 N	Force; motor position	-	≤1ms @7Hz

Without any claim of exhaustiveness, the conclusion that can be drawn is that mechanical change of fabric stress–strain behavior can represent a viable and cost-effective solution for wearable haptics, for both discriminative and affective touch, presenting a high level of portability, with high potential impacts in different application fields. At the same time, fabrics can stimulate skin in a natural fashion, enabling an intuitive cue delivery with minimal effects on haptic perception—this opens up promising avenues in augmented reality, i.e., through the superposition of additional tactile cues during the exploration of real objects, as in reference [63]. Of course, to push forward this paradigm, it is important to further reduce device dimensions and encumbrance, and, at the same time, take into account user's point of view and acceptability.

In an ideal future, we could have devices that are "transparent". In other terms, as we are not aware of the cloths we wear, we should not be aware of the WHS we use. Only in this manner WHS can become a key part of our lives. To paraphrase the words of the well-known American industrialist H. Ford, the true progress is to put technology "on the body" of everyone.

Acknowledgments: This work is partially supported by the EC funded project "SoftPro: Synergy-based Open-source Foundations and Technologies for Prosthetics and RehabilitatiOn" (H2020-ICT-688857), by the EU funded project "WEARHAP: WEARable HAPtics for humans and robots" (no. 601165), and by the ERC advanced grant "SoftHands: A Theory of Soft Synergies for a New Generation of Artificial Hands" (no. 291166).

Conflicts of Interest: The author declares no conflict of interest.

References

1. Aristotle. *A New Aristotle Reader*; Ackrill, J.L., Ed.; Princeton University Press: Princeton, NJ, USA, 1987; pp. 161–205.
2. Hannaford, B.; Okamura, A.M. Haptics. In *Handbook on Robotics*; Siciliano, B., Kathib, O., Eds.; Springer: Heidelberg, Germany, 2008; pp. 719–739.
3. Chinello, F.; Malvezzi, M.; Pacchierotti, C.; Prattichizzo, D. A three DoFs wearable tactile display for exploration and manipulation of virtual objects. In Proceedings of the Haptics Symposium 2012, Vancouver, BC, Canada, 4–7 March 2012; pp. 71–76.
4. Traylor, R.; Tan, H.Z. Development of a wearable haptic display for situation awareness in altered-gravity environment: Some initial findings. In Proceedings of the 10th Symposium on Haptic Interfaces Virtual Environment and Teleoperator Systems, Orlando, FL, USA, 24–25 March 2002; pp. 159–164.
5. Leonardis, D.; Solazzi, M.; Bortone, I.; Frisoli, A. A wearable fingertip haptic device with 3 dof asymmetric 3-rsr kinematics. In Proceedings of the 2015 IEEE World Haptics Conference (WHC), Evanston, IL, USA, 22–26 June 2015; pp. 388–393.
6. Pacchierotti, C.; Chinello, F.; Malvezzi, M.; Meli, L.; Prattichizzo, D. Two finger grasping simulation with cutaneous and kinesthetic force feedback. In Proceedings of the International Conference, EuroHaptics 2012, Tampere, Finland, 13–15 June 2012; pp. 373–382.

7. Lieberman, J.; Breazeal, C. TIKL: Development of a wearable vibrotactile feedback suit for improved human motor learning. *IEEE Trans. Robot.* **2007**, *23*, 919–926. [CrossRef]

8. Prattichizzo, D.; Pacchierotti, C.; Rosati, G. Cutaneous force feedback as a sensory subtraction technique in haptics. *IEEE Trans. Haptics* **2012**, *5*, 289–300. [CrossRef] [PubMed]

9. Pacchierotti, C.; Tirmizi, A.; Bianchini, G.; Prattichizzo, D. Enhancing the performance of passive teleoperation systems via cutaneous feedback. *IEEE Trans. Haptics* **2015**, *8*, 379–409. [CrossRef] [PubMed]

10. Tomasello, M.; Carpenter, M.; Call, J.; Behne, T.; Moll, H. Understanding and sharing intentions: The origins of cultural cognition. *Behav. Brain Sci.* **2005**, *28*, 675–691. [CrossRef] [PubMed]

11. Kuchenbecker, K.J.; Ferguson, D.; Kutzer, M.; Moses, M.; Okamura, A.M. The touch thimble: Providing fingertip contact feedback during point-force haptic interaction. In Proceedings of the 2008 Symposium on Haptic Interfaces for Virtual Environment and Teleoperator Systems, Reno, NE, USA, 13–14 March 2008; pp. 239–246.

12. Yang, G.H.; Kyung, K.U.; Srinivasan, M.A.; Kwon, D.S. Quantitative tactile display device with pin-array type tactile feedback and thermal feedback. *Proc. IEEE Int. Conf. Robot. Autom.* **2006**, *2006*, 3917–3922.

13. Buma, D.G.; Buitenweg, J.R.; Veltink, P.H. Intermittent stimulation delays adaptation to electrocutaneous sensory feedback. *IEEE Trans. Neural Syst. Rehabil. Eng.* **2007**, *15*, 435–441. [CrossRef] [PubMed]

14. Casini, S.; Morvidoni, M.; Bianchi, M.; Catalano, M.G.; Grioli, G.; Bicchi, A. Design and realization of the CUFF—Clenching upper-limb force feedback wearable device for distributed mechano-tactile stimulation of normal and tangential skin forces. In Proceedings of the 2015 IEEE/RSJ International Conference on Intelligent Robots and Systems (IROS), Hamburg, Germany, 28 September–2 October 2015; pp. 1186–1193.

15. Kim, H.; Seo, C.; Lee, J.; Ryu, J.; Yu, S.; Lee, S. Vibrotactile display for driving safety information. In Proceedings of the 2006 IEEE Intelligent Transportation Systems Conference, Toronto, Canada, 17–20 September 2006; pp. 573–577.

16. Prattichizzo, D.; Chinello, F.; Pacchierotti, C.; Malvezzi, M. Towards wearability in fingertip haptics: A 3-DoF wearable device for cutaneous force feedback. *IEEE Trans. Haptics* **2013**, *6*, 506–516. [CrossRef] [PubMed]

17. Shull, P.B.; Damian, D.D. Haptic wearables as sensory replacement, sensory augmentation and trainer—A review. *J. Neuroeng. Rehabil.* **2015**, *12*, 59. [CrossRef] [PubMed]

18. Bianchi, M.; Battaglia, E.; Poggiani, M.; Ciotti, S.; Bicchi, A. A Wearable Fabric-based Display for Haptic Multi-Cue Delivery. In Proceedings of the 2016 IEEE Haptics Symposium (HAPTICS), Philadelphia, PA, USA, 8–11 April 2016; pp. 277–283. [CrossRef]

19. Scheggi, S.; Aggravi, M.; Morbidi, F.; Prattichizzo, D. Cooperative human-robot haptic navigation. In Proceedings of the 2014 IEEE International Conference on Robotics and Automation (ICRA), Hong Kong, China, 31 May–7 June 2014. [CrossRef]

20. Sergi, F.; Accoto, D.; Campolo, D.; Guglielmelli, E. Forearm orientation guidance with a vibrotactile feedback bracelet: On the directionality of tactile motor communication. In Proceedings of the 2008 2nd IEEE RAS & EMBS International Conference on Biomedical Robotics and Biomechatronics, Scottsdale, AZ, USA, 19–22 October 2008; pp. 433–438.

21. Van Wegen, E.; De Goede, C.; Lim, I.; Rietberg, M.; Nieuwboer, A.; Willems, A.; Jones, D.; Rochester, L.; Hetherington, V.; Berendse, H.; et al. The effect of rhythmic somatosensory cueing on gait in patients with Parkinson's disease. *J. Neurol. Sci.* **2006**, *248*, 210–214. [CrossRef] [PubMed]

22. Witteveen, H.J.B.; Droog, E.A.; Rietman, J.S.; Veltink, P.H. Vibro- and electrotactile user feedback on hand opening for myoelectric forearm prostheses. *IEEE Trans. Biomed. Eng.* **2012**, *59*, 2219–2226. [CrossRef] [PubMed]

23. Bark, K.; Wheeler, J.; Shull, P.; Savall, J.; Cutkosky, M. Rotational skin stretch feedback: A wearable haptic display for motion. *IEEE Trans. Haptics* **2010**, *3*, 166–176. [CrossRef]

24. Vuillerme, N.; Chenu, O.; Demongeot, J.; Payan, Y. Controlling posture using a plantar pressure-based, tongue-placed tactile biofeedback system. *Exp. Brain Res.* **2007**, *179*, 409–414. [CrossRef] [PubMed]

25. Vuillerme, N.; Pinsault, N.; Chenu, O.; Fleury, A.; Payan, Y.; Demongeot, J. A wireless embedded tongue tactile biofeedback system for balance control. *Pervasive Mob. Comput.* **2009**, *5*, 268–275. [CrossRef]

26. Hui, T.; Beebe, D.J. An oral tactile interface for blind navigation. *Neural Syst. Rehabil. Eng. IEEE Trans.* **2006**, *14*, 116–123.

27. Mann, S.; Huang, J.; Janzen, R.; Lo, R.; Rampersad, V.; Chen, A.; Doha, T. Blind navigation with a wearable range camera and vibrotactile helmet. In Proceedings of the 19th ACM International Conference on Multimedia, New York, NY, USA, November 2011; pp. 1325–1328.
28. Wall, C.; Weinberg, M.S.; Schmidt, P.B.; Krebs, D.E. Balance prosthesis based on micromechanical sensors using vibrotactile feedback of tilt. *IEEE Trans. Biomed. Eng.* **2001**, *48*, 1153–1161. [CrossRef] [PubMed]
29. Nanhoe-Mahabier, W.; Allum, J.H.; Pasman, E.P.; Overeem, S.; Bloem, B.R. The effects of vibrotactile biofeedback training on trunk sway in Parkinson's disease patients. *Park. Relat. Disord.* **2012**, *18*, 1017–1021. [CrossRef] [PubMed]
30. Van Erp, J. Presenting directions with a vibrotactile torso display. *Ergonomics* **2005**, *48*, 302–313. [CrossRef] [PubMed]
31. McKinney, Z.; Heberer, K.; Fowler, E.; Greenberg, M.; Nowroozi, B.; Grundfest, W. Initial biomechanical evaluation of wearable tactile feedback system for gait rehabilitation in peripheral neuropathy. *Stud. Health Technol. Inform.* **2014**, *196*, 271–277. [PubMed]
32. McKinney, Z.; Heberer, K.; Nowroozi, B.N.; Greenberg, M.; Fowler, E.; Grundfest, W. Pilot evaluation of wearable tactile biofeedback system for gait rehabilitation in peripheral neuropathy. In Proceedings of the 2014 IEEE Haptics Symposium (HAPTICS), Houston, TX, USA, 23–26 February 2014; pp. 135–140.
33. Velázquez, R.; Bazán, O. Preliminary evaluation of podotactile feedback in sighted and blind users. *Conf. Proc. IEEE Eng. Med. Biol. Soc.* **2010**, *2010*, 2103–2106. [PubMed]
34. Yang, T.H.; Kim, S.Y.; Kim, C.H.; Kwon, D.S.; Book, W.J. Development of a miniature pin-array tactile module using elastic and electromagnetic force for mobile devices. In Proceedings of the Third Joint EuroHaptics conference, 2009 and Symposium on Haptic Interfaces for Virtual Environment and Teleoperator Systems, Salt Lake City, UT, USA, 18–20 March 2009; pp. 13–17.
35. Wentink, E.C.; Talsma-Kerkdijk, E.J.; Rietman, H.S.; Veltink, P. Feasibility of error-based electrotactile and auditive feedback in prosthetic walking. *Prosthet. Orthot. Int.* **2015**, *39*, 255–259. [CrossRef] [PubMed]
36. Minamizawa, K.; Fukamachi, S. Gravity grabber: Wearable haptic display to present virtual mass sensation. *ACM Emerg. Technol.* **2007**, *8*. [CrossRef]
37. Johansson, R.S.; Westling, G. Signals in tactile afferents from the fingers eliciting adaptive motor responses during precision grip. *Exp. Brain Res.* **1987**, *66*, 141–154. [CrossRef] [PubMed]
38. Westling, G.; Johansson, R.S. Responses in glabrous skin mechanoreceptors during precision grip in humans. *Exp. Brain Res.* **1987**, *66*, 128–140. [CrossRef] [PubMed]
39. Lederman, S.J.; Klatzky, R.L. Haptic perception: A tutorial. *Atten. Percept. Psychophys.* **2009**, *71*, 1439–1459. [CrossRef] [PubMed]
40. Johansson, R.S.; Westling, G. Roles of glabrous skin receptors and sensorimotor memory in automatic control of precision grip when lifting rougher or more slippery objects. *Exp. Brain Res.* **1984**, *56*, 550–564. [CrossRef] [PubMed]
41. Wolfe, J.M.; Kluender, K.R.; Levi, D.M.; Bartoshuk, L.M.; Herz, R.S.; Klatzky, R.L.; Lederman, S.J.; Merfeld, D.M. *Sensation and Perception*; Sinauer Associates: Sunderland, MA, USA, 2006.
42. McGlone, F.; Wessberg, J.; Olausson, H. Discriminative and Affective Touch: Sensing and Feeling. *Neuron* **2008**, *82*, 737–755. [CrossRef] [PubMed]
43. Carbonaro, N.; Mura, G.D.; Lorussi, F.; Paradiso, R.; De Rossi, D.; Tognetti, A. Exploiting wearable goniometer technology for motion sensing gloves. *IEEE J. Biomed. Heal. Informatics* **2014**, *18*, 1788–1795. [CrossRef] [PubMed]
44. Windmiller, J.R.; Wang, J. Wearable Electrochemical Sensors and Biosensors: A Review. *Electroanalysis* **2013**, *25*, 29–46. [CrossRef]
45. Mukhopadhyay, S.C. Wearable sensors for human activity monitoring: A review. *IEEE Sens. J.* **2015**, *15*, 1321–1330. [CrossRef]
46. Zeng, W.; Shu, L.; Li, Q.; Chen, S.; Wang, F.; Tao, X.-M. Fiber-based wearable electronics: A review of materials, fabrication, devices, and applications. *Adv. Mater.* **2014**, *26*, 5310–5336. [CrossRef] [PubMed]
47. Darden, M.A.; Schwartz, C.J. Investigation of skin tribology and its effects on the tactile attributes of polymer fabrics. *Wear* **2009**, *267*, 1289–1294. [CrossRef]
48. Volino, P.; Davy, P.; Bonanni, U.; Luible, C.; Magnenat-Thalmann, N.; Mäkinen, M.; Meinander, H. From measured physical parameters to the haptic feeling of fabric. *Vis. Comput.* **2007**, *23*, 133–142. [CrossRef]

49. Mäkinen, M.; Meinander, H.; Luible, C.; Magnenat-thalmann, N. Influence of Physical Parameters on Fabric Hand. In Proceedings of the Workshop on Haptic and Tactile Perception of Deformable Objects, Leibniz Haus, Hanover, Germany, 1–2 December 2005; pp. 8–16.

50. Bau, O.; Petrevski, U.; Mackay, W. BubbleWrap: A Textile-Based Electromagnetic Haptic Display. In Proceedings of CHI '09 Extended Abstracts on Human Factors in Computing Systems, Boston, MA, USA, 4–9 April 2009; pp. 3607–3612. [CrossRef]

51. De Rossi, D.; Carpi, F.; Lorussi, F.; Scilingo, E.P.; Tognetti, A. Wearable kinesthetic systems and emerging technologies in actuation for upperlimb neurorehabilitation. *Conf. Proc. IEEE Eng. Med. Biol. Soc.* **2009**, *2009*, 6830–6833. [PubMed]

52. De Rossi, D.; Carpi, F.; Carbonaro, N.; Tognetti, A.; Scilingo, E.P. Electroactive polymer patches for wearable haptic interfaces. *Conf. Proc. IEEE Eng. Med. Biol. Soc.* **2011**, *2011*, 8369–8372. [PubMed]

53. Mueller, F.F.; Vetere, F.; Gibbs, M.R.; Kjeldskov, J.; Pedell, S.; Howard, S. Hug over a distance. In Proceedings of CHI EA '05 CHI '05 Extended Abstracts on Human Factors in Computing Systems, Portland, OR, USA, 2–7 April 2005; pp. 1673–1676. [CrossRef]

54. Rotella, M.F.; Guerin, K.; He, X.; Okamura, A.M. HAPI bands: A haptic augmented posture interface. In Proceedings of the 2012 IEEE Haptics Symposium (HAPTICS), Vancouver, BC, Canada, 4–7 March 2012; pp. 163–170.

55. Sahoo, D.R.; Hornbæk, K.; Subramanian, S. TableHop: An Actuated Fabric Display Using Transparent Electrodes. In Proceedings of the 2016 CHI Conference on Human Factors in Computing Systems, San Jose, CA, USA, 7–12 May 2016; pp. 3767–3780.

56. Sadihov, D.; Migge, B.; Gassert, R.; Kim, Y. Prototype of a VR upper-limb rehabilitation system enhanced with motion-based tactile feedback. In Proceedings of the World Haptics Conference (WHC), Daejeon, Korea, 14–17 April 2013; pp. 449–454.

57. Bianchi, M.; Valenza, G.; Serio, A.; Lanata, A.; Greco, A.; Nardelli, M.; Scilingo, E.P.; Bicchi, A. Design and preliminary affective characterization of a novel fabric-based tactile display. In Proceedings of the 2014 IEEE Haptics Symposium (HAPTICS), Houston, TX, USA, 23–26 February 2014; pp. 591–596.

58. Bergmann Tiest, W.M.; Kappers, A.M.L. Cues for haptic perception of compliance. *IEEE Trans. Haptics* **2009**, *2*, 189–199. [CrossRef]

59. Bianchi, M.; Scilingo, E.P.; Serio, A.; Bicchi, A. A new softness display based on bi-elastic fabric. In Proceedings of the Third Joint EuroHaptics Conference, 2009 and Symposium on Haptic Interfaces for Virtual Environment and Teleoperator Systems, Salt Lake City, UT, USA, 18–20 March 2009; pp. 382–383.

60. Bianchi, M.; Serio, A.; Scilingo, E.P.; Bicchi, A. A new fabric-based softness display. In Proceedings of the 2010 IEEE Haptics Symposium, Waltham, MA, USA, 25–26 March 2010; pp. 105–112.

61. Serio, A.; Bianchi, M.; Bicchi, A. A device for mimicking the contact force/contact area relationship of different materials with applications to softness rendering. In Proceedings of the 2013 IEEE/RSJ International Conference on Intelligent Robots and Systems, Tokyo, Japan, 3–7 November 2013; pp. 484–490.

62. Bianchi, M.; Serio, A. Design and Characterization of a Fabric-Based Softness Display. *IEEE Trans. Haptics* **2015**, *8*, 152–163. [CrossRef] [PubMed]

63. Condino, S.; Viglialoro, R.; Fani, S.; Bianchi, M.; Morelli, L.; Ferrari, M.; Bicchi, A.; Ferrari, V. Tactile augmented reality for arteries palpation in open surgery training. *Med. Imaging Augment. Real.* **2016**, accepted.

64. Catalano, M.G.; Grioli, G.; Farnioli, E.; Serio, A.; Piazza, C.; Bicchi, A. Adaptive synergies for the design and control of the Pisa/IIT SoftHand. *Int. J. Rob. Res.* **2014**, *33*, 768–782. [CrossRef]

65. Godfrey, S.; Bianchi, M.; Bicchi, A.; Santello, M. Influence of Force Feedback on Grasp Force Modulation in Prosthetic Applications: A Preliminary Study. *IEEE Eng. Med. Biol. Mag.* **2016**, accepted.

66. Ajoudani, A.; Godfrey, S.B.; Bianchi, M.; Catalano, M.G.; Grioli, G.; Tsagarakis, N.; Bicchi, A. Exploring teleimpedance and tactile feedback for intuitive control of the Pisa/IIT SoftHand. *IEEE Trans. Haptics* **2014**, *7*, 203–215. [CrossRef] [PubMed]

67. Scilingo, E.P.; Bianchi, M.; Grioli, G.; Bicchi, A. Rendering softness: Integration of kinesthetic and cutaneous information in a haptic device. *IEEE Trans. Haptics* **2010**, *3*, 109–118. [CrossRef]

68. Field, T. *Touch*; MIT Press: Cambridge, MA, USA, 2003.

69. Huisman, G. A touch of affect: Mediated social touch and affect. In Proceedings of the the 14th ACM international conference on Multimodal interaction, Santa Monica, CA, USA, 22–26 October 2012; pp. 317–320. [CrossRef]

70. Calvo, R.; D'Mello, S. Affect detection: An interdisciplinary review of models, methods, and their applications. *Affect. Comput. IEEE Trans.* **2010**, *1*, 18–37. [CrossRef]

71. Posner, J.; Russell, J.; Peterson, B. The circumplex model of affect: An integrative approach to affective neuroscience, cognitive development, and psychopathology. *Dev. Psychopathol.* **2005**, *17*, 715–734. [CrossRef] [PubMed]

72. Valenza, G.; Greco, A.; Citi, L.; Bianchi, M.; Barbieri, R.; Scilingo, E.P. Inhomogeneous Point-Processes to Instantaneously Assess Affective Haptic Perception through Heartbeat Dynamics Information. *Sci Rep.* **2016**, *6*, 28567. [CrossRef] [PubMed]

73. Nardelli, M.; Valenza, G.; Bianchi, M.; Greco, A.; Lanata, A.; Bicchi, A.; Scilingo, E.P. Gender-specific velocity recognition of caress-like stimuli through nonlinear analysis of Heart Rate Variability. In Proceedings of the 2015 37th Annual International Conference of the IEEE Engineering in Medicine and Biology Society, Milan, Italy, 25–29 August 2015; pp. 298–301.

74. Valenza, G.; Greco, A.; Nardelli, M.; Bianchi, M.; Lanata, A.; Rossi, S.; Scilingo, E.P. Electroencephalographic spectral correlates of caress-like affective haptic stimuli. In Proceedings of the 2015 37th Annual International Conference of the IEEE Engineering in Medicine and Biology Society, Milan, Italy, 25–29 August 2015; pp. 4733–4736.

75. Bianchi, M.; Valenza, G.; Greco, A.; Nardelli, M.; Battaglia, E.; Bicchi, A.; Scilingo, E.P. Towards a Novel Generation of Haptic and Robotic Interfaces: Integrating Affective Physiology in Human-Robot Interaction. *IEEE Int. Symp. Robot Hum. Interact. Commun.* **2016**, accepted.

76. Bradley, M.; Lang, P.J. Measuring Emotion: The Self-Assessment Semantic Differential Manikin and the semantic differential. *J. Behav. Ther. Exp. Psychiatry* **1994**, *25*, 49–59. [CrossRef]

77. Bianchi, M.; Valenza, G.; Lanata, A.; Greco, A.; Nardelli, M.; Bicchi, A.; Scilingo, E.P. On the role of affective properties in hedonic and discriminant haptic systems. *Int. J. Soc. Robot.* **2016**. [CrossRef]

MDPI AG

St. Alban-Anlage 66

4052 Basel, Switzerland

Tel. +41 61 683 77 34

Fax +41 61 302 89 18

http://www.mdpi.com

Electronics Editorial Office

E-mail: electronics@mdpi.com

http://www.mdpi.com/journal/electronics